ALASKA HERRING HISTORY

PACIFIC FISHERMAN
ALASKA
HERRING

Figure 0.1. *Pacific Fisherman*, 1935 Yearbook.

ALASKA HERRING HISTORY

THE STORY OF ALASKA'S
HERRING FISHERIES AND INDUSTRY

James Mackovjak

University of Alaska Press
Fairbanks

© 2022 by University Press of Colorado

Published by University of Alaska Press
An imprint of University Press of Colorado
245 Century Circle, Suite 202
Louisville, Colorado 80027

All rights reserved

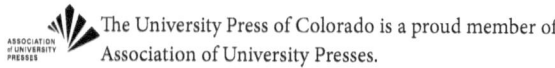 The University Press of Colorado is a proud member of Association of University Presses.

The University Press of Colorado is a cooperative publishing enterprise supported, in part, by Adams State University, Colorado State University, Fort Lewis College, Metropolitan State University, University of Alaska Fairbanks, University of Colorado, University of Denver, University of Northern Colorado, University of Wyoming, Utah State University, and Western Colorado University.

ISBN: 978-1-64642-343-9 (paperback)
ISBN: 978-1-64642-344-6 (ebook)
https://doi.org/10.5876/9781646423446

Library of Congress Cataloging-in-Publication Data

Names: Mackovjak, James R., author.
Title: Alaska herring history : the story of Alaska's herring fisheries and industry / James Mackovjak.
Other titles: Story of Alaska's herring fisheries and industry
Description: Fairbanks : University of Alaska Press, [2022] | Includes bibliographical references and index.
Identifiers: LCCN 2022014575 (print) | LCCN 2022014576 (ebook) | ISBN 9781646423439 (paperback) | ISBN 9781646423446 (epub)
Subjects: LCSH: Herring industry—Alaska—History. | Pacific herring fisheries—Alaska—History. | Pacific herring—Alaska—History.
Classification: LCC SH351.H5 M26 2022 (print) | LCC SH351.H5 (ebook) | DDC 338.3/727452—dc23/eng/20220329
LC record available at https://lccn.loc.gov/2022014575
LC ebook record available at https://lccn.loc.gov/2022014576

Cover illustration: courtesy, NOAA Fisheries

ALASKA HERRING HISTORY

THE STORY OF ALASKA'S
HERRING FISHERIES AND INDUSTRY

James Mackovjak

University of Alaska Press
Fairbanks

© 2022 by University Press of Colorado

Published by University of Alaska Press
An imprint of University Press of Colorado
245 Century Circle, Suite 202
Louisville, Colorado 80027

All rights reserved

 The University Press of Colorado is a proud member of Association of University Presses.

The University Press of Colorado is a cooperative publishing enterprise supported, in part, by Adams State University, Colorado State University, Fort Lewis College, Metropolitan State University, University of Alaska Fairbanks, University of Colorado, University of Denver, University of Northern Colorado, University of Wyoming, Utah State University, and Western Colorado University.

ISBN: 978-1-64642-343-9 (paperback)
ISBN: 978-1-64642-344-6 (ebook)
https://doi.org/10.5876/9781646423446

Library of Congress Cataloging-in-Publication Data

Names: Mackovjak, James R., author.
Title: Alaska herring history : the story of Alaska's herring fisheries and industry / James Mackovjak.
Other titles: Story of Alaska's herring fisheries and industry
Description: Fairbanks : University of Alaska Press, [2022] | Includes bibliographical references and index.
Identifiers: LCCN 2022014575 (print) | LCCN 2022014576 (ebook) | ISBN 9781646423439 (paperback) | ISBN 9781646423446 (epub)
Subjects: LCSH: Herring industry—Alaska—History. | Pacific herring fisheries—Alaska—History. | Pacific herring—Alaska—History.
Classification: LCC SH351.H5 M26 2022 (print) | LCC SH351.H5 (ebook) | DDC 338.3/727452—dc23/eng/20220329
LC record available at https://lccn.loc.gov/2022014575
LC ebook record available at https://lccn.loc.gov/2022014576

Cover illustration: courtesy, NOAA Fisheries

CONTENTS

Dedication *vii*
Foreword *xi*
Acknowledgments *xiii*

Introduction *3*

PART I: HERRING: THE FISH AND ITS UTILIZATION, 1878–1966

1. Alaska Herring: The Basics *9*
2. Early Development of Alaska's Herring Industry *21*
3. Salted Herring: The Early Years *43*
4. Early Alaska Herring Fishery Regulation and Research *64*
5. Alaska's Herring Industry Expands: 1924–1931 *72*
6. A Chronicle of Alaska's Herring Industry: 1932–1948 *84*
7. A Chronicle of Alaska's Herring Industry: 1949–1966 *111*
8. Bait Herring *130*

PART II: ROE HERRING

9. Genesis and Management of Alaska's Roe-Herring Fishery *143*
10. Sitka Sound Roe-Herring Fishery *152*
11. Resurrection Bay and Prince William Sound Roe-Herring Fisheries *179*
12. Lower Cook Inlet and Kodiak Area Roe-Herring Fisheries *211*
13. Togiak Roe-Herring Fishery *217*

14. Norton Sound Roe-Herring Fisheries *254*
15. Food Herring in the Modern Era *266*

PART III: HERRING SPAWN ON KELP

16. Genesis of Alaska's Herring Spawn-on-Kelp Fishery *273*
17. Prince William Sound Herring Spawn-on-Kelp Fisheries, 1981–1999 *285*
18. Southeast Alaska Herring Spawn-on-Kelp Pound Fisheries *296*
19. Togiak and Norton Sound Herring Spawn-on-Kelp Fisheries *308*

Epilogue *318*

Notes *321*
Suggested Readings *389*
Index *391*

DEDICATION

This book is dedicated to Clarence L. "Andy" Anderson, the first commissioner of the Alaska Department of Fish and Game (ADF&G). People who knew Andy said he never got enough credit for the fine work he did.

C. L. Anderson, as he is most commonly known in the fisheries literature, was born in Seattle in 1894 but spent part of his youth in Dawson Creek, British Columbia, and Fairbanks, Alaska. He studied fisheries biology at the University of Washington, in Seattle, graduating in 1917. For the next two years, Anderson was employed by the Bureau of Fisheries to help introduce Alaskans to the Scotch method of curing herring. In 1919, he returned to Seattle to teach fisheries biology, fishing methods, and fish processing at the University of Washington's newly created College of Fisheries. In 1921, on a fellowship from the American-Scandinavian Foundation, Anderson journeyed to Norway to spend a year studying the fishing industry there. Among his interests was the herring fishery at Alesund.

Anderson then returned to the College of Fisheries to teach and pursue a master's degree. Recognized as an expert in herring packing, Anderson worked during the summer of 1923 in a quality-control capacity at the Franklin Packing Company's herring plant at Port Ashton, in Prince William Sound. He received a master's degree in fishery science in 1924. His thesis was on methods of curing herring. Anderson then returned to Alaska to work again for the Franklin Packing Company, and he remained there at least through the end of 1924.[1]

In 1927, Anderson founded and began operating Perfection Smokery, in Seattle. He re-engaged in the Alaska herring business in 1935, when with two partners he purchased the former plant of the Alaska Salmon Meal and Oil Company, near Cordova, and moved it to Thumb Bay, in Prince William Sound. Perfection Fisheries, as it was named, operated both a herring saltery and a herring reduction plant at least through the 1940 season.[2]

In 1943, Anderson sold Perfection Smokery and took the position of chief technologist for the Washington State Department of Fisheries. He became the department's assistant director and in January 1949 was appointed its acting director. In April of that year, the Alaska Board of Fisheries, in recognition of Anderson's broad education and his training and practical experience in fisheries, chose him to head the newly created Alaska Department of Fisheries. The department's budget in 1949 was $70,000, and its staff initially consisted of Anderson, one assistant (Lewis MacDonald, a fisheries biologist), and a secretary. They operated out of a one-room office. In 1957, this department became the Alaska Department of Fish and Game, with Anderson at its helm.[3]

The new state of Alaska gained jurisdiction over its fisheries in January 1960. Among Anderson's imprints on fishery management in Alaska was the division of responsibilities between the Board of Fisheries and the Department of Fish and Game. The board was responsible for establishing the harvest parameters and allocating the harvest among the various gear groups, while the department was responsible for the management of the fisheries, such as seasonal opening/closure dates. As Clem Tillion, Alaska's "Fish Czar" under Governor Walter Hickel in the late 1960s, said, "The brilliant thing that Clarence Anderson left us with was separating the people who protect the resource from the people who allocate the resource."[4]

Anderson retired from the Department of Fish and Game in 1962. Alaska governor Bill Egan praised Anderson as the "Father of the Alaska Department of Fish and Game," an acknowledgment of his long years of work in organizing the department.[5] *Pacific Fisherman* also praised Anderson's work in establishing the department, advancing its program, and creating an esprit de corps that attracted a staff of dedicated young scientists.[6] At present, the Alaska Department of Fish and Game has a staff of 1,700 in forty-seven field offices.

C. L. Anderson's farewell words to the fishing industry:

> Upon my retirement as Commissioner of the Alaska Department of Fish and Game, I wish to express deep appreciation for your cooperation and good will during my 12 years in service of Alaska.
>
> During this period, a Fish and Game Department has been developed from an idea on a few scraps of paper to a well-functioning organization of 175 carefully chosen, well-trained dedicated employees. These now administer the great fish and game resources of Alaska in protection, biological research and management.
>
> Setting up a department to utilize and harvest the vast wealth of Alaska's fish and game resources on a sustained-yield basis has been my primary concern and endeavor since 1949.
>
> Several factors, such as farsighted legislative support, contributed greatly to the realization of these endeavors.

However, without your active interest that manifested itself generously through the years in both moral and material support, I know that the aims for the department could not have been fully realized.

Your continued future support will enable your department to accomplish its true mission of serving all the people of Alaska in the management of the state's fish and game resources.[7]

In 1962, the Alaska Department of Fish and Game honored Anderson by renaming its first substantial research vessel, the seventy-one-foot wooden former seiner *Lucky Boy*, the *C. L. Anderson*.

C. L. Anderson passed away in April 1966. The Clarence L. Anderson Building, on the University of Alaska's Juneau campus, is named in his honor.

Figure 0.2. His wasn't only a desk job. Clarence Anderson planting eyed salmon eggs in spring, probably in Southeast Alaska. (Alaska State Library, image no. P58-93)

FOREWORD

Author Jim Mackovjak takes the reader on an excellent journey through a period of Alaska's history in which a fishery begins, encounters multiple challenges, and adapts to meet the demands of competing interests, changing markets, and the world events that have influenced this important segment of Alaska's fishing industry that continues today.

As a former (very limited) participant and fishery manager, I appreciate the time and effort Mr. Mackovjak has dedicated to accurately compiling the statistics and information about the processing industry players, the communities, and the fishermen involved.

This work is a must-read for those who have an interest in fisheries from any perspective, be it management, processing, marketing, harvesting, or Alaska history in general. The herring fisheries in Alaska will undoubtedly continue to face challenges and will need to continue to adapt. *Alaska Herring History* sets the table by showing us where the fishery has been, and it should be an important text to help guide where it goes from here.

I'd like to add that my father-in-law, Alaska fisheries legend Clem Tillion, who wrote the foreword to Jim's *Alaska Codfish Chronicle*, passed away in October 2021. With him in his casket is, among other items, a jar of pickled herring.

Sam Cotten, former commissioner,
Alaska Department of Fish and Game

ACKNOWLEDGMENTS

Compiling the history of Alaska's Pacific herring fishery was a big, complex project, and I couldn't have completed it without the help, support, and encouragement of numerous individuals and institutions.

To name all who have contributed would make this section read like a telephone book, but first among those I would like to thank is Frank Norris, the former regional historian for the National Park Service in Alaska and my editor. More than anyone, Frank has helped make my writing more comprehensible, and I am forever indebted to him. The care with which he reads my work—evidenced by his ability to catch small inconsistencies spread over hundreds of pages—never ceases to impress me.

Greg Streveler, my neighbor in Gustavus who has vast knowledge of Alaska's natural history and marine environment, carefully read a relatively early draft of the work. His technical comments were invaluable, as were his suggestions regarding how I might reorganize the work.

Jim Balsiger, head of NOAA Fisheries in Alaska; John Jensen, of the Alaska Board of Fisheries and the North Pacific Fishery Management Council; and Sam Cotten, former commissioner of the Alaska Department of Fish and Game, each reviewed a draft of the book and advocated for its publication. Mr. Cotten graced the book by writing its foreword.

Retired ADF&G fisheries manager Jeffrey Skrade shared his vast knowledge of the Togiak roe-herring fishery and reviewed a draft of the manuscript.

Staff at the Alaska Department of Fish and Game, especially biologists Glenn Hollowell, Jim Menard, Bo Meredith, and Geoff Spalinger, provided invaluable information.

Pioneer roe-herring fishermen Beaver Nelson, Ken Moore, and Clyde Curry supplied me with information and photographs and reviewed draft material.

Ann Holmstrand and Gretchen Bersch shared reminiscences and photographs of the early roe-herring fishery in Resurrection Bay.

Rhonda Hubbard, whose father, Ray Anderson, was a pioneer in the roe-herring industry, shared information and photographs.

Tom Swanson and Ed Wyman helped explain the spawn-on-kelp fishery to me.

Historian colleagues Karen Hofstad and Bob King offered constant assistance and encouragement.

Peggy Parker, at *Seafood News*, helped keep me up to date on developments in Alaska's herring fisheries.

The Alaska Department of Fish and Game and the National Marine Fisheries Service (NOAA Fisheries) are admirable for the prodigious amount of material they make easily available on the internet. The Department of Fish and Game's half-century collection of annual regional reports provided an invaluable real-time record of the development of Alaska's herring fisheries.

Thanks to Fritz Funk, Thomas Thornton, and the anonymous individual who peer reviewed my draft material and provided invaluable, critical commentary. Carefully reviewing a 109,000-word manuscript wasn't a quick or easy chore, and I admire their intellectual stamina. Their input benefited this book immensely.

I have no end of respect and appreciation for the many institutions that work to preserve our history, and I want to thank especially the staff at the Alaska Resources Library and Information Service (ARLIS), in Anchorage; the Alaska State Library, in Juneau; and the Egan Library, on the campus of the University of Alaska Southeast, in Juneau. I also want to thank Danielle Devore, of the Alaska State Court Law Library (Juneau), for help with legal documents. To Danielle, I think my name and "request" are synonymous.

Finally, I want to thank the University of Alaska Press for taking on this project. Nate Bauer, the press's associate director, provided just the right amounts of encouragement and caution as he shepherded my project through the acquisition process.

In addition, I want to thank all those who strive to protect herring and the environment that is essential to their survival. Any place in our oceans where herring abound is a better place for it.

ALASKA HERRING HISTORY

INTRODUCTION

Herring Bay? There are four Herring Bays in Alaska. Two are in Southeast Alaska, one is in Prince William Sound, and one is on the Kenai Peninsula, near Seldovia. And the name Seldovia is itself derived from the Russian words *Zaliv Seldevoy*, which translate to "bay of herring."

Additionally, Metervik Bay, near Togiak, in Bristol Bay, is also known as Herring Bay. There are also two Herring Coves, the Herring Islands, and Herring Point.[1] Fish Egg Island, near Craig, in Southeast Alaska, is named for the herring eggs that are deposited there.

> There is reliable information to the effect that schools of herring many miles in extent appear frequently about the fishing shores.—Tarleton Bean, fish expert, 1889[2]

> The most abundant food fish in the waters of the world is the herring, and weight for weight this fish has a greater nutritive value than most meats and other fish, while its low price brings it within the reach of all and makes it pre-eminently the poor man's food.—John N. Cobb, US Bureau of Fisheries, 1920[3]

> Klawock, Alaska, January 1913: Last January at Klawak, on the west coast of Prince of Wales Island, there occurred an unusually enormous run of herring. So numerous were the fish as they crowded into the bay that hundreds of thousands or even millions were stranded and suffocated. When the tide receded they were left in a solid mass over the beach to a depth in places of several feet.—Barton Warren Everman, US Bureau of Fisheries[4]

Alaska's marine waters support at least 384 species of fish. Arguably, the most important of them is the small silver Pacific herring (*Clupea pallasii*).[5] In the North Pacific marine ecosystem, Pacific herring are, in the words of Alaska cultural anthropologist Thomas Thornton and his colleagues, a "foundation and bellwether species."[6] Herring are also characterized as a keystone species because of their vital role in the marine food web.

The family *Clupeidae*, to which the Pacific herring belongs, is the world's most valuable family of food fishes. It includes—in addition to saltwater herring—menhaden, shad, alewives, freshwater herring, and many other species.[7]

> In addition to their direct commercial importance, herring also are of great indirect importance as a food supply for many other commercially important predacious species of fish such as king and coho salmon, cod fish and halibut. They are also extensively preyed upon by whales, seals, sea lions, birds, and by other fishes.—Alaska Department of Fish and Game, 1963–1964[8]

> When the herring vanish, so does everything else dependent on them.—June Allen, *Ketchikan Daily News*, 1993[9]

Pacific herring, because of their typically high oil content, are energy rich.[10] They are classified as forage fish—also called prey fish or bait fish—and are a favored food of top-tier predators, among them whales, sea lions, king salmon, halibut, and marine birds. As such, they occupy a key position in the marine food web, linking the energy and nutrients produced by plankton to mammals, birds, and large-bodied fish.

> Forage fish provide the main pathway for energy to flow from very low trophic levels—plankton—to higher trophic levels—predatory fish, birds, and mammals. They transfer a large proportion of energy in the ecosystem and support or regulate a variety of ecosystem services.—Lentfest Ocean Program, 2012[11]

Herring mass into immense schools that move along coastlines and migrate across open water. In fact, the fish's name is derived from the German term *heer*, meaning "army."[12]

Herring abundance can fluctuate widely, but the reasons behind the fluctuations are poorly understood. Thus, prudent management of the herring resource requires the employment of conservative, ecosystem-based management principles.

Alaska's herring industry is a big, big subject. There have been numerous herring fisheries in the state, ranging along the coast from Kah Shakes, on Southeast Alaska's southern tip, to Dutch Harbor, in the Aleutian Islands, to Kotzebue Sound, above the Arctic Circle. Each fishery is or was at least locally important and often unique. Some, such as the roe-herring fishery at Togiak, were large factors in the global industry. A comprehensive history of Alaska herring fisheries big and small would take volumes, and, in the interest of concision, I have chosen to focus on those I consider most important.

Over the years, Alaska herring have been harvested for a wide variety of purposes, including "reducing" them into fertilizer, fish meal, and fish oil; curing (salting) them; using them for bait; stripping the females of their roe (eggs); and harvesting herring-spawn-laden kelp. Based on those uses, this book is divided into three parts. Part I is a history of the reduction (fertilizer/fish meal/fish oil) and the cured (salted) herring industries, which were largely integrated operations. Also included in part I is a discussion of Alaska's bait-herring fisheries. Part II, in mostly chronological form, is a history of the roe-herring fisheries in Southeast Alaska, Prince William Sound, Kodiak Island, lower Cook Inlet, Togiak, and Norton Sound. Part III is a history of Alaska's herring spawn-on-kelp industry.

Part I
Herring
The Fish and Its Utilization, 1878–1966

1

ALASKA HERRING
The Basics

> Pacific herring live a nervous life. Though their chief function is to feed on abundant zooplankton in the ocean and attain great numbers in the process, in turn, just about every predator within their range feeds voraciously on these little protein converters.—Tom Ohaus, Pacific Fishing, 1990[1]

GENERAL DESCRIPTION

There are minor anatomical and behavioral differences between Pacific herring (*Clupea pallasii*) and Atlantic herring (*Clupea harengus*), but they are so small and so variable that the two species are undistinguishable to the casual observer. Atlantic herring, however, tend to be somewhat larger.

Herring have a blue-green upper body with silvery sides and belly and are devoid of markings. Their scales are large and easily removed. Herring have one short dorsal fin, deeply forked tails, and no spines in their fins. Herring flesh is oily, which adds to the fish's flavor and has made herring a valuable source of industrial oil.[2] The high oil content is also a major reason for the fish's value as a forage species.[3]

Fully grown Pacific herring may be as much eighteen inches long and weigh up to 1.8 pounds, but a nine-inch specimen is typically considered large. In Alaska, Bering Sea herring are genetically distinct from Gulf of Alaska herring and are larger and longer-lived than their Gulf of Alaska cousins.[4]

AGE AND GROWTH

Pacific herring reach sexual maturity at three to four years of age, when they may be eight or even ten inches long, and they spawn each year for the rest of their lives. A herring's typical life span is eight to sixteen years, but occasionally fish reach twenty years. Herring are not fast-growing fish, but they continue to grow each year, albeit at a slower rate as they age. Although the age/size relationship likely varies for a host of reasons, research done in the 1920s determined that herring in Kachemak Bay and the Kodiak-Afognak district did not reach a size suitable for the cured-herring market (about 10.5 inches in total length) until about five years of age, on average. Herring in Prince William Sound required about six years to reach curing size. At Togiak, herring generally entered the roe-herring fishery at age five.[5]

RANGE AND MIGRATION

> The migrations of herring are very erratic; they may desert an old feeding or spawning ground for years, and then return in vastly increased numbers, without anyone knowing the reasons for the disappearance or return.—*Pacific Fisherman*, 1919[6]

Pacific herring inhabit the North Pacific Ocean's coastal waters in an arc that extends from the Sea of Japan northeast to the Arctic Ocean and then southeast to Baja California, Mexico. Gulf of Alaska herring migrate little, generally moving less than 100 miles between their spawning, feeding, and wintering grounds. By contrast, most Bering Sea herring annually migrate to offshore wintering grounds, which can be 1,000 miles or more distant from their coastal spawning and feeding grounds.[7]

About 90 to 95 percent of the total Bering Sea spawning biomass migrates to Bristol Bay to spawn. These fish then migrate southward along the Alaska Peninsula and concentrate in the vicinity of Unalaska Island in the late summer. In the fall, they return to their wintering grounds near the Pribilof Islands. The remaining 5 to 10 percent spawn in Kotzebue Sound, Norton Sound, the Kuskokwim Sound–Yukon River delta area, and the Aleutian Islands.[8]

The movements of individual schools of herring can vary substantially from year to year. The fish may be seasonally abundant at a given location for several years, then—for reasons unknown but perhaps due to changes in ocean conditions—disappear, only to return in subsequent years. This characteristic presents a challenge for fisheries managers: is a sudden scarcity of herring at a specific location the result of natural causes or of overfishing?

DIET AND FEEDING

Herring feed mainly on zooplankton, phytoplankton, nonplanktonic crustaceans, and small fish. They generally feed at night in surface waters, especially where there is upwelling. During the day, the fish remain near the seafloor.[9] Primarily during the summer, when food is usually abundant, herring build up fat stores that will help them survive the lean months.[10]

SPAWNING

In general, Alaska herring spawn earliest at southern latitudes and progressively later as one proceeds north. In southern Southeast Alaska, herring spawning begins as early as the middle of March. In Prince William Sound and Cook Inlet, herring typically spawn in early April. At Togiak, in Bristol Bay, herring typically spawn in late April or early May, whereas in Norton Sound, in the northern Bering Sea, spawning can occur well into July. Typically, herring arrive on their spawning grounds several days before spawning, with the larger, older fish arriving before the smaller, younger fish.

The eggs of Pacific herring are translucent, pale amber in color, and about 1 to 1.5 millimeters in diameter. The number of eggs produced by a female herring at each spawning averages about 20,000 but varies considerably. In races of small fish, individual females produce fewer than 10,000 eggs, while those in large races may produce more than 100,000. These large numbers are important because even under ideal conditions, many eggs fail to hatch, and egg mortality may exceed 99 percent.

Herring prefer to spawn in kelp-rich shallow-water areas of the subtidal zone but will spawn even in the intertidal zone. There are roughly equal numbers of males and females in a spawning aggregation, but they do not pair to spawn. For them, spawning is a group event, with entire schools—often millions of individual fish—releasing their milt and eggs into the water almost simultaneously. The males initiate the event by spraying their milt, following which the females release their eggs, after which the males once again spray milt. This gives the nearshore water a milky appearance, sometimes for miles. The eggs, which are fertilized as they sink toward the seafloor or drift in the current, are coated with an adhesive membrane and adhere to vegetation and the ocean substrate. Egg survival is higher for eggs that adhere to vegetation than for those that sink to the seafloor. After spawning, herring migrate to their summer feeding grounds.[11]

In Southeast Alaska in 1956, herring spawned along 148 miles of coastline. In 1957 and 1958, they spawned along 132 miles and 134 miles, respectively.[12] Herring were relatively abundant in the region during those years. And the schools of spawning herring could be dense. In 1991, the Alaska Department of Fish and Game (ADF&G) estimated that 3,550 tons of herring spawned along each mile of a stretch of coastline on the northern shore of Montague Island, in Prince William Sound.[13] Bob DeJong, an ADF&G biologist who managed the Sitka Sound roe-herring fishery for a number of years, said in 1991 that generally there were about 500 tons of herring for each mile of coastline showing egg deposition.[14]

Anthropologist Edward Nelson, who spent the years 1877–1881 along Alaska's Bering Sea coast, described herring spawning in Norton Sound in June 1881:

> At this time these fish form a continuous line along the beach, passing from south to north in unbroken succession, spawning on the seaweeds and rocks from above low-tide mark to a fathom below it. They enter all the inner bays and swarm about every reef and rocky point. The water boils with them along shore as they struggle about in a dense mass along the short seaweed in spawning, and they can be easily caught in one's hands. The females move slowly among the weeds, and press in the midst of them, depositing their eggs, which adhere to whatever they come in contact with, by means of a gummy secretion with which they are coated. Thrusting my hand under the water for a half minute was sufficient for it to be covered with eggs.[15]

FROM EGG TO FISH

Fertilized herring eggs hatch after about two weeks, depending on the water temperature. The herring larvae, which are about 5.5–7.5 millimeters long, drift and swim in the nearshore coastal currents, consuming their yolk sacs to develop and grow.

After about two weeks, the yolk sac has been consumed, and the larvae, now about ten millimeters long, begin feeding on plankton. If adequate food is not readily available, however, the larvae soon become so weak that they cannot capture food and quickly starve. Another hazard is unfavorable water currents that sweep larvae out to sea or to areas without adequate feed. Add to this predation by other fishes and animals, and it is easy to understand why larval mortality is high.

Two to three months later, the surviving larvae have metamorphosed into juveniles and begin schooling in shallow coastal waters. These schools then migrate to deep water—up to 1,300 feet deep—for the next two to three years. Upon becoming sexually mature, the fish join the adult population. Although some mixing occurs, tagging studies indicate that Pacific herring tend to stay in the same school for years.[16]

ABUNDANCE

Herring are one of the most abundant fish species in Alaska. Population changes can be substantial on both small and large geographic scales. Fluctuations are determined largely by marine conditions, which affect herring survival, growth, and recruitment. Commercial fishing, too, has an impact. In the late 1930s, annual catches that sometimes exceeded 100,000 metric tons were probably too high and may have caused certain stocks to decline.[17]

EARLY NATIVE USES OF HERRING

> For many First Nations and Native American groups from Alaska to Washington, the nutritionally valuable and readily harvested herring and its roe were integral to daily lives and worldviews.
> —Iain McKechnie et al., 2014[18]

> Immense shoals of herring visit the bays and estuaries of Alaska at various seasons of the year, and they form an important item in the food supply of the [N]atives where this fish is found.
> —US Bureau of the Census, 1890[19]

Herring—*yaaw* in Tlingit, *iinang* in Haida, and *uksruktuuk* in Iñupiat—was and to some extent remains an important component of the traditional diet of Alaska's Native people. Herring also have a cultural value, especially in Southeast Alaska.

Over several decades beginning in 1985, a group of researchers analyzed nearly half a million fish bones in 171 coastal archaeological sites—most of which had been occupied within the past 2,500 years—in

Washington, British Columbia, and Southeast Alaska. Their goal was to provide a proxy measure of past herring distribution and abundance. The researchers found that, in the sites they examined, the bones of Pacific herring were the most ubiquitous. But in Southeast Alaska, where eighteen sites were examined, salmon bones tended to dominate, followed by those of herring, Pacific cod, and sculpin.[20]

For Southeast Alaska Natives, Sitka was a center for obtaining herring and herring eggs. Fyodor Litke, a Russian explorer who sailed around the world in 1826–1829, reported that in the spring of 1827 up to a thousand Natives were gathered at Sitka and an equal number were on nearby islands to take advantage of the abundance of spawning herring.[21]

Ivan Petroff, who visited Alaska in 1880 and 1881, wrote in his 1882 *Report on the Population, Industries, and Resources of Alaska* that Natives consumed herring "both fresh and dried, but the larger portion of the catch is converted into oil." In his 1898 report on Alaska's fisheries, Jefferson Moser, captain of the US Fish Commission's oceanographic research steamer *Albatross*, noted that Natives in Southeast Alaska used herring only when they were available in local waters, curing none for winter food.[22]

Moser described the simple yet effective method Natives used to obtain herring:

> In catching them for their own use, a long stick or pole having at the end, and for some distance from it, a large number of sharp-pointed nails, is swept through the water, with a paddle-like motion, like a rake, impaling the fish on the nails. At the end of the movement the pole is brought over the canoe, given a shake which detaches the fish and then thrust into the water again. In this manner a canoe load is quickly made.[23]

Pioneer travel writer Eliza Scidmore, who traveled to Southeast Alaska in 1883 and 1884, wrote that each nail on what she called the Natives' "primitive rakes" caught two or three herring.[24]

Along the Bering Sea coast, by 1881 Native fishermen typically caught herring with a more conventional type of gear: beach seines. The fish were woven into "strings" along lengths of rye grass or draped over drying racks and then sun-dried (figure 1.1). The dried fish was either consumed locally or traded with interior peoples for other items.[25] By about 1906, fishermen at Grantley Harbor, on the Seward Peninsula, were salting herring on a small scale and selling the fish at Nome and various other settlements in the region.[26] Natives along the Bering Sea

Figure 1.1. Herring braided with rye grass and hung to dry, Toksook Bay, 1979. (James Barker)

coast also collected herring spawn on kelp, some of which was dried to preserve it for winter use. The product was boiled before being eaten.[27]

Southeast Alaska Natives also ate herring eggs. Moser described the fish coming to the shores in April in "countless numbers" to spawn, at which time

> the Indians plant hemlock twigs at the low-water mark, where they become covered with spawn, after which they are gathered in canoe loads. The spawn is called "Alaska grapes," and is consumed by the [N]atives in large quantities, either fresh or dried, and cooked as occasion demands, and for winter use. Usually it is eaten with rancid oil, which is the sauce that goes with all their delicacies, even with berries.[28]

Ivan Petroff called herring spawn on hemlock "a favorite article of food in a semi-putrid state."[29] This item became a minor item of commerce in the Native communities in British Columbia and Alaska.[30]

In southern Southeast Alaska, especially around Hydaburg, Natives harvested *Macrocystis pyrifera* (giant kelp) on which herring eggs had been deposited. The product was consumed locally and, when dried, was also a trade item.[31]

Not everyone was pleased with the Native harvest of herring eggs. Among those seeking to prohibit this activity was Carl Spuhn,

Herring

Figure 1.2. Herring and herring roe on hemlock branches. (NOAA Fisheries)

president of the Alaska Oil and Guano Company, which converted herring into oil and fertilizer. In 1910, Spuhn condemned the practice as the "wholesale destruction of herring spawn."[32] John Cobb, of the Bureau of Fisheries, agreed, writing in 1910 that "this practice should be prohibited by law."[33] And in his 1914 report on Alaska's fisheries, E. Lester Jones, also of the Bureau of Fisheries, wrote,

> The present practice of the Indians in southeast Alaska of taking millions of herring eggs every season and drying them for food should be stopped at once, for this not only

Figure 1.3. Freshly landed Pacific herring. (James Mackovjak)

means partial destruction of the future supply of herring, but is quite needless, since these Indians have many other ways of obtaining food.[34]

Despite this criticism, the Native practice of harvesting herring eggs continued and perhaps even increased. Ward Bower, the Bureau of Fisheries agent in Alaska for a number of years, wrote of an unusually large run of spawning herring at Sitka in April 1930. Natives there collected large quantities of herring spawn on hemlock, some of which they shipped to Juneau, Haines, and other locations.[35]

The gathering of eggs by Alaska Natives was never outlawed, but the Alaska Department of Fish and Game eventually regulated it as a subsistence fishery. The practice—now considered a customary and traditional use—continues to this day.

COMMERCIAL FISHERY

[H]erring is the most abundant food fish in Alaskan waters, and that diminution of the supply by the most intensive fishing is only a remote possibility.—Ward Bower, US Bureau of Fisheries, 1924[36]

For commercial purposes, Pacific herring are identical to their Atlantic cousins. Except for herring to be used locally as food or bait in other

Alaska fisheries, Alaska producers of herring products had to compete with products from countries on both sides of the North Pacific and North Atlantic Oceans.

In Alaska, Pacific herring are the only commercially fished species of forage fish. This fishery occurs in coastal waters and was managed by the federal Bureau of Fisheries until 1940, when the bureau merged with Biological Survey to form the US Fish and Wildlife Service. This agency managed the herring fishery until 1960, when (with several exceptions) the State of Alaska took over the management of state-waters fisheries.

The primary locations at which large-scale commercial herring fisheries have occurred are in Southeast Alaska, especially along Chatham Strait and in Sitka Sound; in Prince William Sound, especially in lower Valdez Arm and among the islands north of Montague Island; in lower Cook Inlet, especially in Kachemak Bay and Kamishak Bay; on the Shelikof Strait shore of Kodiak Island; and at Togiak, in Bristol Bay. A relatively small herring fishery occurs in Norton Sound, not too distant from the Arctic Circle.

Today, as has been the case since the inception of the herring fishery in the early 1880s, herring are caught primarily with purse seines. Gillnets—which, based on their mesh size, can select for larger fish—have been employed in some fisheries, as have beach seines. Midwater trawls have also been and continue to be employed in Alaska to catch bait herring.

The method for measuring Alaska's herring catch has evolved over time. The traditional method was by volume. Fish were loaded into a wooden barrel that was on a pivot so it could be easily emptied. Originally, a full barrel was considered to represent 200 pounds of fish, but the size was officially changed to 250 pounds in the late 1930s. Complicating matters, some reduction plants in Alaska didn't even bother to measure the quantity of fish they received. Rather, they calculated the quantity based on the amount of fish meal it produced. In 1956, federal fisheries managers abandoned the barrel measure and began using short tons. Modern fishery managers use either short tons or metric tons.[37]

Alaska's herring harvest peaked in 1937, when 139,000 tons were caught, primarily to supply the territory's reduction plants.[38] The wholesale value of the fishery peaked in 1996 at nearly $100 million, driven almost entirely by a strong Japanese market for herring roe.[39]

COMMERCIAL PRODUCTS

The first major commercial use of herring in Alaska, in the 1880s, was for the "reduction" of raw herring into meal and oil. The meal was used as animal feed as well as fertilizer. The oil was used in the manufacture of soap and a host of other products. This industry, which consumed vast quantities of herring, peaked in the 1930s but persisted until the 1960s. Beginning during World War I, lesser quantities of herring were utilized to make cured (salted) herring, a product that was mostly marketed on the US East Coast. The cured-herring industry was relatively important for a couple of decades but by 1940 had largely faded away.

Beginning in the 1960s, to help meet the Japanese demand for *kazunoko*—the egg skeins of ready-to-spawn herring—the roe-herring ("sac roe") fishery developed in Alaska. This grew to become one of Alaska's largest fisheries. Ancillary to the roe-herring fishery was the fishery for herring spawn-on-kelp (*kazunoko kombu*), which was in great demand in Japan.

Other products, such as kippered herring and frozen herring fillets, have been produced over the years, but in comparatively small quantities. Each will be discussed. Also, relatively small quantities of herring were used to feed foxes, mink, and other furbearers on the fur farms that were located mostly on islands in Alaska. The heyday of the fur farms was the 1920s, during which there were about 700 farms.[40]

The most consistent—and least controversial—use of herring has been as bait to catch halibut, salmon, and shellfish. The bait fishery first developed in the late 1890s, when Seattle-based halibut vessels began fishing in Alaska waters, and it has persisted to this day. The bait fishery is discussed in chapter 8.

CONTROVERSY

> Beyond question, herring should not be used for fertilizer, oil, or fish meal.—E. Lester Jones, Alaska agent, Bureau of Fisheries, 1914[41]

Alaska's herring fishery has not been without controversy. In the early 1900s, people began to question the wisdom of reducing herring to make fertilizer, fish meal, and fish oil rather than using the fish as human food or as bait to catch other fish. And they also questioned the effect large catches of herring had on the highly valuable fish that feed on herring, such as halibut, king salmon, and coho salmon. Moreover, the fisheries that supplied the reduction plants were high-volume, intensive, and nonselective, and they seriously depleted herring populations in some areas.

Nevertheless, the large-scale production of fish meal and fish oil from herring continued until 1966, after which Alaska's last herring reduction plant, at Big Port Walter, in Southeast Alaska, was shuttered. At about the same time, the roe-herring fishery developed. Though the fishery caught male and female herring equally, it utilized only the herring roe.

In the early years of the fishery, when the roe was extracted by workers in Alaska, about 90 percent of the herring catch's weight—the males and the roe-stripped carcasses of the females—was discarded, usually by grinding it and pumping it overboard near processing facilities. This, too, sparked controversy.

At the December 1973 meeting of the Alaska Board of Fish and Game, Carl Rosier, director of the Alaska Department of Fish and Game's commercial fish division, said, "The wasteful practice of utilizing only the sac roe cannot be permitted to continue when there is a worldwide demand for herring flesh." The board agreed and directed the Department of Fish and Game to manage the state's herring fishery for fullest use as food and bait.[42]

Despite Rosier's assertion, the demand for food herring was limited, and the demand for bait herring was relatively static. The market for herring roe, however, was robust and drove the expansion of the herring fishery.

The dumping of herring carcasses largely ended in the mid-1970s, when several processors installed reduction plants to produce fish meal from fish and shellfish offal. By that time, though, most processors had begun freezing whole herring in blocks for shipment to Japan.[43] Even frozen, the male herring were worth next to nothing. For processors, dealing with them was regarded as a cost of being in the roe-herring business.

In 1977, Alaska's Legislature outlawed roe stripping but later made two temporary exceptions for operations in the Bering Sea.[44]

2
EARLY DEVELOPMENT OF ALASKA'S HERRING INDUSTRY

According to John Cobb's report, *The Commercial Fisheries of Alaska in 1905*, the first year for which reliable information regarding Alaska's herring fishery was available was 1878. That year—the same year Alaska's first salmon cannery began operations—individuals at Wrangell, in Southeast Alaska, engaged in the business of catching herring. They extracted the oil from some of their catch and salted and dried the remainder.

The salted fish, which was intended for human consumption, was packed in wooden barrels, likely of the size used for salted salmon and holding about 250 pounds of fish. The Wrangell operation's salted production totaled 37,500 pounds (150 barrels) and had a total value of $900. Cobb reported that 25,000 pounds of herring was salted the following year but did not disclose the location. In 1880, according to Cobb, the Western Fur and Trading Company, at Kodiak, put up 15,000 pounds (500 thirty-pound boxes) of smoked herring as well as about 5,000 pounds of herring salted in barrels.

Cobb reported no food-herring production in Alaska during the years 1881–1890.[1] The herring industry, however, had expanded dramatically in 1882–1884. But it wasn't salted or smoked herring that drove the expansion: it was fish oil and fertilizer made from herring. Food herring, as will be discussed in chapter 3, would be an industry of only minor importance until World War I, when imports of cured herring from Europe were cut off.

REDUCING HERRING

The development of the oil and fertilizer industry was initiated by the Northwest Trading Company, which established itself in Alaska in 1880. Led by Carl Spuhn and J. M. Vanderbilt, the company established stations for trade with the Natives. One station was at Killisnoo, near the Tlingit community of Angoon (Kootznoowoo), on the west shore of Admiralty Island, where the Hudson's Bay Company had once maintained a station. Northwest Trading also constructed at Killisnoo what may have been the first whale-reduction plant on the US Pacific Coast.[2]

The whaling venture didn't work out well. On October 22, 1882, the accidental explosion of a harpoon bomb killed a Tlingit shaman. To ensure compensation for his death, Natives at Angoon took two Northwest Trading Company whaleboat men hostage. In response, the US Navy shelled and burned most of Angoon. This brutal action, intended to teach the Natives a lesson, sparked controversy and led to the eventual end of military control of the district of Alaska. Both whaleboat men survived the ordeal, having been sheltered by several Natives, but the incident probably jeopardized the Northwest Trading Company's ability to remain in the whaling business. Another factor may have been that the price of whale oil, which peaked during the Civil War, had been declining.[3]

Herring, however, were abundant in the area. Pioneer travel writer Eliza Scidmore, who traveled in Southeast Alaska in 1883 on the steamship *Idaho* and in 1884 on the sidewheeler *Ancon*, wrote that "from the end of August into January, the waters of Chatham Strait are black with herring,"[4] that "once in August the mail steamer passed through one school for four hours—the water silvered as far as could be seen, many whales and flocks of gulls attracted by this run of plenty."[5]

In 1882, as something of an experiment, the Northwest Trading Company began reducing herring into oil, likely utilizing the equipment that had been installed to render whale blubber. "Everybody welcomed the advent of the new industry," wrote John Cobb in 1906.[6]

The owners of the Northwest Trading Company were ambitious, and in 1883 the company constructed a salmon cannery at Pyramid Harbor, on the west shore of Chilkat Inlet, near present-day Haines. The cannery, among the first half-dozen constructed in Alaska, was a pioneer endeavor, and it involved considerable financial risk.

Meanwhile, at Killisnoo the results of making herring oil must have been encouraging because in 1884 the company installed machinery to *reduce* herring into oil and fish meal, the latter of which was primarily used as fertilizer. The machinery was the same as that employed in reducing menhaden (*Brevoortia tyrannus*) on the Atlantic coast—an industry that began in the early 1850s—and workmen from a menhaden factory there trained the workers at Killisnoo.

The Killisnoo operation was successful and marked the genesis of the herring reduction industry in Alaska, an industry that would become a large—though not entirely welcome—component of Alaska's fishing industry for more than eight decades.

The reduction process involved grinding the fish, cooking them, pressing the oil out, and drying the resulting material. Typically, reduction plants operated continuously, with semiautomatic feed, long

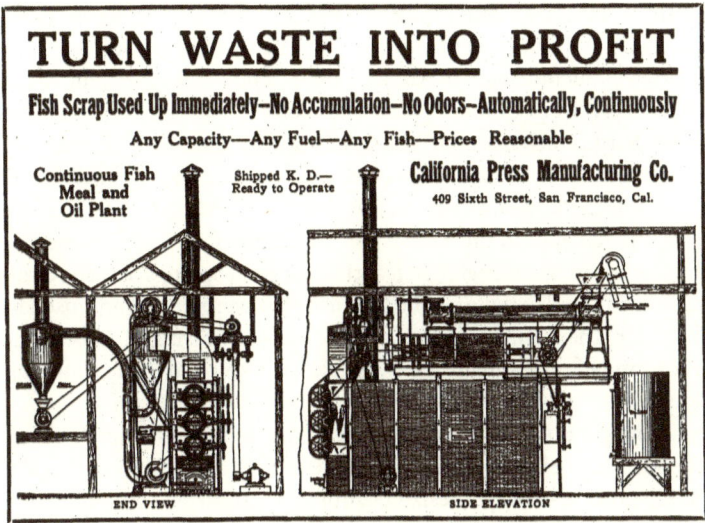

Figure 2.1. Advertisement, California Press Manufacturing Company. (*Pacific Fisherman*, June 1921)

tubular pressure cookers, rotary presses, and large rotary dryers. The larger units could process about four tons of raw fish per hour. Heavy fuel oil and, at least at one location (Killisnoo), coal were used to generate steam for cooking and heat for drying, as well as for the steam engines that powered some machinery. Where available, mechanical power was also provided by Pelton (water) wheels and, later, gasoline and diesel engines.[7]

Though it varied from season to season, 100 pounds of raw herring processed in a typical reduction plant in Alaska yielded about twelve pounds of fish meal and a little less than two gallons of oil.[8] Meal was shipped in 100-pound sacks, while oil was shipped initially in barrels and later in bulk in dedicated tanks aboard the steamships that served the reduction plants. By the 1960s, tug-and-barge operations took over service to the reduction plants. The barges thus employed were fitted with oil tanks.

At Pyramid Harbor, the Northwest Trading Company's salmon-canning venture did not fare well, resulting in the company's bankruptcy in 1888. In an effort to recoup their losses, some company stockholders in 1889 organized the Alaska Oil and Guano Company, which purchased the Northwest Trading Company's herring-reduction operation at Killisnoo.[9] The new company would have the herring oil and meal business in Alaska essentially to itself for more than three decades: until 1917, the Killisnoo plant was, with one small exception, the only herring-reduction plant in Alaska. (The exception was the small, crude Hume Fertilizer Company plant established in 1906 at Scow Bay, near Petersburg, which managed in its short operational life to produce five tons of fertilizer and 805 gallons of oil.) And, at least in 1905, the Killisnoo plant was the largest fish-reduction plant on the US Pacific Coast.[10]

The "Guano" in the Alaska Oil and Guano Company's name was a reference to the nitrogen-rich accumulated droppings of seabirds on islands in the dry regions of the world that were mined and used as fertilizer. The fertilizer (meal) produced from herring in Alaska was a substitute for real guano and was primarily shipped to the Hawaiian Islands and used on sugarcane plantations.

The guano market plummeted in the mid-1910s, after German chemists Fritz Haber and Carl Bosch developed a method—the Haber-Bosch process—for synthesizing ammonia, a nitrogen compound that is a component of synthetic fertilizer. The market for fertilizer produced from herring, however, remained robust at least into the mid-1920s. *Pacific Fisherman* wrote in 1925 that although herring meal was usually classified as edible, most of it was still used for fertilizer.[11]

Before long, however, the value of herring meal as a fertilizer began to diminish. *Pacific Fisherman* wrote in 1935 that only a "minute part" of the fish meal produced on the Pacific Coast was used as fertilizer, and the portion that made its way onto the fields was typically substandard product that had been damaged by dampness or deterioration.

Fish meal, especially that produced from whole fish, was far more valuable when used as an animal feed supplement. In 1927, Olaf Floe, who headed the Northwestern Herring Company, which operated a herring packing and reduction plant at Port Conclusion, in Southeast Alaska, and was president of the Pacific Herring Packers Association, wrote that Alaska herring meal was "an important factor in the development of the Washington and California poultry industry," noting that it was used as well in livestock feed.[12] Poultry farmers knew that a diet of high-protein fish meal greatly shortened the time necessary to grow marketable birds.[13]

The protein content of herring meal produced circa 1950 was 72 percent.[14] Moreover, when used as an animal-feed supplement, herring meal was indirectly food for humans. This helped blunt some of the criticism of the reduction plants' conversion of food fish into meal and oil.

Herring oil manufactured in conjunction with meal was generally considered to be the more valuable product of herring reduction.

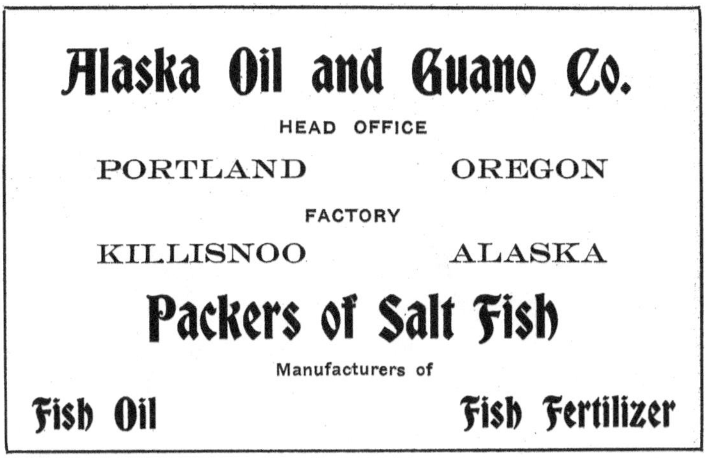

Alaska Oil and Guano Co.

HEAD OFFICE
PORTLAND   OREGON

FACTORY
KILLISNOO   ALASKA

Packers of Salt Fish

Manufacturers of
Fish Oil   Fish Fertilizer

Figure 2.2. Advertisement, Alaska Oil and Guano Company. (*Pacific Fisherman*, December 1909)

Alaska herring oil was primarily used in the manufacture of soap, but it was also used in paints, for tanning, and for many other purposes. Depending upon the application, herring oil competed with other fish oils as well as oil derived from oilseed crops (soybeans, cottonseed, sunflower seed, canola, rapeseed, hempseed, sesame seed, perilla seed, and peanuts).[15] As such, the price of herring oil could fluctuate quickly and significantly. The market price for tallow, a hard, fatty substance made from rendered animal fat, was the prime index for the "fats" market, the category that included herring oil.[16]

In Alaska, the greatest production of meal and oil was during the years 1929–1966, when more than 90 percent of all the herring caught were sent through reduction plants.[17]

The Killisnoo herring reduction operation impressed visitors. After visiting there in 1884, Eliza Scidmore praised the Killisnoo reduction plant and settlement as the "model industrial establishment on the coast." She said it was "well built and tidily kept" and the reduction plant itself was "a model of neatness and order." Skidmore added that "despite the odours," Killisnoo's gardens were worthy of a visit.[18]

In his 1894 report to the secretary of the interior, Alaska's governor James Sheakley reported the Alaska Oil and Guano Company as being "a very well-managed enterprise."[19] And Jefferson Moser, captain of the US Fish Commission's steamer *Albatross* during its 1898 and 1900 voyages to Alaska (both of which included visits to Killisnoo), described

Figure 2.3. Alaska Oil and Guano Company reduction plant. Photo by Vincent Soboleff, who was in Killisnoo from 1886 to 1920. (Vincent Soboleff Photograph Collection, Alaska State Library)

Figure 2.4. Barrels of herring oil at Killisnoo. (Vincent Soboleff Photograph Collection, Alaska State Library)

the herring operation there as "quite extensive, the buildings large, machinery excellent, storehouses roomy, wharves commodious."[20] Coal fueled the plant's boilers.[21]

CATCHING HERRING

From the beginning, fishermen at Killisnoo used purse seines to catch herring. A purse seine is an encircling type of net designed to catch fish that school near the water's surface, such as herring, salmon, and tuna. Aside from the materials used in their construction, purse seines used in the 1890s were similar to those in use today.

As used at Killisnoo, a purse seine was a long wall of tarred cotton webbing hung between a corkline and a leadline. Cork floats were strung along the corkline to float the top of the net, and the leadline was weighted so the net hung vertically. Along the entire length of the leadline were hung purse rings, often of brass, through which the heavy purse line (essentially, a drawstring) was run. This enabled closing—pursing—the seine and prevented the catch from escaping at the bottom. Once pursed, the seine was pulled aboard the boat until the catch was concentrated in the bunt, a section of the seine made of heavier webbing.

Herring fishermen at Killisnoo seined by what was known as the Norwegian method. E. J. Huizer, of the Alaska Department of Fisheries, described the process in a 1952 report:

> In practice, the Norwegian method used two large skiffs, each manned by about eight men, in which a seine 200 fathoms in length was equally divided. The two boats, spaced 20 to 30 feet apart, rowed about in search of a school of herring. When a school was located, the skiffs rowed in opposite directions around it, paying out the seine from each skiff until the body of fish was encircled by the net. Pursing was done by hand, and the fish were brailed directly into a large mother ship, which was standing by.[22]

Brailed is a term for using a large dipnet with an openable bottom (a brailer) to remove fish from a seine. Once the brailer was filled with fish, a steam-powered windlass and the rigging on the tender was used to lift it and maneuver it over the vessel's hold. There, a drawstring that held the bottom of the brailer closed (much like a pursed seine) would be "tripped" by yanking on a line leading to a small releasing mechanism, loosening the drawstring. This opened the bottom of the net to release the fish.

Though generally replaced in the early 1900s by seining from power vessels, as will be described below, the Norwegian method was used by the Killisnoo plant until about 1924 and by the Arentsen & Company saltery and reduction plant at Port Walter until 1927. Typically, the seines were fourteen fathoms (84 feet) deep and 175 fathoms (1050 feet) long, with 1.5-inch (stretched measure) mesh.[23]

At the time of Jefferson Moser's first visit to Killisnoo, in 1898, Alaska Oil and Guano employed three steam-powered vessels and several nonmotorized skiffs in its fishing operations.

The eighty-eight-foot *Dolphin*, at sixty tons* net register, was the largest and was used for cruising through Southeast Alaska's inside waters and catching herring wherever they were found. The vessel carried a crew of eighteen men and their fishing gear—primarily a seine and its associated skiffs. When herring were plentiful, the crew could brail from 800 to 1,000 barrels of fish (approximately 160,000 to 200,000 pounds) from the seine into the boat's hold in an hour. Once the *Dolphin* was loaded, it returned to Killisnoo to discharge the fish, take on fuel and other provisions, and then depart on another fishing cruise.

The other two steamers, the *Favorite* and the *Louise*, at forty-two net tons and five net tons, respectively, fished in the vicinity of the reduction plant.

The three steamers' crews totaled ten Native and twenty white men. Of their wages, Moser wrote: "The white fishermen are paid $50 a month, and board; the Native fishermen get $1.50 a day, or practically $45 a month, for they are paid even if detained, and while board is not stipulated, they practically get it."[24]

At its herring reduction facility, which typically operated from July 1 through December 31, Alaska Oil and Guano in 1898 employed thirty-three Natives, seventeen whites, five Japanese, and three Chinese. Regarding their wages, Moser wrote:

> Laborers generally have $1 a day, but about 10 of them, who have been employed a long time and are faithful, having the more difficult work to do, receive $1.50 per day. Boys are paid 50 cents a day. A good Native fisherman or laborer makes about $200 a season. All wages are paid in cash.[25]

The herring catch processed at Killisnoo during the 1897 season, which was considered poor because of stormy weather, totaled 35,000 barrels, or approximately seven million pounds. From this, Alaska Oil and Guano produced 125,000 gallons of oil, valued at twenty-five cents per gallon, and 780 tons of fertilizer, valued at $27 per ton.[26]

---

* Tonnage is a volumetric measurement based on the amount of coal that could be carried in a vessel's protected cargo space.

Figure 2.5. Steamship *Favorite* coming to the seine. (Vincent Soboleff Photograph Collection, Alaska State Library)

USES OF HERRING EXPAND, ENGENDERING CONTROVERSY

But changes that challenged the use of herring to manufacture oil and fertilizer were afoot. The first was the diversification of the food products manufactured from herring. Perhaps most prominent was the sardine cannery of the Juneau Canning Company, at Juneau, which was established in 1904. Sardines are small herring or herring-like fish that are packed tightly together in a can, usually after part of their natural oil and moisture has been removed. They are often packed in olive oil or a sauce.

During its first year of operation, the company canned 3,173 cases of "mustard sardines"—herring that had their heads, tails, and viscera removed and were then canned in mustard sauce—but the product was unable to compete with those produced on the Atlantic Coast. The Juneau Canning Company also smoked and salted herring. The fish were caught at Taku Harbor, just south of Juneau.

Importantly, the company's manager, Ashton Thomas, who had learned the herring trade in Maine and Nova Scotia, began advocating for a law prohibiting the use of herring and other food fish for the manufacture of fertilizer. By late 1906, Thomas was heading a movement to do so.[27] In his 1906 report on Alaska's fisheries, John Cobb wrote that citizens of Southeast Alaska had sent to Congress "a heavily signed protest, asking for the enactment of a prohibitory law."[28]

About the same time Thomas began canning herring, the US West Coast commercial salmon troll fishery was born at Ketchikan. There, during the winter of 1904–1905, large schools of king salmon were observed feeding on herring. The fish were mostly located in waters where nets could not be easily employed, and fishermen, both white and Native, began catching them by trolling from canoes. Total production that season was 272,000 pounds of king salmon, most of which were packed in ice and shipped to Seattle.[29] The troll fishery expanded quickly and within a few years became one of the most important and profitable fisheries in Southeast Alaska. Troll fishermen viewed the reduction of herring into oil and fertilizer as a threat to the abundance of the king salmon upon which their incomes depended. Millard Marsh and John Cobb, both agents of the Bureau of Fisheries, agreed, writing in 1910 that "there is little question that the serious depletion of the herring schools would correspondingly impair the abundance of king salmon."[30]

Halibut fishermen, for whom herring was a favored bait, likewise opposed herring reduction. In their 1906 report on the fishes of Alaska, Bureau of Fisheries agents Barton Evermann and E. L. Goldsborough noted that herring "has come to be in great demand" as bait for the halibut fisheries.[31] Halibut fishermen had what was later described by an Alaska district fisheries official as "a deep-seated distrust" of the herring reduction industry, and they allied with the troll fishermen.[32]

The editors of *Pacific Fisherman* didn't avoid the issue, noting in 1906 that there was an ecological risk in herring reduction that had not yet been taken into account: "whether the taking of such vast quantities of herring as is required for the extensive manufacture of guano and fertilizer, will not so deplete the supply as to interfere with the continued habitat of other fishes."[33]

At least in the opinion of some, this attitude was manifested in new federal legislation. The 1906 federal Act for the Protection and Regulation of the Fisheries of Alaska contained a provision making it "unlawful for any person, company, or corporation wantonly to waste or destroy salmon or other food fishes taken or caught in any of the waters of Alaska."[34] Many considered the reduction of herring to be wanton waste and argued that it was illegal under the 1906 legislation.

In 1908, *Pacific Fisherman* agreed, writing that there seemed to be sufficient authority in the 1906 legislation "to put a stop to the use of herring, salmon and other food fish for fertilizer purposes." The journal added that if the authority didn't exist, the existing law should be amended to add it. *Pacific Fisherman* reasoned that "it will be but a short time until the demand for all food fishes of this Coast will greatly increase, and every effort should be made to prevent their wanton destruction by the fertilizer companies."³⁵

By 1909, the Bureau of Fisheries' Cobb and Marsh were of the opinion that, because of the great need of herring for food and bait, "the time has arrived when the use of food fishes in the preparation of fertilizer should be prohibited." As a matter of fairness to the Alaska Oil and Guano Company, with its large investment in its Killisnoo plant, Cobb and Marsh thought the company should be allowed at least one full season to adjust its operation, perhaps by reducing offal from nearby salmon canneries.³⁶

There was also political opposition to the manufacture of fertilizer from herring. In 1906, Congress passed the so-called Delegate Act, which authorized a nonvoting Alaska delegate to Congress.³⁷ James Wickersham was elected to that position in 1908. By early 1911, Wickersham was advocating for the passage of a law prohibiting the manufacture of fertilizer from herring in Alaska.³⁸ He found a like-minded colleague in Senator Wesley Jones (R-Washington). In March 1912, Jones introduced legislation that after January 1, 1914, would categorize the manufacture of fertilizer or fish oil from food fishes in Alaska, except offal from fish-processing plants, as wanton waste and make it illegal.³⁹

Jones's proposed legislation was discussed at an April 1912 Senate fisheries subcommittee hearing. In written testimony, Charles Nagel, secretary of commerce and labor, stated that his department—within which the Bureau of Fisheries was situated—was "not in possession of statistics which fully prove serious depletion of the supply of herring." Nagel acknowledged the prevailing anti-reduction sentiment and suggested that "reasonable concession" should be given to the investors in the Alaska Oil and Guano Company, but he was not specific.⁴⁰

The Alaska Oil and Guano Company's main defense of herring reduction had been that the fish it reduced were not suitable for use as food.⁴¹ In a December 1910 letter provided to the subcommittee, Carl Spuhn, president of the Alaska Oil and Guano Company, added that "no scientific data have been yet obtained which justify the assumption that the use of the herring for the purpose of the manufacture of oil and guano is either an injury or detriment to other fishing industries in Alaska."

Spuhn also offered a disingenuous defense of herring reduction:

> The fact is, however, and the experience of all fishermen in Alaskan waters verifies this, that, either from natural causes or from the character of the food which the herring feeds upon, millions of the fish die each year, and it is no uncommon sight

in sailing through Alaskan waters to sail through millions of floating dead herring. Any industry, therefore, which may make possible the catch of the fish before death ensues and the use of its product, should, it is submitted, be encouraged instead of suppressed, because this means a utilization of what would otherwise be so much waste product.[42]

Spuhn's claim of massive herring die-offs commonly occurring was false. Probably calculating that herring reduction was going to be made illegal, Alaska Oil and Guano suggested the company be given ten years to close out its business.

Alaska governor Walter Clark, who had visited Killisnoo on two or three occasions, attended the hearing. Clark recognized that public sentiment was against herring reduction, but he thought the Alaska Oil and Guano Company should be given more than two years, but not more than eight years, to bring its operations to a close.[43]

While the discussion likely had value, federal legislation became a moot point because Senator Jones's bill died in committee. Anti-reduction sentiment, however, remained very much alive, including in the Bureau of Fisheries. In his 1914 report on Alaska's fisheries, federal fisheries agent E. Lester Jones wrote, "While this [Alaska Oil and Guano Company] factory is now the single and isolated case, it seems to me that it should not be allowed to continue operations but should be permanently closed by the Government."[44]

Opposition to herring reduction would continue, but it would manifest itself primarily in the form of territorial taxes on oil, fertilizer, and meal manufactured from herring. The aforementioned Act for the Protection and Regulation of the Fisheries of Alaska (1906) contained a provision requiring fish-processing companies in Alaska to pay license taxes on their production. The tax on fish oil was ten cents per barrel (fifty gallons), while the tax on fertilizer was twenty cents per ton.[45] At the time of its enactment, this tax applied to only one company, the Alaska Oil and Guano Company.

The situation changed in August 1912, when Congress created the Territory of Alaska. The legislation that established the territory specified that the federal government retained control of the fisheries. The territory could, however, levy taxes.[46] In May 1913, Alaska's Legislature levied the first tax on the territory's fisheries; it was a case tax that applied only to canned salmon.[47] Two years later, in April 1915, the legislature expanded the tax to include all salted or mild-cured fish except herring.[48]

In April 1917, Alaska's Legislature used its taxation authority to bolster its desire to discourage nonfood use of the herring resource by adding herring reduction plants to the list of businesses that would be taxed. And the tax was substantial: for "Fish-Oil-Works," companies using herring in whole or in part in the manufacture of fish oil, the tax was two dollars per barrel. For "Meal Plants," companies manufacturing fertilizer or fish meal in whole or in part from herring, the tax was two dollars per ton.[49] These were in addition to the federal taxes already in effect. At the time of its enactment, this tax applied to only one company, the Alaska Fish Salting and By-Products Company, formerly the Alaska Oil and Guano Company.

In 1918, the Alaska Fish Salting and By-Products Company produced 138,012 gallons of herring oil, valued at about $35 per barrel. It also produced 645 tons of fertilizer, valued at about $73.25 per ton.[50] Based on these values—which could over time fluctuate significantly—the territorial tax on the value of oil that year was about 5.7 percent, and the tax on fertilizer was about 2.7 percent.

The following year (1919), the Alaska Fish Salting and By-Products Company filed suit in the Alaska District federal court, claiming the 1917 territorial license tax was unconstitutional, invalid on several counts, and designed to destroy its business. The company hoped to have the tax voided and to recover the taxes it had paid in 1917 and 1918. The court, however, upheld the tax. Alaska Fish Salting and By-Products Company appealed the decision, and in January 1921, the US Supreme Court affirmed the lower court's decision, adding that those who enter business take the risk of taxes being levied or increased.[51]

That same year, Alaska's territorial legislature replaced the existing herring oil/meal/fertilizer taxes with dramatically lower taxes of forty cents per barrel on all fish oil and forty cents per ton on all fish meal.[52] The impetus for the Alaska Legislature's moderation of its fish oil and fish meal taxes was greater acceptance of reduction plants, due to the growth of the cured/salted herring industry beginning in 1917. There was no demand for small herring in the Scotch-cure market (discussed at length in chapter 3), yet herring were mostly caught in purse seines, which didn't discriminate regarding fish size. With the large, desirable herring came quantities of herring that could not be profitably cured—bycatch, in the parlance of modern fishery managers. Turning these fish, as well as the offal and waste from herring salting/curing operations, into oil and meal/fertilizer offered one way to utilize—and profit from—fish that had hitherto been discarded.

In 1918, Wilson Fisheries Company installed a reduction plant at its cannery and saltery at Port Walter, on Baranof Island, in Southeast Alaska, joining the Alaska Fish Salting and By-Products Company in the herring-reduction business.[53] Other firms quickly followed suit, and in

1920 nine companies operated reduction plants in the territory. Seven were in Southeast Alaska, and two were in Prince William Sound.[54] This was the beginning of the integration of the herring-curing and herring-reducing components of Alaska's herring industry.

For its part, the Bureau of Fisheries in 1923 recognized that the herring curing industry, being new to Alaska, faced serious competitive challenges due to its distance from the principal East Coast markets and from European competition. "For these reasons," according to Henry O'Malley, the bureau's commissioner, "the Bureau has been reluctant to impose any obstacles in the way of development of a young industry which is recognized to have great possibilities of development."[55] As *Pacific Fisherman* wrote in 1935:

> In Southeast Alaska and Prince William Sound, the herring curing industry would not exist if it were not for the fish oil and meal manufacturing plants which are operated in conjunction with the salteries. Of the herring taken in these districts, only the largest and fattest fish can be cured profitably. The oil and meal plants utilize the unsuitable fish and they make possible the conduct of a rounded operation without waste of herring.[56]

In a second article that year, *Pacific Fisherman* noted that herring in the Kodiak and Aleutian Islands districts were larger and more uniform in size, which made it possible for herring operations to be conducted on the basis of curing alone.[57] Despite this assertion, three herring-reduction plants were established in the Kodiak Island area in 1935.

Foreign competition, however, was taking its toll on Alaska's herring-curing industry. In 1935, some 60,000 barrels of herring (fifteen million pounds) were Scotch cured in Alaska. Two years later, production fell to 8,500 barrels.[58] It was the beginning of the demise of the herring-curing industry in the territory.

In 1939, members of the US House of Representatives' Special Subcommittee on Alaskan Fisheries journeyed to Alaska to survey the territory's fisheries and to hold public hearings. While in Alaska, they were accompanied by members of the territorial legislature's Joint Committee on Fisheries.[59] Coincidentally, their visit coincided with the first-ever depletion-driven closure of the main herring fishing grounds in Southeast Alaska. And they heard a lot about herring, particularly opposition to reduction plants. Salmon trollers were particularly vociferous.

In June 1940, the subcommittee issued its recommendations. Among them was that "the taking of herring for conversion into oil, meal or fertilizer should be entirely prohibited."[60] The committee's recommendation was not acted upon. World War II had begun, and Congress had more pressing business than Alaska's fisheries.

MODERN PURSE SEINING

By about 1910, herring fishermen at Petersburg and Ketchikan began switching from oar-powered skiffs to power boats to purse seine. The boats were about thirty-two feet long and powered by gasoline engines of about ten horsepower. By 1918, the majority of the herring operators in Southeast Alaska, and all of those in the recently developed herring fishery in Prince William Sound, were using power seine boats.[61] That year, power seine boats in Southeast Alaska had an average net tonnage of seventeen, ranging from eleven to thirty-one tons. They were all powered by gasoline engines and carried a crew of five to seven men. Even larger boats—in the fifty-foot range—were employed by Puget Sound–based seiners who in about 1917 began journeying to Alaska to participate in the expanding herring fishery.

The year 1922 marked the beginning of a tremendous expansion of the summer herring fishery, mostly to supply the high volumes of herring needed to operate the increasing number of reduction plants in the territory. This required an increase in seiners' carrying capacity, and the size of the boats in the fishery grew. A disadvantage of the larger boats was that the crushing weight of big loads in their holds rendered quantities of otherwise suitable herring unfit for curing.

By the late 1920s, seiners in the seventy-to-seventy-five-foot range were common. Crew size had increased slightly, to six to eight men. By 1929, about half of the seiners were powered by diesel engines, which were safer, more reliable, and more fuel efficient than their gasoline counterparts.

The largest herring seiner used in Alaska in 1927 was the seventy-seven-foot-long *Valencia*, which fished in the Chatham Strait fishery. The vessel had a beam of nineteen feet and was powered by a 140-horsepower diesel engine. On one trip that season, the *Valencia* caught 989 barrels (197,800 pounds) of herring, of which nearly 900 barrels (180,000 pounds) was carried in the hold. The balance was carried on deck. After the herring season, the *Valencia* went south to fish for tuna. Other large herring seiners during the off-season fished for sardines along the California coast.

At the peak of the 1930 herring season in Southeast Alaska, when thirteen reduction plants were in operation, sixty-nine seiners fished for herring. Fully sixty-five fished throughout the season.

The "highliner"—the vessel with the largest catch—that year was the *Mary M*, which fished for the Buchan & Heinen Packing Company plant at Port Armstrong, on Baranof Island. The vessel caught 14,450 barrels (2.89 million pounds) of herring. The next-highest catch, around 14,000 barrels (2.8 million pounds), was made by the *Edgar C*, which fished for the Storfold & Grondahl Packing Company plant at Washington Bay, on Kuiu Island. In 1932, the *Edgar C* was the all-Alaska highliner, landing 24,800 barrels (4.96 million pounds) of herring.[62] As of 1950, the record single-haul herring catch was 1,260 tons (2.52 million pounds), by the Canadian purse seiner *Maple Leaf C*. The haul was made near Prince Rupert, and three other boats immediately came to assist with the huge catch. It took twelve hours to brail the fish into the nine packers that transported them to processing plants.[63]

In the spring of 1934, the Chatham Strait Fish Company, which operated a saltery and reduction plant in Prince William Sound, took delivery of the first modern steel seine boat ever built in the United States. The *CSF* was seventy feet long and could carry 900 barrels (225,000 pounds) of herring in its hold. Though its normal crew comprised eight men, the boat had twenty bunks, the extra dozen to accommodate plant workers transported to and from Alaska, which saved on steamship fares.[64]

The purse seines employed by herring fishermen at that time ranged from about 175 to 250 fathoms in length and from about 12 to 30 fathoms in depth. The mesh size was typically 1.5 inches, stretched measure.[65]

In his 1929 report on Alaska's herring fishery, Bureau of Fisheries biologist George Rounsefell described the process of locating herring:

> Most of the purse seining is done at night, but occasional good hauls are made in daylight, especially in the Kodiak-Afognak district. The seine boats arrive at the fishing grounds about dusk and cruise slowly about with a man always on watch. He discovers the presence of a school of herring either by seeing them "flipping" at the surface, or, if it is too dark to see, by hearing the gentle splashing. The herring "flip" best at dusk and just before dawn. Sometimes when the herring are not "flipping" the fishermen resort to "leading." A man rows slowly about in a small skiff, dragging a very fine line, to the end of which is attached a heavy piece of lead. This holds the line taut and perpendicular, so that one can tell when the line is passing through a school of herring by feeling the line jerk as the herring strike against it.[66]

Adding to Rounsefell's description, in his 1965 report on Southeast Alaska's herring fishery, Stephen Rogers, of the US Bureau of Commercial Fisheries, wrote:

> Early in the fishing season, most fishing is done during the day, while in the latter portion of the season (August and September) more night fishing is carried on. A set is made when a sufficiently large school of fish is spotted near the surface. At night, the fish are spotted by the luminescence caused when they disturb the surface of the water.[67]

The act of purse seining for salmon and herring is similar, but because a herring seine was typically deeper, longer, and heavier, the purse line was wire cable instead of the rope used on a salmon seine. A unique characteristic of a purse seiner at this time was its turntable, which occupied the entire after quarter of the boat. Its purpose was to make setting (deploying) and hauling (retrieving) the seine quicker and easier, and it was fitted on one side with a roller.

With the turntable's roller facing aft, the seine was piled on the turntable, with the corkline on one side and the leadline on the other. The vessel's seine skiff, itself a stout vessel, was carried atop the seine for travel to and from the fishing grounds. Once on the grounds, the skiff was usually put in the water and tied close behind the seiner. A line running from one end of the seine's corkline was fastened to the stern of the seine skiff. The ends of the seine's purse line, which was considerably longer than the leadline, were tied or clipped into the corkline at their respective ends.

When a school of herring was located, the captain hollered, "Let 'er go!" and the skiff, with one or two fishermen (skiff men) aboard, was let loose. While the seine boat powered ahead in a great arc, encircling the fish, the skiff men rowed in the opposite direction, basically anchoring one end of the seine while it paid out over the roller on the turntable into the water behind the seine boat. Once the seine was entirely paid out, the seine boat made its way to the skiff to complete the encirclement of the fish.

The two ends of the purse line were then recovered, and the purse winch on the seine boat's deck was set in motion. At the end of each of the two horizontal shafts on the winch was a cylinder-like fitting about eight inches in diameter known as a gypsy (a horizontal capstan). Each end of the purse line was hauled by winding a few turns of line around one of the rotating gypsies and then tightening the free end. The increased friction tightened the line around the gypsy, and the

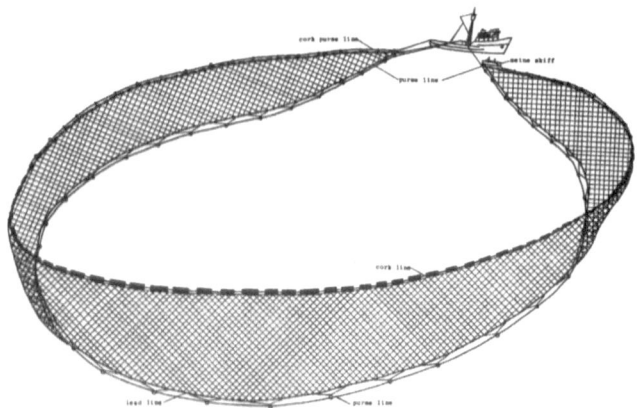

Figure 2.6. Purse seine. (*Commercial Fisheries Review*, November 1953)

purse line was hauled aboard, closing the bottom of the seine. Both ends of the purse line were hauled simultaneously.

Once the seine was pursed, the fishermen, beginning at one end, pulled the seine aboard over the turntable's roller, piling it on the turntable, which had been rotated so the roller faced outboard on the side of the vessel the seine was on. As more and more of the seine came aboard, the herring were ultimately concentrated in the seine's bunt (known as the money bag), from which the fishermen brailed them into the vessel's hold. Once the seine was emptied of fish, the fishermen piled the rest of it on the turntable, which they then rotated so its roller faced aft. This completed, the seiner was ready for another set.[68] However, when a herring seiner's hold was full and sea conditions allowed, herring were also carried on deck. Being deckloaded, of course, precluded making additional sets.

In the years immediately following World War II, echo sounders (depth finders, or fathometers) became common on herring seiners. In addition to showing the depth of the water, the machines were used to locate deep-running schools of herring.[69] In his 1946 report on Alaska's fisheries, Ward Bower, the US Fish and Wildlife Service's Alaska agent, noted that at Kodiak, "all of the herring seiners were equipped with depth finders which were used in locating the herring, thereby greatly increasing the catches."[70] Over the years, echo sounders became increasingly sensitive and sophisticated.

Figure 2.7. Herring seiner, probably in Kodiak Island waters, with seine piled on turntable. The boat is low in the water, so its hold is likely filled with herring. (Lowell Wakefield collection, Kodiak Historical Society)

In the early 1950s, nylon began replacing cotton and linen in fishing nets. The synthetic material was stronger and required far less maintenance than cotton and linen, which needed to be dried after use and were often treated with preservatives. Similarly, plastic floats began replacing cork floats.

In the mid-1950s, seiners began using power blocks to haul their seines. A power block is a large hydraulically driven rubber-covered V-section sheave about 28 to 36 inches in diameter. When the seine was ready to be hauled, the block was opened, and an end of the net was placed on the sheave. The block was then closed, hauled to the top of the boom, and power was applied. The V-shape of the sheave, aided by molded cleats, gripped the net and pulled it out of the water and over the sheave. From there it dropped down to the aft section of the deck, where crewmen piled it so it was ready for the next set.[71] One advantage of using power blocks is they reduced the size of the crews carried on seine boats. They also eliminated the need for turntables. Almost immediately, power blocks became standard equipment on all seine boats, from skiffs to tuna seiners. The power blocks in service today hew closely to the original design.

Purse seine vessels used to seine herring in Alaska today are salmon seiners that serve double duty in the herring fishery. State regulations limit salmon seiners to fifty-eight feet in length.

Figure 2.8. Purse seiner *Tidings* with deckload of herring. The vessel's hold is likely full, too. (Lowell Wakefield collection, Kodiak Historical Society)

WHALES VERSUS HERRING

In November 1906, *Pacific Fisherman* published an article titled "Whale Slaughter Harms Herring Industry." A whaling station had been established on Vancouver Island several years before, and herring were no longer as plentiful in Puget Sound as they had been. According to fishermen, the herring entered shallow coastal waters to escape the whales, which preyed upon them. As a result of there being fewer whales, the herring now stayed in the open ocean, where they were not accessible to fishermen.

There were plans to establish a whaling station in Southeast Alaska, and *Pacific Fisherman* thought the government should prevent this from happening.[72] It was all for naught, because the following year (1907), the Tyee Company established a station at Murder Cove, on the southern end of Admiralty Island.[73]

Alaskans apparently shared the belief that whales drove herring into shallow waters. But in his 1913 report on Alaska's fisheries, federal fisheries agent Barton Warren Evermann countered that belief, writing that "it seems more reasonable to suppose that the whales follow the herring than to believe that they drive them."[74]

Alaska's Territorial Legislature was not convinced. In the spring of 1919, the legislature introduced a memorial asking Congress to prohibit the killing of whales in Alaska waters, stating that "it is the general opinion of people who, for years, have closely observed conditions on the coast of Alaska that the disappearance of herring from shores is wholly, or partly, due to the extermination of whales now going on in Alaska." *Pacific Fisherman*, however, pointed out that herring abundance is very erratic, and that the catch of herring in 1918 was about four times that of 1917. The memorial was postponed indefinitely.[75]

Nevertheless, the opinion that the abundance of herring along the coast depended upon there being enough whales to drive the fish into shallow waters prevailed. But, as Bureau of Fisheries herring expert George Rounsefell pointed out in 1928, "the herring in the inside waters do not come in from the outer coast. They are permanent residents in their own particular waters, and any number of whales cannot add to their abundance."[76] The issue seemed to have been put to rest.

In fact, because humpback whales (*Megaptera novaeangliae*), which range throughout Southeast Alaska, feed extensively on herring, an abundance of whales likely diminished the availability of herring to fishermen. Conversely, industrial whaling in Southeast Alaska (1907–1922) substantially reduced the whale population and its predation on herring.

Case in point: Roy Chapman Andrews, of the American Museum of Natural History, traveled twice to Alaska to document whaling operations, spending time aboard a Tyee Company whaler. Andrews later wrote that when the Tyee Company began operations in 1907, "finback and humpback whales were there in the hundreds, [but] they were soon all killed," and the whalers shifted their hunt to the open ocean, particularly the waters around Cape Ommaney.[77]

In 1911, the Tyee Company abandoned its Murder Cove site and reorganized as the United States Whaling Company.[78] The new company built a whaling station at Port Armstrong, near the southern tip of Baranof Island—and closer to Cape Ommaney. The station reportedly processed 181 humpback whales during its eleven years of operation (1912–1922).[79]

This reduction in the whale population in the region likely contributed to a temporary boom in herring abundance that enabled the rapid growth of the region's herring-reduction industry. An indicator of this may have been that, upon ceasing operations, the Port Armstrong whaling station was leased by the Buchan & Heinen Packing Company, which converted it into a herring-reduction plant.[80] The volume of herring subsequently removed by overfishing, however, greatly exceeded the volume that had been removed by whales, and herring stocks declined precipitously.[81]

Due to the nearly two decades of legal whaling in the region and to unregulated offshore whaling, humpback whales became nearly extinct in Southeast Alaska and were listed under the Endangered Species Act. Humpbacks, however, have recovered substantially, and the distinct population segment of humpback whales that frequents Alaska waters is no longer listed as endangered.[82] Regarding humpbacks and herring, the Alaska Department of Fish and Game states that "the recovery of populations of predator species, such as humpback whales, may impact herring populations."[83]

At Sitka, the arrival of humpback whales (and sea lions) in Sitka Sound each March is a welcome sight because their presence usually signals that schools of spawning herring are entering the sound and that the opening of the roe-herring fishery is imminent.

And in Prince William Sound, researchers during the years 2006–2014 mapped the movements of humpback whales as indicators of herring movements. In twenty-two five-to-seven-day seasonal surveys, the researchers mapped whale locations, collected prey items near feeding whales, and used hydroacoustics to estimate prey density. They determined that seasonal movements of whales into the sound were driven largely by the movements of adult herring.[84]

3
SALTED HERRING
The Early Years

> Salted and smoked herring are staple articles of diet in Europe. For some reason, the herring has never attained a corresponding popularity in the United States and Canada. Perhaps this may be accounted for by the enormous variety of large food fishes with which American waters abound. Or it may be that the standard of living of the average person in America is higher than in Europe and, therefore, fresh or canned fish is demanded.—Donald K. Tressler et al., 1923[1]

Salting, either with dry salt or in brine, was, with drying, the only widely available method of preserving fish until the nineteenth century, when canning was developed. Salted fish of various kinds remain popular in parts of the world. Salted herring is a generic term for herring packed in salt or brine (pickle). Dry-salted herring is simply herring that is packed in salt, usually in wooden barrels. This product is usually for human consumption but may also be for bait, in Alaska typically for halibut bait. Herring that is preserved in brine is referred to as being cured and is almost always packed in special wooden barrels. Properly curing herring is an exacting, labor-intensive process.

There were few barriers to entering the herring-salting trade in Alaska. Once a supply of herring had been located, it was simply a matter of obtaining barrels, salt, and a place to work.

By 1897, the Alaska Oil and Guano Company, the herring-reduction operation at Killisnoo, in Southeast Alaska (see chapter 2), had expanded to include salted herring, producing 950 half-barrels of the product that year. (By the measure of the time, a half-barrel contained 100 pounds of fish.) The barrels were made on site from local Sitka spruce, as was the common practice of salmon salteries in the region.[2] The company continued to salt modest quantities of herring, and by 1900 its product, especially rich in oil, had attracted the attention of the eastern market, which historically had obtained salted herring from Europe. An eastern interest that year placed a large order for the salted herring, but Alaska Oil and Guano was for some reason unable to fill it.[3] East Coast buyers, as will be discussed later, tended to have specific requirements for the size and quality of herring packed, the method used in packing them, and even the barrels the herring were packed in.

Although Alaska Oil and Guano and several other companies had salted small quantities of herring prior to 1900, the herring-salting industry in Alaska may be said to have commenced about that time at Petersburg. It was a rocky start. In the fall of 1899, the Icy Strait Packing Company, led by Peter Buschmann, a Norwegian immigrant experienced in fisheries, was constructing a salmon cannery at Petersburg. While doing so, the company salted 1,500 barrels (300,000 pounds) of herring that were mostly caught in Wrangell Narrows.[4] The fish, however, found little interest among buyers, and the operation was deemed a financial failure.

During the next decade, individual fishermen salted herring, but the activity failed to attract any considerable amount of capital. The fishermen salted the fish on scows that they towed from place to place, going where the herring were.[5] When not engaged in fishing operations, the scows were likely moored at Scow Bay, on Wrangell Narrows just south of Petersburg, which had the advantage of being along the steamship route between Juneau and Seattle, and thus provided the fishermen with a means of getting their product to market.

In 1903, Alaska governor John Brady characterized the new entrants into the industry as Swedes and Norwegians who were salting herring in a manner that suited the demands of Scandinavian immigrants in the northwestern states.[6] Alaska Department of Fisheries biologist E. J. Huizer described herring produced by the Norwegian method (Norwegian cure) as "poorly gutted, carelessly graded and packed, and heavily salted, all of which resulted in an unattractive pack."[7] The Scotch cure, which will be discussed below, involved a far more exacting process.

Simultaneous with the early effort at Petersburg, the Yakutat & Southern Railway Company constructed a herring saltery at Yakutat, a Native community on the Gulf of Alaska, east of Prince William Sound. Herring at Yakutat were known to be especially large, but unfortunately, they failed to appear for the next four years, and the saltery packed salmon instead.[8]

But the industry had potential. In 1905, John Cobb, of the US Bureau of Fisheries, noted, "There is room for a very great development of the herring industry. For many years, salmon absorbed all the attention and capital, but since the slump in profits in the latter business during the last four years, more attention has been directed to herring."[9]

The salted-herring industry in Alaska nevertheless remained relatively stable until 1910, when British Columbia experienced a partial failure of its herring run, which happened again the following year. These failures presented an opportunity for salting herring in Alaska, and the industry quickly expanded. Notable was the construction in 1910 of a large saltery at Ketchikan. Overall, production of salted herring for human consumption increased from a mere 45,600 pounds in 1910 to 2 million pounds in 1911. Most of the product was dry salted—brined, then packed in an excess of salt in wooden boxes—and sold in eastern Asia, which had previously been supplied by British Columbia salteries.

More Alaskans entered the field in 1912, and their total production was more than 13.7 million pounds. Unfortunately for the Alaska operators, herring returned to British Columbia in great numbers that year, and the demand for Alaska herring cooled. In 1913, production in Alaska—which for the first time included herring salted in Prince William Sound—declined to 8.7 million pounds. In 1914, the market for dry-salted herring fell precipitously, and most of the operators went out of business. After 1914, there was only one year, 1918, when the quantity of dry-salted herring exceeded a million pounds. The market for dry-salted herring remained modest in later years, and 1941 was the last year herring were commercially dry-salted as a food item.[10]

"FEEDY" HERRING: THE IMPOUNDMENT SOLUTION

During July and August, herring feed heavily in order to grow and put on fat reserves that enable them to survive the leaner months. A large portion of the herring's summer diet is small red copepods, as many as 3,000 of which have been found in a single herring's stomach. This "red feed," as it is called, is fat and oily and is the main source of the herring's abundant oil.

But feedy herring—fish whose stomachs were full of feed—presented a problem: their flesh, especially the belly flesh, decomposes rapidly when the fish are killed. This renders the fish unsuitable for curing or even for bait. In September and October, however, the herring apparently change their diet. The red substance is not noticeable in their stomach, and the fish are suitable for curing. Thus, much of the herring curing during the early years occurred in the fall.[11]

In the summer of 1912, Ashton Thomas, who at that time had a bait-herring operation at Ketchikan, found a solution to the red-feed problem: Thomas impounded purse seine-caught herring in an enclosure made of webbing (netting). After three or four days, the herring had digested and eliminated the red feed, and the fish was suitable for both food and bait. The impounding process facilitated the curing of summer herring, which, because they tended to be fat, made a superior pack.[12]

Impounding, which Thomas called a "poor-man's cold storage," also provided operators with an inventory of herring that could be held for long periods of time and used as production needs and efficiency demanded.[13] At Unalaska in 1928, herring were held in pounds (enclosures) for nearly a month after fishing ended.[14] According to Thomas, herring properly handled in purse seines could be moved as far as five miles. Of impounding, he said that "in the development of the industry, there has been no greater step made."[15]

Herring pounds were made of cotton webbing that was heavier than the webbing used in seines. Because these fixed pounds had to stay in the water for a long time, the webbing was heavily tarred. Strips of webbing about eighty fathoms (480 feet) long and fitted with floats on top and weights on the bottom were formed into a square and secured in a protected location near the fishing grounds that, because the pounds had no bottoms, had to be shallower at high tide than the webbing was deep. At Kodiak in 1932 and 1933, herring impounded for bait escaped when strong tides lifted the pound.

Of the fishing operations, once a seine boat had a load of herring in its seine, it blew its whistle to summon a towboat to tow it to the pound. The seine boat was fitted with a bridle fastened to its bow and stern, and the towboat towed the seine boat sideways to the pound, the seine with its herring dragging behind.

Upon reaching the pound, the top edge of the seine was attached to the top edge of a side of the pound, and the two corklines held below the water. Crewmen then pulled the seine into the boat, spilling the herring into the pound.[16]

In 1927, Lee Wakefield, of the Wakefield Fisheries Company, which had salteries and reduction plants in Southeast Alaska and Prince William Sound, experimented with floating herring pounds. Wakefield's pounds were relatively small, about thirty feet square and twenty feet deep, but had a couple of big advantages: they could be carried on boats and put out when a catch had been made, and, since the floating pounds had bottoms, fish that died could be salvaged and utilized in a reduction plant. Wakefield told *Pacific Fisherman* in 1928 that he would no longer use fixed pounds.[17]

In his 1929 report on Alaska's herring fishery, Bureau of Fisheries biologist George Rounsefell pointed out the conservation concerns over impounding:

> When the wind and tide are unfavorable, or when the haul is made too far from the pound, there is great danger of the herring being smothered by being forced into dense masses

during the towing. In some cases, the pounds have been placed in water too shallow and the receding tide has left the herring stranded. Occasionally storms drag a pound ashore, smothering the herring. Even with the best of care, a small percentage of the impounded herring will soon die from infection where the scales have been rubbed off against the web.[18]

Herring pounds nevertheless remained important tools for keeping a supply of live herring, and they were utilized in Southeast Alaska until at least the early 2000s for holding bait herring. Moreover, the herring spawn-on-kelp fishery that still occurs in Southeast Alaska impounds—and later releases—herring.

Returning to the development of the salted-herring industry, in his 1913 Alaska fisheries report, Bureau of Fisheries Alaska agent Barton Warren Evermann recognized a major shortcoming of the herring-curing industry:

> Too often there has been much carelessness in handling herring. Strictly fresh fish of uniform size, properly eviscerated and thoroughly cured are absolutely essential to a wholesome preserved product. Stale fish, irregular sizes, indifferent curing and packing, and short weights have been the causes for the poor demand hitherto prevailing for Alaska salt herring.[19]

Several years later, the outbreak of World War I in Europe, by cutting off the US supply of herring cured in Scotland and Norway, presented an opportunity for Alaska herring operators. As shall be seen, the industry did indeed enjoy a boom over the next several years, but hard times followed the end of the war.

The first expansion of Alaska's herring industry during the war was primarily a diversification. In 1916, Ashton Thomas, who had canned smoked herring at Juneau in the early 1900s and had since 1914 operated a bait-herring business, organized the Alaska Herring & Sardine Company in partnership with Lee Wakefield. The company constructed a fish-processing plant at Port Walter, a prime location for exploiting the usual abundance of herring in the Cape Ommaney area. As a rule, the fish there were of good size and ordinarily remained in the vicinity for a considerable time each year.

Thomas and Wakefield equipped their plant with two lines for canning kippered (smoked) herring and one line for canning salmon. Also incorporated in the facility was a reduction plant to utilize fish offal and the herring that were too small to kipper. The company also produced Scotch-cured herring.

Herring for kippering were typically delivered to the plant at night or early in the morning, then spread in a thin layer on the floor and sprinkled with a thin layer of salt. There they remained until the cannery workers were ready to process them. As needed, the fish were placed on a table, where their heads were cut off and their viscera removed. After a brief soaking in brine, the fish were taken to the smokehouse and hung by the tails on sticks studded on both sides with rows of sharpened nails and subjected overnight to smoke made by burning alder—a common tree in coastal Alaska. Once kippered, the fish were packed in oval one-pound cans (typically five to eight fish per can) that were then sealed and placed in boiling water for about two hours. When cool, the cans were ready for labeling, boxing, and shipping.[20]

Production commenced that summer. To ensure their product met quality standards, Thomas and Wakefield brought in Scottish women who were experienced in the herring trade. By the end of the season, the plant had produced nearly 20,000 cases of kippered herring.

In late 1917, Lee Wakefield purchased Ashton Thomas's interest in the company and then merged with Wilson & Company, a large Chicago packing house, to form Wilson Fisheries Company.

Production of kippered herring at the Port Walter plant continued to grow, and in 1919 the pack was some 66,000 cases. Unfortunately, despite an extensive advertising campaign for what the company called "The New American Breakfast Dish," efforts to find a satisfactory domestic market for canned kippered herring failed. After putting up some 3,600 cases in 1920, the Wilson Fisheries Company abandoned its kippered-herring effort.[21] The company, however, would continue to be an important factor in Alaska's cured-herring industry.

In the spring of 1917, a second company, the Alaska-Pacific Herring Company, which was owned by US, British, and Norwegian interests, erected what *Pacific Fisherman* called "a magnificent plant" at Port Walter, not far from the Alaska Herring & Sardine Company plant. The plant was equipped to salt herring and to can both salmon and herring. Production in 1917 included 8,052 barrels of herring salted according to both the Norwegian and Scotch methods and 20,816 cases of canned herring.[22]

Production in 1918 included 10,600 barrels of herring—the largest pack made at any one establishment in Alaska that year—and 20,500 cases of canned herring. As was the situation at the Alaska Herring & Sardine Company plant, some of the company's 200 employees were from Scotland and were familiar with the Scotch cure. In early 1919, the Alaska-Pacific Herring Company was acquired by the Southern Alaska Canning Company, which was primarily interested in canning salmon.[23] Like the Wilson Fisheries Company, it soon abandoned its herring-canning effort but would continue to be an important factor in Alaska's cured-herring industry.

THE SCOTCH CURE

The Scots set the standard for cured herring. Herring cured by the Scotch method had to be carefully graded by size, properly gutted, salted lightly, and neatly packed into special barrels. As such, the product commanded a higher price than did herring cured by the comparatively crude Norwegian method.

The principal markets for Scotch herring were New York, Boston, Philadelphia, and Chicago.[24] It was the Jewish populations, particularly the older Jewish populations, in these cities who were the customers for Scotch-cured herring. And they were very discerning customers.

The market, though, was substantial. *Pacific Fisherman* reported that in the years leading up to World War I, some fifty million pounds of cured herring had been annually imported into New York from Scotland.[25]

With the advent of the war, however, Scotland was unable to supply the US market, and buyers looked elsewhere. Alaska had herring, but it wasn't until 1917, when the United States entered World War I, that a concerted effort was made to produce Scotch-cured herring in the territory. That effort was led by the Bureau of Fisheries, which was interested in increasing food supplies—especially from underutilized species—and in lightening the drain on meat supplies needed by the US Army and US allies during the war.[26] In *Pacific Fisherman*'s view, "Those in a position to pickle herring in this country owe it as a duty in the present emergency greatly to increase their packs."[27]

The United States entered World War I in April 1917, and within a fortnight the bureau secured the services of August H. D. Klie, a partner in a wholesale herring firm in New York City and a recognized authority on the Scotch-curing method.[28] The bureau then sent letters to all the companies and individuals in Alaska who were involved in the fisheries, requesting their cooperation to better utilize the territory's herring resource. Included with the letter were directions for curing herring by the Scotch method, a process heretofore unknown in Alaska. It was not only the abundance of herring in Alaska that drove the decision to promote the Scotch cure but also the cool climate. The peculiar flavor of Scotch-cured herring is caused by the blood that remains in the fish. Blood spoils quickly if the temperature is permitted to rise much above 50° Fahrenheit.

In early May, the bureau sent Klie to Alaska to instruct Alaskans in the preparation of herring by the Scotch-cure method and to provide marketing advice.[29] Klie was assisted by students at the fisheries school of the University of Washington, one of whom was Clarence Anderson, who in 1949 would be appointed director of the newly created Alaska Department of Fisheries.[30]

Klie and his assistants arrived in Ketchikan in mid-May and then proceeded to Wrangell and Petersburg.[31] In early July, Klie completed his work and returned to Seattle. His assistants were dispatched along Alaska's coast from Southeast Alaska to Kachemak Bay, in lower Cook Inlet, to furnish assistance to those who might want it.[32] Complicating matters, Alaska herring salters had traditionally packed their fish in 200-pound-capacity barrels, but the Scotch-cure trade required special barrels of 250-pound capacity.[33] Fortunately, the Bureau of Fisheries was able to arrange for a supply of Scotch-cure barrels.

Klie judged his and his team's effort a success. As noted above, two large firms, the Alaska Herring & Sardine Company and the Alaska-Pacific Herring Company, both at Port Walter, at the south end of Baranof Island, began packing herring according to the Scotch method, and Klie estimated that Alaska would produce 25,000 barrels of Scotch-cured herring during the 1917 season.

His estimate was overly optimistic. Production of Scotch-cured herring totaled 7,622 barrels, but almost twice that amount, 13,576 barrels, were packed according to the Norwegian method. Likely this was because the Norwegian method required less labor and did not require special barrels. Moreover, Klie's educational effort had started relatively late in the season.[34]

Nevertheless, the amount of herring cured in Alaska in 1917 was about eight times as much as had been cured in 1914.[35] Though Prince William Sound operators contributed only about 550 barrels to the total, the sound would soon become one of the centers of the Alaska herring fishery.[36] (Small quantities of herring had been cured in Prince William Sound as early as 1913.[37])

The 1917 production, though smaller than Klie had hoped for, was a solid start and, in preparation for the 1918 season, Klie published a detailed 2,400-word description of the Scotch-curing process in the January 1918 issue of *Pacific Fisherman*.[38] Basically, the process—carried out totally with hand labor—comprised three steps: gibbing, rousing, and packing. What follows is an abbreviated version of Klie's description, enhanced slightly with details from Robert Browning's *Fisheries of the North Pacific* (1980).[39]

Herring ready for curing were, while being sprinkled with salt, shoveled into a wooden box about ten feet long, five feet wide, and three feet deep with openings in the sides and the bottom to allow blood and other liquids to drain off. Once drained, and as needed, the herring were moved to the gibbing tables.

To gib a herring is to use a knife similar to a paring knife with a blade about two inches long to remove its gills and gutbag (stomach, liver, intestines, etc.) but not its milt or roe. This work was performed by women —often referred to as "herring chokers." Gibbers often wrapped their fingers with bandages to protect them. As they worked, the gibbers sorted herring by size and quality into baskets (typically four of them) placed conveniently at each gibber's side.

The next step was rousing. From the gibbers' baskets, herring were dumped into wooden tubs, each large enough to hold enough herring—about 250-260 pounds—to pack a regular Scotch-style barrel. Here, the herring were carefully sprinkled with salt until they were well covered on all sides. August Klie considered this work "the secret of Scotch curing" and the "most important part of the process."[40] Typically, about one barrel of salt was used to salt three barrels of herring.

Next came packing. The roused fish were carefully layered belly up in a prescribed manner in barrels, a handful of salt sprinkled between the layers. Once a barrel was full, it was covered and left upright for two days, allowing the fish to settle. The barrel was then filled as full as possible with fish of the same selection and cured on the same day the fish in the barrel had been cured. The barrel was then sealed and placed on its side for about ten days. After this period, the barrel was stood on end and opened, and a bunghole was bored through a wide stave near the barrel's center. The "pickle" (brine) that ran out of the bunghole was caught in a pail and then poured over the top of the open barrel, washing the fish. This was repeated three or four times.

Drawing the pickle off caused the fish to settle further, and several additional layers of herring—again, from the same day's pack—were added to fill the barrel. These fish had not been washed, so before they were layered into the barrel, they were placed in a rousing tub and washed in their own pickle. Generally, one extra barrel of herring was required to bring five barrels up to their advertised net weight.

The bunghole was then plugged, and new pickle (or the original pickle to which additional salt had been added) was poured into the top of the barrel. The barrel was then closed and its top stenciled with the packer's name, the station at which the fish were packed, the selection, etc.

Once this was complete, the barrel was rolled on its side, bunghole up, and its bunghole was opened. Pickle was added until the barrel was full, and then the bunghole was plugged. The barrel was now ready for shipment.

The 1918 season would set a record: nearly 89,000 barrels of herring were cured in Alaska. Of these, the equivalent of some 39,000 barrels were Scotch cured, and some 50,000 barrels were Norwegian cured. The Alaska-Pacific Herring Company plant at Port Walter made the largest pack of any single establishment: 10,600 barrels, most of which was Scotch cured.

Part of the reason for the increase was the expansion of the herring curing industry in Prince William Sound and lower Cook Inlet. (Herring were first cured at Halibut Cove, in lower Cook Inlet, by 1914.) Though many were small operations, fifty-eight companies, including fifteen at Halibut Cove, cured herring in Alaska in 1918.[41] John Cobb considered the herring in lower Cook Inlet to be "the largest and finest found in Alaskan waters."[42]

Herring at Halibut Cove were generally taken with gillnets, which are long, fairly shallow nets with a mesh size that allows fish to push their heads but not their bodies through. A fish that has pushed into a gillnet is prevented from backing out by its gill covers, which flare out when open. As with a seine, a corkline fitted with floats keeps the gillnet suspended, and a weighted leadline keeps it spread vertically.

When fished, gillnets are typically stretched at right angle to the shore and are either anchored or allowed to drift with the current. An advantage of employing gillnets in the herring fishery was that by using mesh of the proper size, the taking of small fish—fish that were not suitable for curing—was minimized. This was especially important at locations that lacked reduction facilities, such as at Halibut Cove.

The gillnets used at Halibut Cove were about fifty fathoms (300 feet) long and three fathoms (eighteen feet) deep and were anchored while fishing. Because herring tended to move close to the water's surface at night, gillnets were typically set in the evening and the herring "picked" from it in the morning. Fishermen picked the net from a large skiff by lifting one end of the net into the skiff, then working their way along it, shaking their catch into the skiff as they went, in the process resetting the net. Though it would likely not catch many herring during the day, keeping the net in the water saved the effort of resetting it later, and it also prevented other fishermen from taking the spot.

When the 1918 pack began to arrive in Seattle, agents of eastern US buyers examined it to ensure it met the buyers' standards, opening individual barrels in the process. They found that several packers had been careless in their work. The problems included sloppy packing, undersalting, oversalting, and improper grading. The inferior product was barely marketable and made potential buyers skeptical of herring cured in Alaska.[43]

John Cobb described the situation in a November 1918 *Pacific Fisherman* article:

> The chief trouble with the domestic pack has been that the packers, many of whom are fishermen with no shore plants, or grossly inadequate ones, have frequently insisted upon curing the fish in the manner which appealed most to themselves, and with practically no regard to the wishes of the ultimate consumer. Slipshod methods of dressing, cleaning and curing the fish have also prevailed, while negligence, to put it mildly, has been shown in packing the fish. The chief evil has been that many packers have put big and little fish together in the same barrel. Buyers of herring should know the approximate, if not exact, number of fish to the barrel, and if the packer has not properly graded his fish according to the standard sizes and packed them in separate barrels, the buyer will naturally offer the fisherman a much lower price than if they had been properly graded, as before he can offer them to the jobbing houses, he must empty every barrel, grade the fish properly, and then repack, marking on the outside of each barrel the approximate number of fish in it, and all this costs money and requires time.[44]

H. F. Moore, deputy commissioner of the Bureau of Fisheries, was of a like mind. After praising his agency's work in 1917 as having "laid the foundation for the development of a great herring industry in the territory," he cautioned that "a few careless or unscrupulous men eager for immediate profit may utterly ruin all chance" of the development of this industry. He added that "the Jewish housewife of New York is a better judge of herring than many of the men who are packing them."[45]

Despite these warnings, quality issues would continue to plague Alaska's cured-herring industry, in large part because—as stated previously—barriers to entry into the industry were minimal: almost anyone with some barrels, salt, and a source of herring could get into the business.

There were conservation concerns as well. These were tied to the use of purse seines—gear that did not discriminate among the sizes of fish caught—to catch herring in Southeast Alaska and Prince William Sound. Given the limitations on the size of herring that were suitable for salting/curing, considerable quantities of fish were discarded.[46]

Besides the waste of herring, the prevailing belief was that accumulations of dead fish in the water tended to drive herring away.[47] Add to this an aesthetic factor: the sight of dead fish brought up from the bottoms of the bays by the churning of the propellers of steamships coming to the plants was offensive to tourists and residents alike.[48] The solution, of course, was to incorporate reduction plants into herring salting/curing operations so undersize fish and fish offal could be utilized.

BARRELS

As noted above, the Scotch cure required special barrels. In a 1926 advertisement, the Western Cooperage Company, a barrel manufacturer in Seattle, wrote, "Experienced operators have known for a long time that the quality of their salt fish pack depends to a great extent on the quality of their containers."[49] At least for Scotch-cured herring, this statement was accurate. Scotch-cure expert August Klie wrote in 1918, "Herring packed Scotch style in good domestic barrels seldom bring the same high price in the American markets as those packed the same way in the Scotch-style barrel or half barrel."[50] And *Pacific Fisherman* warned in 1919 that barrel quality mattered: "A few leaky barrels in a shipment, with the spoilage of their contents, will entail losses which would more than offset the additional cost of proper containers."[51]

Although some herring salting/curing operations, such as the Alaska Oil & Guano Company, at Killisnoo, felled Sitka spruce trees to manufacture barrels on site, most curers purchased barrels. The first cooperage in Southeast Alaska, Henry Imhoff Cooperage, was established at Ward Cove, near Ketchikan, in 1879. A 1922 company advertisement listed salmon barrels but said nothing about barrels for herring. Henry Imhoff Cooperage operated until Henry Imhoff's death in 1928.[52]

The capacity of a standard Scotch-style whole barrel was thirty-two US gallons, while the capacity of a half barrel was sixteen US gallons. The barrels were considered to hold 250 pounds and 125 pounds of fish, respectively.[53] In 1917, at least one company, the Seattle-based Western Cooperage Company, began manufacturing Scotch-cure barrels (figure 3.1).[54] Based on the amount of advertising it did, it seems that the company

Figure 3.1. Barrel advertisement, Western Cooperage Company. (*Pacific Fisherman*, July 1924)

supplied most of the barrels for Alaska Scotch-cure operations. It advertised barrels manufactured from "well-seasoned," "hard" Douglas-fir.[55] August Klie rated the Pacific Coast–manufactured barrels as "equal to or better than the regular Scotch barrel, as the fir used is almost entirely free of knots and the staves rarely break or crack."[56] In 1921, the Pacific Cured Fish Association approved a standard barrel for Alaska Scotch-cured herring. The association was concerned about complaints of barrels being underweight, and the approved barrel was of a size that, when full, would contain not less than 260 pounds of herring.[57]

SALT

> Salt continues to be to a large extent the cornerstone of the fish business.—*Pacific Fisherman*, 1916[58]

In his 1918 description of curing herring by the Scotch method, August Klie included several warnings. One was "Don't use cheap or poor salt."[59] Salt used in Alaska's herring-curing industry was solar salt produced by the evaporation of seawater. It was produced on southern San Francisco Bay, where abundant sunshine and wind from May through October made conditions ideal for the process.

About forty tons of seawater were required to produce one ton of salt. A typical salt operation described by *Pacific Fisherman* in 1916 included seven shallow reservoirs encompassing some 1,600 acres and delineated by fifteen miles of levees. *Pacific Fisherman* praised the salt produced there as "without doubt the best and most suitable for the curing and salting of fish."[60] Specifically to accommodate the needs of the fishing industry, in 1925 the San Francisco–based Arden Salt Company constructed a 300-foot salt dock on Seattle's waterfront.[61] The cured/salted herring industry was likely one of the company's biggest customers.

POST–WORLD WAR I YEARS

The 1918 cured-herring pack in Alaska had indeed set a record, but early in 1919, Ashton Thomas, now associated with the Franklin Packing Company, which operated a large herring-salting plant at Port Ashton, in Prince William Sound, explained the challenge that the industry faced. While noting that the industry had made substantial progress, Thomas wrote that "the progress has been almost entirely due to war conditions," and, now that the war was over, the industry was confronted with "a very serious situation" that threatened it with "entire extinction."

The situation involved the difference between the cost of transporting herring from Alaska to New York, the principal Scotch-cured herring market in the United States, and from Scotland to New York. It cost approximately $6.50 to ship a barrel of herring from Alaska to New York, versus approximately $2.50 from Scotland to New York. What Alaska producers needed to ensure the survival of their industry, according to Thomas, was for Congress to levy a duty on imported herring.[62]

Alaska's Territorial Legislature agreed, and in April 1919 it passed a memorial stating that the salted-herring industry was of the "greatest importance" to the territory and that "with the ending of the world war it will be impossible for this industry to be sustained without Governmental aid to offset the advantage European producers have in the New York market on account of low trans-Atlantic freight rates." The memorial then asked Congress to levy an import duty of not less than $3 per barrel on salt herring, this being "the only hope of the survival of this important industry."[63]

Some wanted the duty to be higher: a New York agent for an Alaska packer suggested an import duty of $7.50 per barrel. August Klie, now out of the service of the Bureau of Fisheries and buying herring under his own account, echoed John Cobb's earlier observation that Scotch-cured herring was "poor man's food" and an import duty of $7.50 per barrel would make it a luxury item and demand would immediately fall. Moreover, American importers would strongly oppose the duty.

Klie suggested that "as long as we can again depend on Scotland for our entire supply of herring . . . it will be advisable to discourage the curing of Alaska herring for our eastern markets in America." He suggested that because of the improvements in the quality of Alaska's cured herring, the West Coast demand for herring would increase, and it was here Alaska packers should focus their efforts.[64]

A duty on cured herring was not forthcoming, and the herring industry entered the 1919 season with trepidation. In addition to the duty-free competition from Europe, the market for Alaska herring, despite Klie's assertion, had declined because of quality issues with the 1918 pack—specifically, the poor selection of fish to be cured and the faulty packing of those fish. Given this grim situation, some packers elected not to operate in 1919, and employment in the industry declined by more than half.[65] The cured-herring pack, too, declined by more than half, to 42,800 barrels. Despite the lower production, packers seemed to acknowledge the superior marketability of the Scotch-cured product. It accounted for fully 80 percent of production, in contrast to the 44 percent it had been the year prior.[66]

Not surprisingly, given the history of Alaska's cured-herring industry, there were some size/quality issues with the 1919 Alaska pack. But there was good news, too: Alaska herring, properly selected and packed, had come to be recognized in the New York market as the best in the world, and commanded a premium over the best Scotch herring. Commanding top prices were large *matjes*—early, fat Scotch-cured herring that were free of milt and roe.[67] Regarding the meaning of "large," according to W. J. Emlach, a herring packer from Scotland, it was "waste of barrels" to pack fish less than 10.5 inches long.[68]

Whether the premium could overcome the high cost of transportation was another question. Moreover, there remained the question of what to do with the lower grades of herring. Reducing the herring to meal and oil was a solution, but most herring-curing stations did not have reduction plants.

Alaska herring production in 1920 was greater than ever before, but unlike 1919, when approximately 52 percent of the harvest was used to produce food products, nearly 80 percent of the 1920 harvest was used as bait or in the production of oil and meal/fertilizer.

Meanwhile, the shift toward producing more Scotch-cured herring continued. In 1920, the product accounted for approximately 94 percent of food-herring production.

Of the fifteen companies that handled herring in Southeast Alaska in 1920, six were exclusively involved in bait herring, seven were primarily operators of reduction plants, and two—the smallest of the operators—were engaged wholly in the preparation of food products. The lack of interest in producing food products from Southeast Alaska's herring was basically because the fish tended to be too small for the Scotch-cure market. As the Bureau of Fisheries herring expert George Rounsefell opined in 1926, "Herring in [Southeastern] Alaska are not good for anything but reduction into meal."[69]

The situation was decidedly different in southcentral Alaska (Prince William Sound and lower Cook Inlet). There, all the thirteen operators primarily produced food products. Two of the companies installed reduction plants—the first in Prince William Sound—to utilize the waste from curing operations. And there was a lot of "waste." That season, the two plants reduced more than ten million pounds of herring into meal and oil—most of it herring that was unsuitable for curing.[70]

Pacific Fisherman noted that although fewer herring had been cured in Alaska during the 1920 season than in 1919, the business was in much better shape because operators now understood what the market demanded and had made a real effort to produce accordingly. Despite this improvement, however, profits for even the best operators were minimal because the railroads had increased the freight rate between Seattle and New York, and European competition had lowered the price of cured herring. Moreover, Prohibition, which outlawed the manufacture, transport, and sale of alcohol within the United States, had taken effect in January. This reduced the market for the cheaper grades of cured herring, which were often consumed by beer drinkers in taverns.[71]

These challenges were considerable, and in March 1921, Alaska herring curers organized into the Pacific Cured Fish Association, whose goal was to "keep in closer touch with developments bearing on their industry, crystalize the ideas of the people engaged in it, and take orderly and effective action when action should be needed." Lee Wakefield, now of the Franklin Packing Company, which had a herring operation in Prince William Sound, was elected president.[72]

Perhaps the first official act of the association was to send a letter to Washington Congressman John Webster asking that a duty of two cents per pound (equal to $5 per Scotch-cure barrel) be levied on imported herring. The herring packers believed they could overcome the industry's remaining disadvantages by lowering per-unit costs through higher-volume production (economies of scale), producing a high-quality product, and urging transportation companies to lower freight rates. Claiming that "the supply of fresh herring is inexhaustible," the association said it would "do more for the development of Alaska than any other part of the fishing industry."[73]

In May 1921, the cured herring industry suffered a small setback: a territorial tax. It may have been simply a matter of fairness because since 1906 there had been a federal tax on pickled and salted salmon, and since 1915 all salted and mild-cured fish—except herring—had been subject to a territorial tax (see chapter 2).[74] The 1921 territorial tax, which applied to all salted or mild-cured fish except salmon and codfish (which were taxed at a higher rate), was 2.5 cents per 100 pounds, equal to 6.25 cents per Scotch-cure barrel.[75]

But all was not grim, at least for Alaska herring curers. During the latter part of August 1921, it was reported that Scotland's herring pack was very light and of poor quality. This unexpected situation increased the demand for Alaska herring to the extent that New York buyers bid against each other, and prices rose to exceptional levels. Alaska packers had already provisioned themselves for the volume of herring they anticipated they would produce during that season, but they did what they could to expand production.

The result was that Alaska's herring pack was the greatest it had been since 1918. A total of 64,600 barrels were cured, of which some 41,000 barrels were packed in Prince William Sound, where, as *Pacific Fisherman* reported, "the fish were in better condition than ever before, and large fish were fairly abundant."[76]

Importantly, Alaska operators had finally learned their lesson about quality, and the product was well received. It was for most operators a satisfactory season, but the failure of the Scotland fishery in 1921 was unprecedented, and there was no reason to believe it would be repeated.[77]

Nevertheless, 1922 began with herring packers preparing for what they hoped would be the largest pack on record.[78] *Pacific Fisherman* noted that the larger companies were now operated by experienced men and, providing suitable fish could be secured, they should have no difficulty in putting up a first-class product. The journal cautioned, however, that some of the new firms in the business were lacking in experience and had chosen locations of "doubtful suitability."[79]

One location, though, was especially desirable: the area in and around Sawmill Bay, in Prince William Sound. There, seven companies outfitted to pack herring. Additionally, about a half-dozen crews were going to thoroughly prospect Kodiak Island for herring opportunities. (Outfits such as these typically packed their fish on barges.)[80] Territory-wide, twenty-two plants cured herring in 1922, nine more than in 1921. In addition to these plants were small operators who cured herring on barges and at shore stations.[81]

Meanwhile, the Pacific Cured Fish Association's lobbying effort for a duty on imported herring achieved some success. In 1922, Congress passed the Fordney-McCumber Tariff, which was designed to help America's factories and farms. Signed into law in September, the bill included an import duty of one cent per net pound on pickled or salted herring ($2.50 per standard Scotch barrel). This duty, however, was not enough to offset the freight-rate advantage enjoyed by European producers.

Those who had hoped for a record pack were not disappointed. By mid-July, herring were plentiful in both Prince William Sound and in Southeast Alaska, and the quality in practically all localities was exceptionally fine.[82] The total pack in 1922 was 145,326 barrels (thirty-six million pounds), more than double that of the previous year. It exceeded the previous record pack (1918) by approximately 55,000 barrels. Scotch-cured herring accounted for more than 99 percent of the 1922 pack.[83]

The one discouraging element was the failure of the seven firms that prospected Kodiak Island's coast to locate suitable fish. In total, they packed about 10,000 barrels, not nearly enough to offset their expenditures.

Nevertheless, Ashton Thomas, who had been involved with Alaska herring for more than two decades, retained his optimism. He said one year's experience at Kodiak was not enough to draw definite conclusions, adding that, in his opinion, the quality of herring at Kodiak was the best in Alaska.[84] Thomas's optimism was warranted: within several years the Kodiak and Afognak Islands area became an important center for Alaska's herring industry.

The 1922 cured-herring pack, while a record, had a serous downside. It had glutted the market, and in the early months of 1923 about 25,000 barrels had not yet been sold and remained in storage in Seattle.

Seriously exacerbating this situation was the handling of herring barrels at Seattle. Scotch-cured herring is a relatively delicate product. Ideally, the herring should be kept in cold storage if being held for any length of time, and, to ensure the herring are adequately in contact with the pickle, the barrels should be stored on their bilges (sides) and rotated every few days. This didn't always happen in Seattle, and an appreciable quantity of fine-quality, well-packed Scotch-cured herring went bad and was dumped. Packers caught in this situation suffered heavy losses, and this tempered their enthusiasm for the upcoming season. Commissioner of Fisheries Henry O'Malley, who had long been an advocate for expanding Alaska's cured-herring industry, acknowledged the product's limited market and advised packers to be "very cautious to avoid over-production."[85]

Overproduction, however, would not be an issue in 1923. Herring were scarce in Southeast Alaska all season, and their quality was low. By the end of July, most packers in the region had shut down. Prince William Sound, too, was plagued by a lack of suitable-quality herring.[86] Leif Buschmann, superintendent of the Franklin Packing Company's Port Ashton plant, spoke of the "poorest herring runs in the history" of the district.[87]

By midsummer, the surplus from 1922 had been consumed. Moreover, about that time it became apparent that only a small amount of herring from Scotland would be available for the US market. New York buyers began scrambling for Alaska herring, and Alaska packers worked to meet the demand. One or two of the plants in Southeast Alaska that had been closed were reopened and operated until the end of the year. In Prince William Sound, several plants operated until Thanksgiving, and a couple did so until about the first of December. At Halibut Cove, crews were still packing herring in January.[88]

When all was said and done, all the cured herring of desirable quality found a ready market at remunerative prices, yet the market was still not sated. In its summary of the 1923 herring season, *Pacific Fisherman* warned,

> The high prices of the past season are likely to be misleading as to the condition of the industry, and it is well to remember that they resulted entirely from the exceptional shortage of fish, not only from Alaska but also from Scotland and other sources. Anything like a normal production would result in much lower prices.[89]

For the duration of the Scotch-cure herring industry in Alaska (which persisted until 1951), the prices it could command depended on the volume and price of the Scottish pack.

Alaska production in 1923 totaled 81,466 barrels, of which all but 2,222 barrels were Scotch cured.[90]

CURED HERRING IN NORTON SOUND

Norton Sound is a seventy-mile-wide arm of the Bering Sea that indents Alaska's northwest coast for about 125 miles. Golovin, a community on the Seward Peninsula, about seventy miles east of Nome and about 150 miles south of the Arctic Circle, deserves special attention because it was the location of the northernmost commercial herring fishery in Alaska. Herring at Golovnin Bay and Golovnin Lagoon tend to be large and fat. Spawning populations typically arrive in late May or early June. Once spawning is complete, the herring leave, but some return to feed in late summer.[91]

A 1929 Bureau of Fisheries report stated that a herring fishery had been conducted at Golovnin Bay since before 1909, but this seems to have been for local use.[92] Fishermen likely employed beach seines to catch the fish. Transportation costs were a chronic problem for Alaska herring packers, and they were especially challenging for packers at Golovin, who had to ship their product through Nome to get it to Seattle. Most of the packers were Seattle-based and resided in Golovin only during the herring-packing season.[93]

The commercial fishery at Golovin seems to have begun in 1916, when four small packers there produced 559 barrels (about 112,000 pounds, by the measurement of the time) of Norwegian-cured herring.[94] The following year (1917), five packers operated at Golovin, and they more than doubled the 1916 pack, producing 1,275 barrels of Norwegian-cured herring.[95] Interest in herring curing at Golovin continued to increase, and in 1918 ten packers produced 5,169 barrels of herring, more than quadrupling the previous year's pack. The lead packer was the Pioneer Mining & Ditch Company (later the Golovin Fishing Station), which cured 2,600 barrels of herring, 500 of which were the first-ever Scotch cured at Golovin.[96]

Most of the workers in Golovin's herring industry were Iñupiat. Tragically, on October 20, 1918, passengers infected with the Spanish influenza virus stepped off a steamship at Nome. The deadly virus quickly spread across the Seward Peninsula and then throughout Alaska. In November alone, 831 deaths were attributed to the influenza. Nearly two-thirds of the deaths of Alaskans during the epidemic occurred in the Nome area. With much of their labor force deceased, herring packers found it necessary to import workers from the lower 48 for the 1919 season.[97]

The 1919 herring-packing season at Golovin was disappointing. Severe storms limited fishing to about fifteen days and caused the loss of many boats and lighters. Production was 2,555 barrels, of which 1,655 were Norwegian cured and 900 were Scotch cured. A considerable quantity of cured herring was sold to people in Nome who had come to appreciate the product.[98]

Interest in Golovin herring was waning because of the cessation of World War I, which allowed the resumption of cured-herring imports from Europe, and the high cost of operating at this remote location.

In 1921, only one packer, the Arctic Whaling & Fishing Company, participated in the fishery. The company packed 562 barrels of Scotch-cured herring.[99] In 1922, the Arctic Whaling & Fishing Company was joined by John Winthers, and together they packed 500 barrels of Scotch-cured herring.[100] The following year, the Arctic Whaling & Fishing Company was again the lone herring packer at Golovin, packing 352 barrels of Scotch-cured herring.[101]

The Arctic Whaling & Fishing Company continued packing modest quantities of herring at Golovin until 1927, when the Golovin Bay Packing Company was organized and purchased the company's plant and equipment. The new company packed modest quantities of herring until 1935. The following year, fishing for herring in Golovnin Bay was a complete failure, there being no catch whatsoever, and in 1937 a fire destroyed the company's plant.[102] Several small operators packed herring at Golovin until 1941, when the advent of World War II caused a labor shortage. Also, shipping space was at a premium, and freight rates had increased because of the war risk.[103]

4
EARLY ALASKA HERRING FISHERY REGULATION AND RESEARCH

> There may be said to be three general classifications of all natural resources on which our economy depends. These are separated into the inexhaustible, as the power of falling water; the non-replaceable, as the mineral deposits, and the replaceable, as the forests, game and fishes. Only in the latter group can continued unwise or excessive exploitation result in a total loss of the resource, yet judicious use results in a yield continuing indefinitely. Our denuded forests lands, our decimated wildlife, and our many declining fisheries show clearly the need for more reasonable methods of utilization.—E. H. Dahlgren, Alaska herring researcher, 1940[1]

REGULATION

In the early 1920s, there was great concern that, in the words of Commissioner of Fisheries Henry O'Malley, Alaska's salmon fishery was "becoming steadily depleted, and further threatened by increased operations." O'Malley recommended to the US secretary of commerce that "fishery reservations," in which the fisheries could be more closely regulated, be established.[2] As a result, President Warren G. Harding in November 1922 used an executive order to create the Southwestern Alaska Fishery Reservation.[3] The reservation was subsequently divided into three districts: Bristol Bay, Cook Inlet, and Kodiak-Afognak. Within this reservation, commercial operations would be subject to regulations and restrictions issued by the secretary of commerce. The first regulations, issued in October 1923, simply required that commercial fishermen and fish processors obtain permits. Of the seventy-one permits issued in 1923, six were issued to herring operations in Cook Inlet. At least five of the Cook Inlet operations were at Halibut Cove.[4]

During the summer of 1923, President Harding visited Alaska, accompanied by, among others, his secretaries of commerce, interior, and agriculture. Also in Alaska that summer were Commissioner of Fisheries O'Malley, who spent almost all of July and August investigating the territory's fisheries, and Representative Wallace White (R-Maine), chairman of the House Committee on Merchant Marine and Fisheries, who was in Alaska likewise to investigate the fisheries.[5] Regarding the herring fisheries, 1923 was an unusual year: there was a shortage of herring in all districts, and the normally important fall season in Prince William Sound was almost a complete failure.[6]

President Harding died shortly after leaving Alaska, but in his last public address, given in Seattle on July 27, he warned that "if Congress cannot agree upon a program of helpful legislation, the [fishery] reservations and their regulations will be further extended by Executive order."[7] It took nearly a year, but on June 6, 1924, Congress, led by Rep. White, passed the Act for the Protection of the Fisheries of Alaska.[8] The legislation, commonly referred to as the White Act, provided the first solid direction for the management and conservation of Alaska's fisheries. It also rendered the Southwestern Alaska Fishery Reservation unnecessary, and on the following day President Calvin Coolidge revoked the executive order that had created it.[9] Among its provisions, the White Act authorized the secretary of commerce to

> set apart and reserve fishing areas in any of the waters of Alaska over which the United States has jurisdiction, and within such areas may establish closed seasons during which fishing may be limited or prohibited as he may prescribe. Under this authority to limit fishing in any area so set apart and reserved, the Secretary may (a) fix the size and character of nets, boats, traps, or other gear and appliances to be used therein; (b) limit the catch of fish to be taken from any area; (c) make such regulations as to time, means, methods, and extent of fishing as he may deem advisable.[10]

On June 21, 1924, less than three weeks after the White Act became law, the Bureau of Fisheries promulgated the first regulations for Alaska's herring fishery. The regulations (1) established herring fishing seasons, except for bait or local food purposes, in the Kodiak Island and Cook Inlet areas; (2) prohibited seining for herring in Halibut Cove and Lagoon; (3) prohibited the maintaining of a herring pound or the dumping of offal and dead herring in the waters of Halibut Cove and Lagoon; (4) restricted gillnets used to catch herring in the Kodiak Island, Cook Inlet, and Prince William Sound areas to having a stretched measure of not less than three inches; and (5) prohibited in the Kodiak Island, Cook Inlet, and Prince William Sound areas the placing of any net or other device across the entrance of any lagoon or bay that would at any time prevent the free passage of herring into or out of said lagoon or bay.[11]

Later that year, the bureau established seasonal fishing closures in Cook Inlet and Prince William Sound. In Cook Inlet, fishing for herring—except for bait and local food purposes—was prohibited from January 1 to May 31. In Prince William Sound, fishing for herring—again, except for bait and local food purposes—was prohibited from January 1 to June 24 and from November 1 to December 31.

Southeast Alaska was spared any regulation until September 4, 1924, when the bureau prohibited commercial fishing for herring in the waters of Kootznahoo Inlet, near Killisnoo, for the remainder of the year.[12]

Regulations implemented in 1925 prohibited fishing for herring in all waters closed throughout the year to salmon fishing, and they also extended to all districts the rule against enclosing lagoons and bays.[13]

W. J. Imlach, whose Imlach Packing Company in 1925 operated three herring plants, derided the idea of being regulated.[14] According to Imlach, "From 1919 to 1923 we in the herring industry had no laws or regulations imposed on us, and no Bureau of Fisheries officials came to visit us. These were the most successful years of the herring industry."[15]

Others in the industry saw at least the seasonal regulations as beneficial, if only because they prevented unscrupulous operators from packing herring that were not in their prime, thus jeopardizing the reputation of Alaska's cured herring. And, importantly, the Bureau of Fisheries exhibited flexibility and worked to accommodate the industry. For example, the Cook Inlet herring season was scheduled to close on January 1, 1925, but the herring at Halibut Cove were still in good shape at that time, so the bureau postponed the closure until fish quality began to deteriorate.[16]

To comply with the regulation that prohibited dumping offal and dead herring at Halibut Cove, at least some operators loaded the offal and dead herring onto one or more trap scows, which were normally used to transport salmon from fish traps to canneries. The scows had no propulsion of their own, were about sixty feet long, had a capacity of about 120,000 pounds, and carried the fish in wooden pens on deck. The scows were periodically towed out of the cove into Cook Inlet and their cargo dumped overboard.

A half century later, Alaska fisheries managers had made no significant changes in how herring offal and dead herring could be disposed of. In the early 1970s, Sitka Sound Seafoods, located along Sitka's waterfront, had a roe-stripping operation (see chapter 10). The roe-stripped female carcasses and the male herring were loaded aboard a scow, which was towed to Eastern Channel, just south of Sitka, and dumped overboard.[17]

Another new regulation took effect prior to the opening of the 1927 herring fishing season in Alaska: a weekly thirty-six-hour period (from 6:00 p.m. on Saturday until 6:00 a.m. on Monday) during which fishing for herring, except for bait, was closed. The closures applied in the Kodiak, Cook Inlet, Prince William Sound, and Southeast Alaska management areas.[18]

RESEARCH

Until 1925, about the only information available regarding Alaska's herring reduction industry was the amount of meal and oil it produced. Little was known about the volume of fish processed or the biology of the herring populations that were being targeted. By 1925, the fishery had become so large that there was concern that the herring resource was being overexploited.[19]

Early in 1925, *Pacific Fisherman* interviewed Clarence Anderson about the state of Alaska's herring industry. In 1917, Anderson, then a fisheries student at the University of Washington, had assisted August Klie in introducing Alaska to the Scotch method of curing herring. He later became an instructor at the University of Washington's College of Fisheries (established in 1919) and in 1922 journeyed to Norway to observe the herring fisheries there. Now, Anderson was employed at Lee Wakefield's Franklin Packing Company, which operated a saltery and reduction plant at Port Ashton, in Prince William Sound, had a scow-based herring-curing operation at Izhut Bay, on Afognak Island, in the Kodiak management area, and operated a packing plant at Seldovia, on the Kenai Peninsula.[20]

Based on his experience in Norway, where the stock of herring seemed to be healthy despite there being no regulation of the fishery, Anderson was skeptical of the regulation of Alaska's herring fisheries. He suggested that any regulations should be based on a careful study of the fisheries and said what Alaska's industry needed was an accurate life history of herring, including their migrations. Anderson noted that herring tend to follow the feed, and a lack of information on herring feed made it impossible to determine why the fish had not appeared in their usual locations in 1924.[21]

Anderson's observations and concerns must have carried considerable weight, because not long after his interview, the Bureau of Fisheries initiated the first comprehensive biological study of Alaska's herring. The bureau contracted twenty-one-year-old George Rounsefell to conduct the study. Rounsefell had been born and raised in Ketchikan, had attended Stanford University, and for the past year had studied sardines at the California State Fisheries Laboratory. He would spend most of the next decade studying Alaska's herring.

The purpose of Rounsefell's study was to understand and, if possible, to forecast fluctuations in herring abundance, to discover whether fluctuations were due to natural causes or to depletion, and, if due to depletion, how this condition might best be remedied.[22] Writing in *Pacific Fisherman* in December 1926, Rounsefell said one of his most important questions was whether all the herring in Alaska belonged to a common stock or were members of separate colonies or races.[23]

Rounsefell departed for Alaska in June 1926 to conduct a preliminary study, and that summer he visited all the important herring centers and collected data for a general study of herring ages and races.[24] The following year, to help discern trends in the fishery, Rounsefell distributed "herring-catch books" to all the herring plants in the territory. In these, buyers recorded the size of each catch and the exact location where it was made.[25] (The Bureau of Fisheries continued to collect this fundamental and important information through 1966.[26]) Rounsefell spent time during 1926 in southcentral and western Alaska, including Prince William Sound, Kachemak Bay, and Kodiak Island.[27]

Pacific Fisherman considered Rounsefell's work important. "The only real safeguard, either for the industry or the herring fishery, must be a knowledge of the facts upon which reasonable regulation can be based," the journal wrote in 1926.[28]

Rounsefell's *Contribution to the Biology of the Pacific Herring, Clupea pallasii, and the Condition of the Fishery in Alaska* was published in 1929. Among his conclusions were: (1) the populations of herring at Craig, Chatham Strait, Stephens Passage, Prince William Sound, Kachemak Bay, Shuyak Strait, Shearwater Bay, Old Harbor, Chignik, the Shumagin Islands, Unalaska, and Golovnin Bay were biologically distinct stocks; (2) the rates of growth among the populations varied significantly (for example, at six years of age, Unalaska herring were 6.5 centimeters longer and 2.8 times heavier than Stephens Passage herring); (3) there was some evidence of depletion in Southeast Alaska, but the available data did not offer sufficient proof; (4) the data indicated severe depletion in Prince William Sound; and (5) depletion had probably occurred in Cook Inlet and in Shuyak Strait. Rounsefell concluded that it was "deemed necessary that additional protection be applied to the herring in the Prince William Sound, Cook Inlet, and Shuyak Strait areas."[29]

The following year, in a report on fluctuations in the supply of herring in Southeast Alaska, Rounsefell was more certain that herring in the region had been depleted, warning of the "depletion that is threatening the commercial extinction of the herring fisheries of southeast Alaska." The production of herring oil and meal/fertilizer in Alaska was centered in Southeast Alaska, and fishing there beginning in the mid-1920s had been intense. Regulations issued in 1924 and each year thereafter restricted fishing somewhat, but they failed to stem the depletion. Rounsefell explained how fishermen had depleted herring stocks:

> An intensive fishery was maintained on the older and better-known fishing grounds until they no longer produced sufficient raw material. Then the fishery sought new grounds, usually at a greater distance from the plant. If the older grounds had now been entirely abandoned, the situation might not have become so alarming. However, this did not occur. The fishermen continued to seek for herring on the old and well-known fishing grounds long after they had ceased to produce a fair return. In going to and returning from newer and more productive grounds they traversed and fished the older grounds. In periods of stormy weather or seasonal scarcity the older grounds, being nearer to the plants and usually more sheltered than the newer, were fished intensively. As a result of these conditions, each fishing ground, once depleted, remained depleted, without any chance to recover, long after it had ceased to be of any real value to the fishery.[30]

According to Rounsefell, one of the major factors contributing to the depletion was the increase in the size and number of purse-seine boats since 1922.[31] In 1930, the year of his report, the number of purse-seine boats engaged in Southeast Alaska's herring fishery peaked at sixty-nine.[32]

To stem the depletion, Rounsefell recommended closing some areas of Southeast Alaska—including much of Chatham Strait—to herring fishing for five years. While this may have seemed to be harsh medicine, as Rounsefell pointed out, the areas were already so depleted that their closure would have minimal impact on the region's total catch. Among his other recommendations were lengthening the weekly closed period in certain areas and shortening the fishing seasons in others. Rounsefell also recommended that "the use of herring of over

10½ inches in total length measured from the tip of the snout to the end of the tail fin for reduction purposes be regarded as wanton waste." His intent was "aimed at stopping the tremendous waste of large fat herring from area 20 [lower Lynn Canal] that have in the past been used chiefly for reduction."[33]

None of Rounsefell's recommendations were implemented for the 1931 herring season. In his 1931 Bureau of Fisheries report on Alaska's fisheries, Ward Bower wrote of herring in Southeast Alaska being "abundant, particularly toward the latter part of the season in the lower Chatham Strait region."[34]

In 1932, however, the Bureau of Fisheries closed Seymour Canal, Gambier Bay, Pybus Bay, and the adjoining waters of Frederick Sound to herring fishing except—with specific restrictions—for herring to be used as bait.[35] These waters were among those Rounsefell had recommended closing.

As an additional measure to stem potential overfishing, in 1939 the Bureau of Fisheries began establishing area quotas (limits) in Alaska's herring fisheries. As will be discussed in chapter 6, the first quotas to be imposed were for Prince William Sound and the Kodiak Island area.

ATTEMPTED REGULATION THROUGH LEGISLATION

In late 1931, James Wickersham, Alaska's nonvoting delegate to Congress, introduced legislation to prohibit the reduction of food fish in Alaska. Wickersham had been voicing his objection to herring reduction since at least 1911, and the bill he introduced in 1931 was exactly the same as had been introduced by Dan Sutherland, his predecessor, in 1927 and again in 1929. Neither bill had made it out of committee.[36]

The Pacific Herring Packers Association had little fear that Wickersham's proposed legislation would pass, but it nevertheless voiced its opposition to it, pointing out that the herring-reduction industry provided employment at a time when jobs were scarce, and that since the market for cured herring had been declining for years, the herring that went to reduction plants were not needed for food.[37]

The association had an ally in Commissioner of Fisheries Henry O'Malley, who warned that the proposed legislation, if it became law, would "destroy the extensive industry which has developed in Alaska of manufacturing oil and meal from herring"—precisely what the legislation was intended to do. Regarding the accusations that the reduction plants were depleting the runs of herring, O'Malley maintained that depletion was not a serious problem, stating, "As in any fishery, seasonal fluctuations occur, and in some places in Alaskan waters herring may be less abundant than in previous seasons; at the same time, they may be more abundant in other parts of the territory."[38]

Like those of Dan Sutherland, Wickersham's bill died in committee.

5
ALASKA'S HERRING INDUSTRY EXPANDS
1924–1931

> The herring fishery of Alaska has undergone a tremendous development in recent years. Gaining an impetus during the World War, it has increased until during the four years 1924 to 1927 an average of 160,000,000 pounds have been taken annually from the waters of Alaska. This ranks next to the take of salmon, the average annual catch of which during the same period was 358,000,000 pounds.—George Rounsefell, 1929[1]

Although the cured-herring component of Alaska's herring industry had diminished somewhat after the record year of 1922, the reduction industry—driven by rising prices for herring oil—boomed. In 1921, oil was selling for about twenty cents per gallon, with, as *Pacific Fisherman* wrote, "the material going to the soap kettle in the absence of other demand."[2] During the summer of 1923, however, an increase in the demand for cottonseed oil and tallow brought attention to fish oils, and the price of herring oil increased dramatically, the product selling for forty-seven to fifty cents per gallon.[3] That year, the territory's fourteen reduction plants—which some Alaskans called "stink plants" because of the odors they omitted—produced 953,000 gallons of herring oil, more than twice the amount produced in 1922.[4] Oil production increased again in 1924, to 975,000 gallons. At the same time, the production of meal, some 4,000 tons, nearly doubled from the previous year.[5]

Notwithstanding that the 1924 herring season in Alaska was characterized by a marked scarcity of suitable fish, particularly in Prince William Sound, production of cured herring that year totaled 77,151 barrels, only about 4,000 barrels fewer than in 1923 but about half the record pack of 1922.[6]

For Alaska's cured-herring industry, 1925 was largely a successful year. Though Prince William Sound experienced another failure, herring were fairly abundant in all the other districts, and they were of good size and quality. The pack, 137,639 barrels (34.5 million pounds), was the second largest on record.[7]

Pacific Fisherman celebrated Alaska's herring industry with two feature articles in 1925. "Remarkable Growth of the Alaska Herring Industry," published in June, noted that the industry had for the past several years "ranked among the really important branches of the Alaska fish business."⁸ "Remarkable Development of the Alaska Herring Fishery," published in September, praised "the rapid rise of the herring fishery, which within a few years has grown from insignificant proportions to be one of the most important and promising branches of the fishing industry" as having been "a spectacular feature in the development of Alaska."⁹ Similarly, in his 1925 report, Bureau of Fisheries Alaska agent Ward Bower noted that "the greatest expansion of Alaska fishery operations in 1925 occurred in the herring industry."[10]

This expansion was attributable chiefly to the increased output of oil and meal/fertilizer in Southeast Alaska, itself driven by the large increase in the price of oil. From 1925 to 1927, Alaska's herring-reduction plants annually consumed over 100 million pounds of raw fish, peaking at 150 million pounds in 1926.[11] That year, Alaska's herring reduction plants set a record. Production of herring oil totaled 2.97 million gallons; production of meal/fertilizer, some of which was derived from salmon cannery offal, was 12,041 tons.[12] *Pacific Fisherman* estimated the herring-reduction industry in Alaska had "probably reached a higher stage of development than anywhere in the world."[13]

Bureau of Fisheries scientist Elmer Higgins, however, worried that the industry had developed to "alarming proportions," that the "manufacture of these by-products, at first a side line to utilize the wastage of the salteries, has now become the chief object of the fishery." Higgens noted that the decline in abundance of herring in certain regions had aroused the fear of depletion among herring fishermen, as well as halibut and salmon fishermen who needed a consistent supply of bait herring.[14]

My Herring-Choker
Dan McNeil, 1924

You may rave about your Burmah girl adown by Mandalay
Or your jazz-bo baby stepping in some classy cabaret
Or some bathing beauty posing beside the sun-kist sea,
But your jazzy janes and dizzy dames don't make no hit with
    me,
There's a queenly herring-choker who has got me hooked for
    fair,
With a silvery crown of herring scales a'shining in her hair.

She don't wear them for adornment, for she's not on dress
    parade;
They're a badge of honest labor for this busy winsome
    maid—

Coarse rubber boots and oil-skins hide her figure's dainty grace,
There's guts and gills and gurry spattered o'er her pretty face,
But she keeps the waste-chute full of gib and a herring in the air
So, no wonder now and then a scale gets tangled in her hair.

Oh, those gleaming flecks of silver, shimmering with each vagrant breeze
Like the flakes of moonbeams sifted through the branches of the trees,
Like the pearly shafts Aurora shoots on drowsy Night's broad screen,
Set my pulse to racing wildly—hitting up on all sixteen
For this queen of herring-chokers has my poor heart in a snare;
Just like those lucky herring scales that nestle in her hair.

She has hardly time to smile at me when I come sailing in,
For she's as busy as a beaver while there's herring in the bin.
But she'll collect a wad of green-backs that will shame a bale of hay
When she cashes in her tickets for her twenty barrels a day.
And when she steps out at the 'Lympic in her furs and silk so rare,
I know I'll miss those herring scales agleam within her hair.

And when our wedding day comes 'round, that joyous day of days,
And she leads me to the alter, I'll be in a blissful haze.
Though she'll be gowned like a princess, to me she'll seem more fair
If I can slip a few bright herring scales into her marcelled hair.[15]

    A notable change in the structure of the industry in 1925 was that eastern distributors had purchased several Alaska herring packing companies.[16] This was a clear indication that the distributors had a high degree of confidence in the market for Alaska-cured herring. It was also an effort to secure a reliable supply of product that was prepared to their own standards.[17] A second, and more important, change was the great increase in the employment of large floating salteries.

Figure 5.1. Herring chokers, Franklin Packing Company, Port Ashton, Prince William Sound. (*Pacific Fisherman*, January 1924)

FLOATING SALTERIES AND REDUCTION PLANTS

Alaskans had been curing herring on scows and vessels in a small way since the turn of the century, but they began using large vessels in 1924, when Ottar Hofstad purchased the 130-foot three-masted former codfish schooner *Esther* and outfitted it to salt herring, and the Nassau Fish Company outfitted the five-masted schooner *Bianca* for the same purpose. The advantage of large floating salteries was that they could move from one district to another to take advantage of the seasonal herring runs—to go where the fish were.

The *Esther* began its first season of herring curing in Prince William Sound, then moved to Kodiak Island, and ended the season in Cook Inlet. The *Bianca* also began the season in Prince William Sound, then moved to Redfox Bay, on Afognak Island. Unfortunately, the *Bianca* was lost during its return voyage from Alaska, but the *Esther* continued to salt herring until 1927, when Hofstad converted it to a floating salmon cannery.[18]

The year 1925 saw a great increase in the number and size of floating salteries. That year, to replace the *Bianca*, the Nassau Fish Company purchased the *Muriel*, a 1,600-ton, 245-foot-long wooden vessel. About the same time, the Utopian Fisheries Company, which had formerly operated several shore stations, purchased the *Donna Lane*, sistership to the *Muriel*. Both vessels were converted into salteries. Wooden vessels were preferred for salting operations because of salt's corrosive effect on iron and steel.

Pacific Fisherman described the alterations made to the *Muriel*, which were similar to those made to the *Donna Lane*:

> A house has been built over almost the entire main deck, which will be used for gibbing and packing. Fish bins have been built fore and aft above the deck, with considerable space between, and gibbing tables along the bins on the outside, leaving a good deck space outside the tables. A conveyor will run along under the tables to remove the offal, which may be taken to a reduction plant if there is one in the vicinity. Fish will be raised from boats alongside by a conveyor and delivered to a deck built above the bins, where they can be spread out to avoid crushing and heating; this being protected from sun and rain by a roof. During the curing and repacking process, the fish will be held in the 'tween-decks section, where there is more than ample room for the purpose; and the hold space is sufficient to accommodate all salt and cooperage required for the season . . .
>
> All conveyors, etc., will be operated by electricity, supplied by three large generators which are part of the regular engine-room equipment, and there is also a donkey engine forward for heavy hoisting. A complete radio outfit is carried . . .
>
> Both vessels have large and well-equipped passenger accommodations in the after section, as well as crew's quarters forward, so the entire crews can be comfortably carried.[19]

That same year, the *Muriel* was renamed ZR3.

Also in 1925, the North American Fisheries Company purchased the 1,035-ton, four-masted schooner *Rosamond* and converted it into a herring saltery, housing over the entire main deck in the process. The *Rosamond*, which was over 200 feet long, forty-one feet wide, and seventeen feet deep, could easily accommodate 10,000 herring barrels in its hold, as well as salt and other equipment, and carried more than 30 crewmembers. The vessel operated in Alaska's herring fishery at least until 1931.

Additionally in 1925, the Libby, McNeill & Libby Company added the 160-foot schooner *Salvator*, and the Latouche Packing Company outfitted the 116-foot Puget Sound freighter *Wakina* as a floating saltery. Unfortunately, the *Wakina* caught fire near Nanaimo, British Columbia, during its maiden voyage to Alaska and was completely destroyed. Fortunately, the crew escaped unharmed.[20]

The floating salteries in 1925 mainly operated in the Redfox Bay area of Afognak Island, and all had successful seasons.[21]

In 1926 and 1927, the Puget Sound Reduction Company used its barge, the *Fort Union*, as a herring reduction plant in the vicinity of Port Armstrong, in Southeast Alaska. In 1927, the floating fleet in Southeast Alaska was augmented by the addition of the steamships *Peralta* (reduction plant) and the 253-foot *Lake Miraflores* (saltery and reduction plant).[22]

By 1928, the Nassau Fish Company had added the 232-foot schooner *La Merced* as a floating saltery, and the Aurora Fish Company had outfitted the 186-foot schooner *Alice Cooke*, likewise, as a floating saltery. The *Alice Cooke* was later owned by the Kalgin Packing Company. It was destroyed by fire in November 1931 while being towed from Latouche to Cordova, in Prince William Sound.[23]

The *Donna Lane*, outfitted in 1928 with a freezer plant that was used to freeze bait herring as well as halibut and some salmon, became something of an institution in Alaska's herring fishery. It operated in the fishery until 1936, there being a pronounced curtailment in the production of cured herring the following year.[24]

Probably the last vessel to be converted into a saltery was the 132-foot former codfish schooner *John A*. In 1936, the vessel was purchased by the Chatham Strait Fish Company, which had a herring operation in Prince William Sound, and was converted into a saltery. The *John A* operated that year at Bluefox Bay, on Afognak Island, in conjunction with the George Hogg & Company reduction plant. Herring offal from the *John A*'s operation and herring that were damaged or too small for curing were reduced at the plant.[25]

Stepping back a few years, cured-herring packers began 1926 with a substantial inventory of carryover from the large pack of the previous season, and there was considerable concern regarding the market for the upcoming season's pack, especially if it, too, was large.[26] Exacerbating the situation was the fact that a record eighty companies engaged in curing herring in Alaska that year. The packers' concern was unwarranted. In the Afognak area, where shore stations and floating salteries were prepared for a large pack, herring failed to materialize in anything like the expected quantities. Likewise, the herring run in Prince William Sound was considerably below expectations. Fishing in Cook Inlet, in the words of *Pacific Fisherman*, was "not notably great."[27] And even though herring were abundant in Southeast Alaska, the fish were considerably smaller than usual, which caused herring packers to curtail their operations. Overall, the 1926 season was a practical failure. The production of cured herring in Alaska totaled 63,149 barrels, the smallest Alaska herring pack since 1920 and less than half of 1925's pack.[28]

For the operators of the thirteen herring-reduction plants in Southeast Alaska (most of whom also operated salteries), 1926 was a record year, with a production of 10,850 tons of meal/fertilizer and 2.9 million gallons of oil. Ward Bower, of the Bureau of Fisheries, estimated that more than 90 percent of the herring caught in Southeast Alaska that year were used in the manufacture of meal/fertilizer and oil.[29]

HERRING PACKERS ORGANIZE

By the mid-1920s, the herring-salting industry (formerly represented by the Pacific Cured Fish Association) and the herring-reduction industry were to a large extent integrated. One of the larger integrated operations was the Baranof Packing Company, at Red Bluff Bay, on the Chatham Strait shore of Baranof Island, in Southeast Alaska. Its four reduction units had a total processing capacity of twelve tons of herring per hour. In 1925, the operation employed more than seventy people, including twenty-two women in the curing department. Six seine boats, manned by forty-six fishermen, supplied herring. Six high-pressure Pelton (water) wheels powered the operation.[30]

During 1926, seventeen of the larger companies in Alaska—eleven in Southeast Alaska and six in Prince William Sound—operated both salteries and reduction plants. Only three companies—two in Southeast Alaska and one in Prince William Sound—operated solely reduction plants. In addition, there were a substantial number of small independent salteries—eight at Halibut Cove alone.[31]

In October 1926, Alaska herring packers gathered in Seattle and began organizing the Alaska Herring Packers Association to facilitate, in the words of *Pacific Fisherman*, "the general protection and betterment of the industry."[32] The move was hastened by the herring industry's desire to defend against what it called "inflammatory and misleading propaganda regarding the activities of the industry and the herring situation in general."[33] This likely referred to the perennial accusations that the manufacture of oil and meal/fertilizer from herring amounted to wanton waste and should be outlawed.[34]

The membership elected Olaf Floe as interim chairman. Floe was head of the Northwestern Herring Company, which operated a saltery and reduction plant at Port Conclusion, in Southeast Alaska.[35] Several meetings later, the group, perhaps attempting to gain more clout by being more inclusive, incorporated as the Pacific Herring Packers Association. The organization's goals were:

> To gather statistics and to disseminate information with reference to the herring industry; to secure the enactment and enforcement of laws to properly regulate, protect and encourage the herring industry; to promote a higher standard of education among the members of the association with reference to both the scientific and practical features of the herring industry; to encourage the production, marketing and branding of products which will in all respects conform to the government requirements as to purity and weight; to encourage the exclusive use of the most sanitary methods for handling, curing and packing herring; to investigate and supply its members with all available information relating to the utilization of waste products, and promote practices tending to the permanent benefit of the industry; and by all lawful and proper ways and means to improve and perpetuate the herring industry on the Pacific Coast, and to promote the welfare of the members.[36]

Following incorporation, Lee Wakefield, of the Baranof Packing Company (saltery and reduction plant at Red Bluff Bay, in Southeast Alaska) and the Franklin Packing Company (saltery and reduction plant at Port Ashton, in Prince William Sound), was elected president.[37] The organization maintained an office in Seattle and was active until at least 1939.[38]

Choosing Seattle as its headquarters made sense. By the time the association was formed, all the larger Alaska herring operators had their headquarters in either Seattle or Portland and had a presence in Alaska only during the herring-fishing season. The companies sent their crews and supplies to Alaska—often aboard company seiners and tenders—in time to prepare for the arrival of herring, and they left soon after fishing ended. Moreover, the herring operations were at remote locations that had few ties to Alaska communities. As Alaska's Legislature later said, "the herring industry contributes nothing to the economy of Alaska."[39]

In retrospect, another notable event during this period was the Storfold & Grondahl Company's entry into Alaska's herring industry in 1925. The company, which *Pacific Fisherman* said was a well-known mild-curer of salmon, also operated a cold-storage plant, and in early 1925 it made preparations to put up a small pack of Scotch-cured herring at its plant at Washington Bay, in Southeast Alaska. The following spring, Storfold & Grondahl installed a reduction plant at its facility.[40] This plant would operate until 1965.

Overall, the 1927 herring fishery was a disappointment. *Pacific Fisherman* attributed the poor production to the failure of herring runs in the Kodiak, Cook Inlet, and Prince William Sound areas; "spasmodic" fishing in Southeast Alaska; and the weekly closed periods.[41] That year, fully twenty-five herring reduction plants operated in the territory—eighteen large plants in Southeast Alaska and seven smaller plants in Prince William Sound.[42] The output of meal/fertilizer (7,237 tons) and oil (2 million gallons) was the smallest in three years, and cured-herring production (57,292 barrels) was the smallest since 1920.

There was, however, an exception to the disappointing season: the bait fishery, which increased its production more than threefold from the previous year, from 2.2 million pounds in 1926 to 8 million pounds in 1927.[43]

The 1928 herring fishery was largely another disappointment. *Pacific Fisherman* printed a grim summary of that year's herring fishery on southcentral Alaska's traditional fishing grounds:

> Prince William Sound, which was for some years an important factor in the production, has had an almost continuous series of failures since 1923, and this year has been the poorest on record. Kachemak Bay, also a large producer for several years, has had its third successive failure; and Red Fox Bay [Redfox Bay, Afognak Island], another former important source of supply, has had poor returns for three years, with a complete failure this season.[44]

But the news wasn't all bad: in Southeast Alaska, production of both cured herring and oil and meal/fertilizer exceeded that of the previous year.[45]

Another encouraging development was the herring fishery's expansion into the eastern Aleutian Islands—specifically, the Unalaska Island area. Occasional small packs of salted herring had been made there, and in 1927 a big run occurred. At the beginning of August 1928, the Utopian Fisheries Company, operator of the floating plant *Donna Lane*, became discouraged by poor fishing in Prince William Sound and at Kodiak and sent two seiners to Unalaska. The seiners encountered a tremendous run of herring. The *Donna Lane* was immediately dispatched. When it arrived about a week later, the seiners had impounded a considerable quantity of herring.

News of the abundance of herring at Unalaska spread quickly, and the schooners *La Merced* (Nassau Fish Company) and *Alice Cooke* (Aurora Fish Company), both floating salteries, soon arrived. Additionally, at least seven packing companies hastily dispatched crews and supplies to Unalaska. The sudden demand for passenger and cargo space swamped the existing transportation capacity, and a special steamship run was made to handle the traffic.

Once at Unalaska, the packers hastily improvised processing facilities, primarily on a dock at Dutch Harbor. Everyone scurried to make the most of an opportunity that might save their season.

About sixteen purse-seiners fished at Unalaska that season. Because the herring there schooled close to the beach, practically no effort was required to locate them. Enormous hauls were made with little difficulty. Though the run was over by the end of August, there were by then some 18,000 barrels (4.5 million pounds) of herring held in impoundments. This inventory allowed packing to continue for nearly a month.

There was a refreshing element of cooperation among the packers at Unalaska that season. They established a perimeter outside the harbors within which offal was not permitted to be dumped. This made for a more wholesome environment around the packing plants, and it was also thought to help maintain the fishery, the prevailing belief among many herring fishermen being that herring did not return to locations where the seafloor had been fouled.[46]

Overall, the Unalaska herring fishery was a huge success, dubbed by *Pacific Fisherman* as "the outstanding feature of the 1928 herring season."[47] With a pack of more than 26,000 barrels, it accounted for about 38 percent of the entire herring pack in Alaska in 1928.[48] This new source of herring was especially important given the scant runs during recent years on the traditional southcentral Alaska fishing grounds.

Another notable event in 1928 was that despite there having been fifteen small operators that packed herring at Halibut Cove, this was the last year herring were packed in commercial quantities there.[49] The herring runs in 1926 and 1927 had been practical failures in Kachemak Bay, and 1928 was no different.[50]

Twenty-five reduction plants operated in Alaska in 1928, and they produced 10,728 tons of meal/fertilizer and 2.5 million gallons of herring oil. Though not a record, it was a considerable increase over the previous year's production. In Southeast Alaska, the movements of herring were irregular, and none of the fifteen reduction plants operated to capacity. This stimulated conversations regarding possible consolidations within the region's herring-reduction industry.[51] In

1929, a dozen reduction plants operated in Southeast Alaska, three fewer than in 1928.[52]

The 1926 record production of meal/fertilizer and herring oil was broken in 1929, when sixteen reduction plants (including one floater) produced 13,807 tons of meal/fertilizer and 3.5 million gallons of herring oil.[53] Production in 1930 was similar: 13,654 tons of meal/fertilizer and 3.5 million gallons of herring oil. The Crab Bay, Prince William Sound, plant of the Chatham Strait Fish Company alone produced 1,454 tons of meal/fertilizer and 445,981 gallons of herring oil, the largest production ever recorded for a single reduction plant in Alaska.[54]

But expansion couldn't go on forever. In 1929, the price of herring oil was forty-five cents per gallon, but with the advent of the Great Depression, the price dropped to twenty-five cents per gallon in 1930, then to seventeen and a half cents in 1931.[55]

That year, seven herring reduction plants operated in Alaska, two in Prince William Sound and five in Southeast Alaska. Significantly, Killisnoo Fisheries (formerly the Alaska Oil & Guano Company and the Alaska Fish Salting & By-Products Company), the pioneer reducer of herring in Alaska, was not among them, ending more than four decades in the business.

As one packer said early in the 1931 season:

> The packers have no hope of making money this season. We are just trying to get by. With the best of luck, we may break even. Under normal conditions, we hope to lose no more money than we would by closing down entirely. We operate for the purpose of keeping our plant in condition and employing as many people as possible.[56]

But the situation would get worse: in 1932, the price of herring oil dropped to an unheard-of ten cents per gallon.[57]

Meanwhile, in 1929 the cured-herring sector packed the smallest amount since 1918, when Scotch curing began in earnest. The 1929 herring catch, however, was actually a record, but the fish were too small for the cured-herring market. The exception was at Unalaska, which accounted for nearly 77 percent of the territory-wide 26,436-barrel pack of Scotch-cured herring. And the fish there were of very high quality, resulting in a pack that was at the time considered to be the best in all Alaska.[58] (The herring caught at Unalaska spawn at Togiak, which is known for especially large fish.)

The situation was considerably different in early 1930. First, the economic depression had caused the market price of cured herring to drop, which tempered the packers' enthusiasm for the upcoming season. This was especially true of the Unalaska fishery. A combination of lower prices for cured herring and the great expense of conducting operations at this remote location did not bode well for operating profitably. "Gloomy" was how *Pacific Fisherman* described the outlook for Aleutian Islands herring.[59] Nevertheless, the floating salteries *Donna Lane*, *Alice Cooke*, and *Rosamund* journeyed to Unalaska to cure herring, and at least five companies set up shore operations there.[60]

Very welcome in 1930 was an abundance of herring in Prince William Sound and Kodiak Island area waters. Importantly, the fish were larger than in previous years and of excellent quality. The total pack of cured herring in Alaska in 1930 was 31,406 barrels, about 15 percent more than the previous year. Nevertheless, because of the onset of the Great Depression, the pack was worth about 12 percent less than in 1929.

Despite the early misgivings about the Unalaska fishery, it produced approximately 46 percent of the 1930 pack. Southcentral Alaska and Southeast Alaska produced 37 percent and 17 percent, respectively.[61]

6
A CHRONICLE OF ALASKA'S HERRING INDUSTRY
1932–1948

This chapter provides a timeline of Alaska's herring industry from 1932 through 1948, a period that marked the demise of the cured-herring industry in Alaska.

> Never in his experience with Southeast Alaska herring has A. Buchan of the Buchan & Heinen Packing Co. noted so many small, immature herring in the Chatham Strait district as during the past season. The fish were evidently of two year classes, as they fell into two distinct size groups, and they were particularly abundant about Port Armstrong.
> Mr. Buchan ordered that his plant's seine boats pass up the small fish, as they were not particularly valuable and he felt they should be permitted to grow without interference as the foundation of the fishery in future years.
> —*Pacific Fisherman*, November 1933[1]

1932

The 1932 herring fishery in Alaska featured a great abundance of fish in the Kodiak area and good showings in both Prince William Sound and Southeast Alaska. Only in western Alaska were herring comparatively scarce, a problem that was compounded by their small size.[2] Territory-wide, cured-herring production totaled 51,225 barrels, an increase of nearly 62 percent over the previous year and the largest since 1928.[3] Seven reduction plants—five in Southeast Alaska and two in Prince William Sound—produced 9,609 tons of meal and 18.8 million gallons of oil, increases of about 21 percent and 16 percent, respectively.[4]

1933

For Alaska companies involved in Scotch-cured herring, 1933 began on an encouraging note: the quality of recent herring imports from Europe was low, and even though depreciated European currencies had lowered the price of the fish, the US market balked at purchasing it.[5]

These factors opened the market for Alaska herring, at least in the short term. By this time, most of the territory's producers of cured herring were larger companies that paid close attention to the quality of their products. Consistently high quality greatly facilitated the marketing of Alaska herring.

And there was more good news: in March 1933, President Franklin Roosevelt signed the Cullen-Harrison Act, which was enacted just months before Prohibition ended and authorized the sale of 3.2 percent beer, which was thought to have too low an alcohol content to be intoxicating. Tavern patrons often snacked on herring and other cured or smoked fish while drinking beer.[6] *Pacific Fisherman* celebrated the legislation with a poem, though neglected to name its author.

Hoch Der Herring [High the Herring]

The herring wiped a streaming eye
And dashed away a tear;
"I fear our fate in Chatham Strait—
"I see they've brung back beer."

From Kodiak to Golovin
The herring shook with fear;
The terror ran to Akutan—
For they have brung back beer.

But up spake Old Harengus,*
"My children be of cheer;
"A nobler fate does await
"Since they have brung back beer.

"We will not go for meal and oil;
"They'll treat us kindly here.
"They'll cure in fine smoke and brine,
"Now they have brung back beer.

"In gilded grills, on shining bars,
"We herring will appear;
"For though we die, we'll glorify
"Their weakling three-two beer."[7]

---

* "Old Harengus" is derived from the scientific name for Atlantic herring, *Clupea harengus*.

In August 1933, the Bureau of Fisheries issued a regulation that prohibited the dumping of herring offal or dead herring in the waters of any bay in which herring spawned. Until this time, regulations had prohibited such dumping in specific bays or waters—Halibut Cove, for example—but the new regulation applied to all of Alaska.[8] This assuaged the previously mentioned prevailing concern of the many herring fishermen who believed that herring did not return to locations where the seafloor had been fouled.

Despite the fact that the 1933 Scotch-cured herring pack was light and inventories were modest, by September the prices being offered for it had declined. According to *Pacific Fisherman*, the prices were "pitifully low," but many packers, in need of cash, were forced to sell. The industry attributed the low prices to what *Pacific Fisherman* called "a small and closely knit buyer's block" that could dictate prices.

Alaska packers who still had fish and could afford to hold onto them refused to sell at what were essentially going-out-of-business prices.[9] In October, they organized the Alaska Herring Marketing Association, a cooperative, in an effort to facilitate the orderly marketing of the year's cured-herring pack by pooling their herring inventories. The impact was an immediate, sharp firming of the market price for cured herring.[10] Despite its success, the Alaska Herring Marketing Association seems to have been dissolved soon after it accomplished its goal.

On a somewhat humorous note, herring men that fall awarded the Southeast Alaska Herring Championship belt to Andrew Buchan, of the Buchan & Heinen Packing Company, which operated a saltery and reduction plant at Port Armstrong. The "trophy" belt had dubious sartorial value: it was a rusty barrel hoop studded with herring scales. And it came with a price: the winner paid for the banquet at which it was awarded.[11]

Though herring were relatively abundant in Prince William Sound and Southeast Alaska, the run at Kodiak was light. Production of Scotch-cured herring in Alaska declined slightly from the previous year, but the production of oil and meal/fertilizer increased somewhat, in part because nine reduction plants (six in Southeast Alaska and three in Prince William Sound) operated in the territory, two more than in 1932. Moreover, the market for herring oil had improved substantially during the past year. In 1932, the price of oil was ten cents per gallon; 1933's production sold for from twelve cents to fifteen cents per gallon.[12]

1934

The outlook for 1934, at least for packers of cured herring, was, according to *Pacific Fisherman*, "brighter than that which has greeted the herring packers at the first of any year for some time past." The reason for this optimism was the recently depreciated value of the US dollar, which made imports from Scotland less likely because curers there would have to sell their fish at a lower price in British pounds (the currency in Scotland) to keep the price in US dollars the same.[13]

In early May, as packers were preparing for the upcoming season, longshoremen on the West Coast went on strike. The packers did what they could to move their supplies and workers north. For gibbers at several Southeast Alaska salteries, this worked out well. The salteries chartered two yachts to transport the women north. As *Pacific Fisherman* wrote, "The girls travelled in style rarely, if ever, enjoyed by herring chokers in the past."[14]

Despite the strike, Alaska herring plants were able to begin operations on schedule, but the question remained of how oil and meal produced in early June at reduction plants along Chatham Strait would be moved to market. Fortunately, in mid-June a truce was declared regarding Alaska shipping, and freighters were dispatched to move the product.[15] The longshoremen's strike ended in mid-July without materially affecting Alaska's herring-packing season.

Herring were abundant in Southeast Alaska, Prince William Sound, and at Unalaska. The fish in Southeast Alaska and Prince William Sound, however, were generally too small for curing. At Kodiak, the herring were exceptionally fine and large, but the run was light. Only at Unalaska was there an increase in cured-herring production. The territory-wide production of 29,637 barrels of Scotch-cured herring represented a decrease of more than 40 percent from the previous year. In part because the quantity was limited, the packers found a ready market and good prices for their product.

By contrast, the production of oil and meal, 3.7 million gallons and 13,851 tons, respectively, was the largest in the history of the industry. Nine reduction plants, each with an associated saltery, operated in the territory. Six were in Southeast Alaska, and three were in Prince William Sound.[16]

The season was capped in October, when Andrew Buchan, of the Buchan & Heinen Packing Company, once again won the coveted Southeast Alaska Herring Championship belt. Buchan's win seems to have been attributed to his company's Port Armstrong plant producing nearly half a million gallons of herring oil.[17]

1935

The most significant development of the 1935 Alaska herring fishery was the establishment of three modest-sized herring-reduction plants on Kodiak Island. *Pacific Fisherman* attributed the plants being established there to a "liberalized attitude of the Bureau of Fisheries," although no relevant regulations had changed.[18]

The introduction of reduction operations enormously increased demands on the Kodiak area's herring stocks. From 1930 to 1934, the annual herring catch had averaged about 5.3 million pounds. From 1935 to 1942, the annual catch would average 54 million pounds—a more-than-tenfold increase.[19]

Though fishing was hindered by poor weather, the addition of the three plants at Kodiak, plus three new reduction plants in Prince William Sound and one in Southeast Alaska, increased production such that a new record was set.[20] In all, nineteen reduction plants operated in the territory in 1935, producing 15,818 tons of meal, about 8 percent more than in 1934, and 3.9 million gallons of oil, about 2 percent more than in the previous year. Significantly, the price of oil had increased to thirty cents per gallon, and the total value of the oil produced ($1.1 million) was 76 percent greater than that of 1934.[21]

Even more significant was the near doubling of the production of Scotch-cured herring. The territorial total was 59,719 barrels, the greatest since 1928. The pack was generally of good quality, but its increased size put such a volume of fish on the market that the price declined slightly. It helped that European competition that year was not substantial.[22]

The cherished Southeast Alaska Herring Championship belt was awarded that year to Olaf Floe, head of the Northwestern Herring Company, which operated a saltery and reduction plant at Port Conclusion, in Southeast Alaska.[23]

1936

The year 1936 began with considerable uncertainty. The market for herring oil was weak, its price having dropped several cents per gallon from the previous year. At the same time, the fishermen who crewed the herring seine fleet were demanding a pay increase. In 1935, the year the fishermen had organized into the Herring Fishermen's Union of the Pacific, plant operators had paid each crewman 6.5 cents per barrel

Figure 6.1. Unloading herring from purse seiner, Buchan & Heinen Packing Company, Port Armstrong, Southeast Alaska. (*Pacific Fisherman* yearbook, January 1929)

(200 pounds) of herring delivered. For the 1936 season, the fishermen wanted seven cents per barrel. (The packer furnished the boat, net, and fuel and oil for the boat; the fishermen furnished their own food.[24]) In addition to the fishermen's demands, herring plant workers in 1936 organized into the Fish Reduction and Saltery Workers' Union and were making demands of their own.[25] These unions would complicate processors' relations with fishermen and employees in upcoming years.

And then there was the perennial issue of how much cured herring would be available from Europe. Compounding the supply issue was Iceland, which in early 1934 had begun exporting substantial quantities of Scotch-cured herring to the United States and seemed poised to increase exports.

The reduced price for oil and uncertainty in the cured-herring market, coupled with the fishermen's demands, jeopardized reduction operations to the point that many companies considered sitting out the season.[26] In late April, to resolve the fishermen's pay issue, Prince William Sound packers suggested utilizing a sliding scale based on the price of herring oil. The fishermen's union rejected the offer, holding out for seven cents per barrel. Not long after, Kodiak packers agreed to the fishermen's price, and packers in Prince William Sound and Southeast Alaska soon followed, although three firms in Southeast Alaska elected to not operate their salteries and reduction plants.[27]

The fact that the packers' rejected offer was based on the price of oil illustrated the importance of oil to the packers' operational profitability. The quality of Alaska herring oil had been steadily improving, and by 1936 the average shipment was almost free of moisture, impurities, and objectionable odors.[28] This undoubtedly increased the marketability of the product.

It may not have been an auspicious time to get into the reduction business, but the Oceanic Fisheries Company, which operated the saltery/freezer ship *Donna Lane*, installed a reduction plant on Raspberry Island, off the north shore of Kodiak Island. The plant would utilize offal from the *Donna Lane*'s operations as well as herring that were unsuitable for curing.[29]

In Southeast Alaska, herring suitable for curing were scarce, and the Scotch-cure pack was the smallest since the method had been introduced in the territory in 1917. The output of meal and oil in the region was the lowest it had been since 1924.

Herring were scarce in Prince William Sound at the beginning of the summer but were very abundant in the Kodiak area, where the run was said to have been the best in recent years. A substantial number of seiners from fish-starved Prince William Sound began fishing at Kodiak, bringing their catches back to Prince William Sound. In his annual report, Ward Bower, the Bureau of Fisheries agent in Alaska, somewhat sarcastically noted that "the long haul to Prince William Sound in unusually warm weather did not improve the quality of the fish." Practically all of the herring thus transported were reduced to oil and meal. Fortunately for Prince William Sound fishermen and packers, a large run of herring arrived there in September, and plants operated at full capacity for several weeks.[30]

The territory-wide production of Scotch-cured herring in 1936 was 45,740 barrels, a decline of 23 percent from the previous year but close to the average annual pack during the past decade. Nevertheless, the business was in a precarious state. Foreign competition was robust, production costs were out of line with probable returns, and the market was limited and, to a large extent, controlled by a few East Coast firms.

Pacific Fisherman opined that 1936 "probably dealt more harshly with the Alaska [cured] herring operators than with any other division of the general industry."[31] Particularly troubling were imports. Alaskans had grown accustomed to stiff competition from Scottish packers, but now exports from Iceland were increasing and compounding the problem. And the quality of the Icelandic product, which was often found wanting, had improved. Scotch-cured Icelandic herring found increasing favor in the US market. Price was also a factor: during 1936, the Icelandic product was delivered to New York at prices that were said to be barely half of what Alaska packers needed to break even. This dire situation did not bode well for the future of Alaska's herring-curing industry.

The situation with herring oil and meal, on the other hand, was pretty much status quo. The seventeen reduction plants that operated in Alaska in 1936 produced 3.8 million gallons of herring oil, a decrease of 1 percent in quantity and 15 percent in value over the previous. Meal production, 15,051 tons, decreased about 6 percent in quantity but increased 46 percent in value.[32]

On the bright side, Olaf Floe won his second Southeast Alaska Herring Championship belt.[33]

1937

Alaska herring curers thought the 1936 season was bad—"intolerable," according to *Pacific Fisherman*—but 1937 was worse.[34]

Curers entered the new year with a large portion of their 1936 pack unsold and, given the availability of cheap imports from Iceland, little possibility of selling it at even break-even prices.[35] The buyer of last resort was a Great Depression–era institution, the US Federal Surplus Commodities Corporation, which purchased about 6,250 barrels of Scotch-cured herring. The corporation paid a below-cost price for the fish, but this at least allowed the packers to salvage something for their efforts.[36]

Given the situation, packers were hesitant to make firm plans for the 1937 season. The consensus was that profitably curing herring would be virtually impossible except at the most favorable locations, and even at those locations, only limited quantities of the best fish should be cured.[37]

Alaska's territorial legislature tried to help. Even when it was enacted in 1922, the federal tariff on cured herring, $2.50 per standard barrel, was not enough to offset the freight-cost advantage enjoyed by European packers. Alaska's territorial legislature generally had little sympathy for reduction plants, but it recognized the cured-herring sector as an important, potentially growing component of the territory's fishing industry. In March 1937, Alaska's Legislature memorialized Congress to double the duty on cured herring.[38] Congress, likely pressured by East Coast buyers, chose not to do so.

Besides the dismal market for cured herring, herring packers, including those that operated reduction plants, faced demands by the recently reorganized Fish Reduction, Salters, and Gibbers Union for higher wages and better working conditions. The union's demands were formalized on March 1, and an agreement—among its provisions were substantial wage increases—was reached with the plant owners in late May.[39]

Fishermen, too, demanded an increase in pay. Although they had agreed upon a crew share of seven cents per barrel for the 1936 season, they now wanted eleven cents per barrel. This high price, said some, would make it impossible for herring-reduction plants to operate.

Plant operators countered with an offer of nine cents per barrel. The fishermen asked for ten cents. Meanwhile, the price of herring oil was dropping. The plant operators held firm with their nine-cent offer, which the fishermen's union, by a narrow vote, accepted. The crew share of nine cents per barrel was the highest it had ever been in the Alaska herring industry.[40]

Burdened by imports of cheap foreign herring and higher labor and fish costs, only seven packers risked curing herring in 1937. It was a bad decision, for even those that were most favorably situated lost money. Despite the fact that, except for at Unalaska and Akutan, herring were generally abundant in the territory in 1937, production of cured herring sank to the lowest level in the industry's history. Of the total of 8,414 barrels of Scotch-cured herring produced, fully 70 percent was attributed to the Kodiak area. In Southeast Alaska—the birthplace of herring curing in Alaska—the pack was a mere 292 barrels.

Pacific Fisherman summed up the position of Alaska's cured-herring industry in January 1938: "Although it is generally acknowledged that the quality of Alaska herring is superior to that from foreign countries, the domestic product labors under such a serious cost handicap that competition is virtually impossible."[41]

Except for those in western Alaska, each of the companies that cured herring also operated a reduction plant. For this component of their operations, 1937 was a banner year. The twenty reduction plants harvested more than 138,000 tons of herring, from which they produced a record 19,636 tons of meal and 5.5 million gallons of oil.

And though the price of meal had declined somewhat, the price of oil had risen handily. Oil production increased 49 percent over the previous year, but the value of the total production increased 122 percent.[42] The overall 1937 Alaska herring catch of 138,898 tons stands as a record to this day.[43]

Even under the best of conditions, fishing is a dangerous occupation. Tragedy struck the herring industry on the night of September 29, 1937, when the Storfold & Grondahl Packing Company's sixty-foot purse seiner *Limit* foundered and was lost with all hands. The vessel had been fishing at Larch Bay, on the southwestern shore of Baranof Island. Thought to have been heavily laden with herring, the *Limit* was on its way to the company's plant at Washington Bay. As it rounded Cape Ommaney, the vessel likely encountered a strong northerly wind blowing down Chatham Strait. Since few items from the boat—among them three hatch covers—were found, it was assumed that the *Limit* was instantly overwhelmed and sank before the crew could do anything to save themselves. Among the eight individuals who perished that night was Olaf Storfold, captain of the *Limit* and brother of John Storfold, president of the Storfold & Grondahl Packing Company.[44]

At the end of the season, at least Olaf Floe, of the Northwestern Herring Company, had reason to celebrate: for the third year in a row (and at a banquet he sponsored and paid for), the indomitable Floe was awarded the prestigious, if rusty, Southeast Alaska Herring Championship belt.[45]

1938

> The growing intensity of this [herring] fishery gave rise to fears of depletion, and in certain parts of the Territory there was ample evidence of reduced abundance.... Facts were obtained on which to base sound restrictive measures, and regulations are now in force which

will restore this fishery to a level of maximum productivity.—Message from President Franklin Roosevelt to Congress, January 1938[46]

So far as cured herring is concerned, the Alaska industry is pretty well resigned to the fact that it is about through, save for supplying the scanty needs of the Pacific Coast and a few inland markets in the western half of the United States.—*Pacific Fisherman*, April 1938[47]

Through the Herring Fishermen's Union of the Pacific, fishermen in 1937 successfully negotiated a high ex-vessel price for herring. That fall, the union combined with several other fishermen's unions to form the coastwise United Fishermen's Union of the Pacific.[48]

In the spring of 1938, the unionized fishermen and plant workers began negotiations with an informal Alaska herring packers' committee. Fishermen asked for a crew share of ten cents per barrel. But the market for herring oil, the fats market, had weakened substantially, and the committee offered six cents per barrel. For plant workers, the committee wanted a 10 percent pay reduction for workers earning between $100 and $120 per month, and a 15 percent reduction for workers earning more than $120 per month.[49]

In late May, just before the fishing season was scheduled to open (June 1 in Southeast Alaska, June 8 in Prince William Sound, and June 15 at Kodiak), the fishermen and plant workers were still at loggerheads with the packers. Meanwhile, the price of oil was continuing to fall.[50] The standoff was finally resolved on June 27, when the fishermen agreed to seven cents per barrel and plant workers accepted pay reductions of approximately 10 percent.[51]

Packers scrambled to get their operations running. Fortunately for them, the Bureau of Fisheries had recognized in early June that the prolonged labor negotiations would delay the herring season and curtail the catch. To compensate for the lost fishing time, the agency eliminated the weekly closed periods in Prince William Sound and at Kodiak.[52] The weekly closed period in Southeast Alaska was allowed to stand, apparently out of concern that the herring stocks there might have been depleted. This proved to be a prudent decision, because through the end of August, the herring run in Southeast Alaska was the poorest in the history of the industry. Although fishing improved somewhat in September, overall, the catch was, in the words of *Pacific Fisherman*, "extremely poor."[53] And the situation would get worse.

The total pack of Scotch-cure herring in Alaska in 1938 was 8,945 barrels. In addition, 1,746 barrels of hard-salted herring were produced. Significantly, no herring were cured or salted in Southeast Alaska. *Pacific*

Figure 6.2. Beach seining herring at Kodiak, probably 1938. (NOAA Fisheries collection)

Fisherman attributed the lack of interest to the herring themselves: the fish lacked the sort of "special characteristics" that would have enabled them to compete with imported herring. The journal pronounced Southeast Alaska's outlook to be—at least so far as cured and salted herring were concerned—"very dark."[54]

The territory-wide meal production of 16,095 tons represented a decrease of about 14 percent over the previous year, and the production of oil, 4.5 million gallons, a decrease of about 19 percent. More than half of the total production of meal and oil, about 55 percent and 53 percent, respectively, occurred in Prince William Sound, which experienced its best season ever. Eight reduction plants operated in the sound. Production of meal and oil in Southeast Alaska, where five reduction plants operated—three fewer than in 1937—represented less than a quarter of the territory-wide production.[55]

1939

> The needs of the western portion of the country; the desirability of some Alaska herring of special characteristics; and the trade's interest in maintaining some vestige of a domestic industry probably will prevent complete extinction of the Alaska curing industry.—*Pacific Fisherman*, January 1939[56]

In their 1939 book, *North Pacific Fisheries*, Homer Gregory and Katherine Barnes classified the herring fishery as a "minor fishery."[57] That year, some 1,060 fishermen and workers were engaged in the herring industry, and the total value of its production—meal, oil, cured herring, and bait herring—was $2.1 million. By contrast, Alaska's canned salmon industry employed 24,921 fishermen and workers, and the total value of its production was $34.4 million.[58]

A threat was hanging in the air when herring packers began preparing for the 1939 season. In early 1937, Commissioner of Fisheries Frank Bell had warned that expansion of herring reduction operations in Prince William Sound and the Kodiak Island area would necessitate the imposition of measures to limit the catch.[59] Herring reduction in Prince William Sound began in 1920, when two reduction plants were installed, and was initiated at Kodiak in 1935, when three reduction plants were installed. In his report on Alaska's fisheries in 1937, Ward Bower noted the "marked expansion in the manufacture of meal and oil" as an "outstanding feature" of the herring industry that year. Six reduction plants operated in Prince William Sound and four in the Kodiak Island area.[60]

Herring were abundant in both areas in 1938, but the run in Southeast Alaska was a disaster. In response, reduction plant operators began shifting away from Southeast Alaska. The Northwestern Herring Company shuttered its reduction plant at Port Conclusion after the 1938 season and the following year established a plant at Drier Bay, in Prince William Sound. Likewise, the Chatham Strait Fish Company, which had not operated its reduction plant at New Port Walter since the 1935 season and was already operating a reduction plant in Prince William Sound, in 1939 established a plant at Zachar Bay, on Kodiak Island. Within these two areas, a dozen herring-reduction plants would operate in 1939: seven in Prince William Sound and five in the Kodiak Island area.[61]

The Bureau of Fisheries feared this expansion would lead to overfishing, as had occurred in Cook Inlet and parts of Southeast Alaska, and in late March the agency issued regulations that limited the herring catch to 200,000 barrels in the Kodiak Island area and 350,000 barrels in Prince William Sound. These limits (quotas)—a new method for regulating herring fisheries in Alaska—were based on average catches during recent years.

As the 1939 season progressed, however, it became apparent that the herring runs in both areas were above average. The Bureau of Fisheries accordingly revised the quotas to permit an additional catch of 100,000 barrels in each area. In the Kodiak Island area, the new quota was reached in late August. In Prince William Sound, fishing slowed toward the end of the season, and the year's catch totaled about 422,000 barrels.[62]

In Southeast Alaska, where there was evidence of severe depletion of herring stocks, the Bureau of Fisheries issued a regulation at the beginning of the year that prohibited all commercial fishing for herring, except for bait purposes, in the vicinity of Cape Ommaney. The Cape Ommaney herring population—a single stock that originates in the Sitka-area spawning grounds—had long been the mainstay for herring packers in Southeast Alaska, contributing about 80 percent of the total catch from 1927 to 1938. The bulk of the catch was taken within five miles of the cape during the summer feeding migration.[63] Only four herring-reduction plants operated in Southeast Alaska in 1939, the fewest since 1922.[64]

The Pacific Herring Packers Association later claimed herring were abundant in the closed area and that had they been able to fish where herring were plentiful, the catch would have been one of the biggest in history.[65]

More changes were afoot. In April 1939, Congress passed legislation that reorganized the federal government's administrative structure. In accordance with this legislation, President Franklin Roosevelt on July 1 transferred the Bureau of Fisheries from the Department of Commerce to the Department of the Interior.[66] The following year, Roosevelt merged the Bureau of Fisheries with the Biological Survey to form the US Fish and Wildlife Service. These changes had little, if any, effect on the management of Alaska's fisheries. Ward Bower, who had been the federal government's chief Alaska fisheries agent since 1925, would continue in that position until his retirement in 1947. In 1944, the Territory of Alaska became Region 6 of the Fish and Wildlife Service, with its headquarters in Juneau.[67]

By 1949, the service began regularly holding hearings to discuss and gain input regarding Alaska herring fishery regulations, including regulations proposed by the herring industry and other fishing-industry groups. The hearings were held in Seattle, typically in late fall, and were sometimes heated and divisive.[68]

Notwithstanding the curtailment of fishing in Southeast Alaska, the 1939 production of herring meal and oil was the second largest in the history of the industry, exceeded only by that of 1937. The sixteen reduction plants produced 16,626 tons of meal and 4.8 million gallons of oil.[69]

The production of Scotch-cured herring in 1939, 11,874 barrels, was substantially more than had been expected at the beginning of the year. The increase was attributed to the beginning of World War II in Europe in September. As had been the situation during World War I, hostilities reduced the amount of cured herring that could be exported to the

US, stimulating production in Alaska. Packers at Kodiak accounted for most of the territory's production. Production of hard-salted herring, primarily bloater stock—large, ungutted herring that were salted in barrels or wooden boxes and shipped to smokeries, where they were lightly smoked—was 2,458 barrels, almost all of which was produced in the Unalaska area.[70]

1940

The new year brought more bad news to Southeast Alaska's herring packers: the newly formed Fish and Wildlife Service had determined that the herring stock had been so seriously depleted that the area would be closed to herring fishing except by gillnets and for bait purposes.

The restriction should have come as no surprise. Although operators and fishermen professed that the supply of herring was virtually unlimited and that annual fluctuations in the catch were attributable to variations in fish behavior, the chronic scarcity of herring in the region could not be attributed solely to normal population fluctuations. Though poor recruitment was a factor, the main cause of the depletion was years of overharvest to feed the region's reduction plants. Clearly, the fishing pressure on Alaska's herring resource needed to be more closely managed, lest herring in other areas of Alaska suffer the same fate as those in Southeast Alaska.

Packers in the region strongly objected to the closure, and in March, five companies with reduction plants in Southeast Alaska proposed a joint operation of a single plant supplied by a maximum of six seine boats. Their proposal included a proviso that no waters would be closed to fishing and there would be no curtailment of fishing during the season. The companies argued that fishing success—or lack of success—would be the best way to determine herring abundance. The Fish and Wildlife Service rejected the proposal.[71]

Packers countered the rejection with a proposal for the joint operation of a single plant and a 60,000-barrel quota. They hoped to garner sufficient revenue from the joint operation to cover their plants' maintenance costs and to prove that herring were abundant. Not long after, all the packers except Arentsen & Company withdrew from the proposal. The company's persistence was rewarded: in early August, the Fish and Wildlife Service agreed to allow the Arentsen & Company reduction plant at Big Port Walter to operate under a 30,000-barrel quota, with a provision limiting the weekly catch in certain areas.

Fishing opened on August 27, and five seine boats headed to the fishing grounds. It was all for naught; the fishermen caught nothing, and the reduction plant was closed on October 9.[72]

The outlook for Prince William Sound and the Kodiak area was considerably brighter. The quota at Kodiak was 300,000 barrels, the same as it had been in 1939, and in Prince William Sound, it was 350,000 barrels, about 17 percent less than the catch in 1939.[73]

But the ability to catch these quotas became doubtful when negotiations between packers and fishermen over the price of herring extended into the summer. It was July 25—the latest date ever—before they agreed on a price of seven cents per barrel. Complicating matters, most of the seiners were still in Seattle, where they had been awaiting the results of the negotiations, so they wouldn't be on the fishing grounds until early August. Because of the season's late start, two reduction plant operators in the Kodiak area elected not to operate.[74]

As a result of the severe restriction on fishing in Southeast Alaska and the delay in the start of the fishing seasons in Prince William Sound and the Kodiak area, the total volume of herring products produced was the lowest it had been since 1924. In neither Prince William Sound nor in the Kodiak area did fishermen reach the established quotas.

Alaska's salteries, however, had quickly geared up their operations to take advantage of the war-induced shortage of cured herring, and they produced 17,918 barrels of Scotch-cured herring in 1940, about 70 percent of which was cured in the Kodiak area. But, given their experience during World War I, the packers understood that once the war ended, they would again face stiff competition from Europe. As had been the case since the genesis of the cured-herring industry in Alaska, high transportation costs largely precluded Alaska packers from competing with Scotland and Iceland in the East Coast market. For the past four years, Alaska cured-herring packers had limited their production to an amount sufficient to supply markets in the western states.[75]

Owing to the late start of fishing and the closure of fishing in Southeast Alaska, production of meal and oil in 1940 was the lowest since 1927 and less than half the amount produced in 1939. Meal production totaled 8,143 tons, and oil production totaled 2.1 million gallons.[76]

In 1940, *Pacific Fisherman* published a three-part series on Alaska herring research by Edwin Dahlgren, a researcher at the Bureau of Fisheries. Dahlgren had worked with and succeeded George Rounsefell in studying Alaska's herring. As part of his study, Dahlgren continued Rounsefell's herring-tagging program, which had begun in 1932.[77] Among Dahlgren's conclusions:

- There was "indisputable evidence" that the Pacific herring population comprised a series of independent populations, each with their own spawning area and feeding grounds.
- Within Southeast Alaska, there were several herring stocks. For example, the herring caught in the outside waters of lower Chatham Strait and Iphigenia Bay (on the outer coast of Prince of Wales Island, between Coronation Island and Noyes Island) were of a different stock than herring caught in the inside waters—specifically, Behm Canal, Ernest Sound, and upper Frederick Sound. Thus, the decline of inside-waters herring populations could not be attributed to "decimation of the 'outside' stocks by the commercial fishery."[78]

Probably because of favorable marine conditions, the 1931 herring year class (herring born in 1931) was very strong in Southeast Alaska, Prince William Sound, and the Kodiak area. The 1932–1934 year classes, however, were weak in all three areas. The 1931 year-class fish entered the fishery in 1934 and sustained it at a high level until 1936, after which there were few juvenile fish to replace fish that had been caught or had been lost to natural mortality.

Dahlgren also discussed the effect of abundance and availability on herring catches. Under ideal conditions, a relatively small school of fish might be easily available to fishermen, and large catches could be made. Conversely, under unfavorable conditions, even a very large school of fish could be difficult to locate and catch, the result being small catches. Poor fishing conditions, however, typically prevailed during only part of the season. Dahlgren reasoned that if catches over a period of years showed a consistent upward or downward trend, availability could not be the only responsible factor. This reasoning, coupled with the landings data that had been collected over more than a dozen years, seemed intended to refute fishing industry claims that herring were abundant even though fishermen hadn't been able to catch them.

Dahlgren ended his articles with an admonition: "Man must temper his demands on the natural resources in accordance with nature's ability to replace his depredations."[79]

1941–1942

The 1941 herring quotas for Prince William Sound and for the Kodiak area were 250,000 barrels each, somewhat lower than the previous year. However, a provision in the regulations allowed unrestricted fishing outside designated quota areas.[80] The Fish and Wildlife Service opened Southeast Alaska's herring fishery, but it did so cautiously, with a quota of only 50,000 barrels.[81]

On April 18, three and a half months earlier than in 1940, fishermen and packers agreed on a share price of nine cents per barrel, two cents more than in 1940. This timing allowed packers to take full advantage of the season.

At Kodiak, fishing outside the quota area was exceptionally good during the early part of the season, and 60,000 barrels were taken near Old Harbor, on the southeastern shore of Kodiak Island. Fishing within the quota area was good as well, and fishermen caught the full quota.

In Prince William Sound, fishing was hindered by contrary weather and periodic scarcities of fish. Though the herring season officially ended October 15, all the operations in the sound were closed by the end of September. The catch was about 33,000 barrels short of the quota.

The limited herring fishery in Southeast Alaska was a disappointment. Only the Storfold & Grondahl reduction plant, at Washington Bay, operated. Three seiners fished for the company, but their catch was 24,200 barrels, a little less than half of the quota.

Despite the modest production in Southeast Alaska and the curtailed fishing in Prince William Sound, the territory's production of meal and oil was substantially greater than in 1940. Meal production totaled 11,764 tons; oil production totaled 2.9 million gallons. For the first time ever, production of meal and oil in the Kodiak area exceeded that produced in Prince William Sound.

Production of Scotch-cured herring, a total of 8,894 barrels, continued its downward trend. The Kodiak area accounted for almost the entire production. No herring were cured in Southeast Alaska, and only a nominal amount was cured in Prince William Sound. One saltery, Hoveland and Nesskaug, operated at Unalaska and produced 683 barrels of Scotch-cured herring and 1,028 barrels of bloater stock. At Golovin, the Native Cooperative Fisheries produced thirty barrels of bloater stock.

On December 7, 1941, the Japanese bombed Pearl Harbor. War had come. In January 1942, *Pacific Fisherman* touted the value of herring to the war effort: "The Alaska herring runs constitute a resource of immense wartime importance, capable of providing food in important volume, or the oil and meal which can play so important a part in supporting industrial and agricultural production."[82]

For the war effort, Secretary of Interior Harold Ickes, who titled himself "Coordinator of the Fisheries," recognized the need for "every possible pound of fresh, frozen, canned and otherwise preserved fishery product." In Alaska, the department focused on securing a supply of canned salmon for the military.[83] (In nearby British Columbia, by contrast, a sizable canned-herring industry developed during the war years.[84])

In anticipation of poor herring runs, the Fish and Wildlife Service reduced the quotas in all three major herring fishing areas. In the Kodiak area, the quota dropped from 250,000 barrels in 1941 to 150,000 barrels in 1942. In Prince William Sound, the quota dropped from 250,000 to 75,000 barrels, and Southeast Alaska's quota was reduced from 50,000 barrels in 1941 to 2,000 barrels per month (except for fish caught by gillnets or for bait purposes), an amount intended to satisfy local needs.

As expected, Alaska's herring industry protested the quota reductions as not consistent with the war effort. "Right at a time when our government is asking for fishery production, and particularly emphasizes the urgent need of fish oils which contain glycerin, these drastic new regulations come out, practically shutting down the herring industry of Alaska," stated W. W. Wilde, secretary of the Pacific Herring Packers Association.[85] The industry's protests proved futile.

Three reduction plants operated in 1942, two at Kodiak and one in Prince William Sound. It being wartime, labor was scarce, and the companies negotiated a fishermen's crew share of sixteen cents per barrel, the highest ever.

At Kodiak, the Apex Fish Company was the only individual company to operate a reduction plant. Three other companies, the Chatham Strait Fish Company, Southwestern Herring, Inc., and the Oceanic Fisheries Company, operated as a joint venture under the name Kodiak Operators. The venture operated the Oceanic Fisheries Company reduction plant at Port Vita, on Raspberry Island, which was supplied by six seine boats.

In Prince William Sound, the one reduction plant that operated, that of the Oceanic Fisheries Company, at Thumb Bay ("Port Oceanic"), on Knight Island, was operated as a joint venture of five companies.

Fishing at Kodiak was outstanding during the first couple of weeks, but it tapered off substantially. The fleet came up several thousand barrels short of the area's 150,000-barrel quota and was back in Seattle by the end of September. In Prince William Sound, the herring run was almost a complete failure; the fishery produced only about 14,000 barrels of fish, less than one fifth of its 75,000-barrel quota.

Production of herring meal and oil sank to an all-time low. Meal production totaled 3,540 tons, and oil production was 832,518 gallons. Scotch-cured herring production, too, sank to an all-time low: 1,370 barrels. Three companies at Kodiak accounted for the cured-herring production.[86]

On the bright side, the take of herring for bait, 7.4 million pounds, was almost a million pounds more than the previous year. That said, in October 1942, *Pacific Fisherman* summed up the season: "All-in-all, the Alaska season was extremely disappointing and some stiff losses were experienced."[87]

1943

On March 1, 1943, Harold Ickes, secretary of the interior and coordinator of fisheries, met in New York City with consultants from the fishing industry. "Our fisheries can produce more food per man-hour and per dollar invested capital than any other element in the food business," said Ickes. He estimated that in 1943 the United States would need to produce some seven billion pounds of seafood products to supply the armed forces, the US allies, and civilian needs. The estimated national production, however, would be only about half that amount. Ickes wanted the maximum production that could be attained without seriously impairing fishery resources, and he was willing to relax some conservation regulations to further that goal.[88]

For Alaska's herring industry, this fundamentally meant increases in the herring quotas in the Kodiak area and Southeast Alaska. The Kodiak-area quota was initially increased to 200,000 barrels, but an additional 25,000 barrels were added about the first of August. In Southeast Alaska, the quota was increased to 100,000 barrels. The quota in Prince William Sound was not increased, but the waters in which the quota could be caught were expanded to embrace the entire area.[89]

The result of the liberalized policy was a 130 percent increase in the herring catch, from 146,409 barrels in 1942 to 337,293 barrels in 1943. Likewise, meal production more than doubled, to 6,330 tons, and oil production nearly doubled, to 1.6 million gallons. Alaska herring were said to be on the lean side that year, so the increase in meal production was not reflected in oil production. The Kodiak area, where four reduction plants operated, accounted for the lion's share of the production. Sixteen seiners fished in the area, two of which delivered fish to the single reduction plant that operated in Prince William Sound. Four seiners supplied the single reduction plant that operated in Southeast Alaska.[90]

Regarding Scotch-cured herring, the market was fairly strong, and some packers provisioned themselves for a substantial pack. However, the herring at Kodiak, where nearly all the curing took place that year, were not of suitable quality, and the season was branded a disappointment. Nevertheless, the territory-wide pack of 4,512 barrels was more than twice that of 1942.[91]

Herring for bait decreased from 1942, to 6.2 million pounds. Several cold-storage plants in Southeast Alaska froze bait herring.[92]

Pacific Fisherman offered an upbeat summary of the season:

> In its relatively liberal attitude toward the Alaska herring industry in 1943 [the Fish and Wildlife Service] undoubtedly was influenced in some degree by appreciation of the national need for fish oil and fish meal. Flush supply of these products was recognized as highly necessary in time of war, and the industry was permitted to undertake to meet the call, although existence of the fishery's stock was not jeopardized by permitting wide-open fishing.[93]

1944

In early 1944, with World War II still raging and the need for fish oil and meal strong, the Fish and Wildlife Service increased the Alaska herring quota by 75 percent, from 400,000 to 700,000 barrels. Incorporated within this increase was a doubling of the Southeastern area quota, to 200,000 barrels. The Prince William Sound quota was likewise 200,000 barrels, 125,000 more barrels than the previous year. The quota in the Kodiak area was raised to 300,000 barrels, up from 225,000.[94] Add to this the fish that could be caught in nonquota waters.

In Southeast Alaska, two reduction plants operated: the Storfold & Grondahl Packing Company plant, at Washington Bay, and the Buchan & Heinen Packing Company plant, at Port Armstrong. In Prince William Sound, one plant operated, that of the Chatham Strait Fish Company, at Crab Bay. The operation represented a consolidation of packers in the sound. In the Kodiak area, four reduction plants operated: Southwestern Fisheries Company, Iron Creek; Apex Fish

Company, Port Wakefield; Oceanic Fisheries Company, Port Vita; and Chatham Strait Fish Company, Zachar Bay.[95]

The herring catch in Southeast Alaska, about 140,000 barrels, was considered satisfactory after several years of skimpy catches, and it was a sign that the abundance of herring had markedly increased. According to *Pacific Fisherman*, more fish could have been caught had the seine fleet been willing to fish more diligently during the last couple of weeks the plants remained open.

For the one reduction plant that operated in Prince William Sound, the season was exceptional. This was in part because early in the season fishermen found a great abundance of herring outside the quota area. Boats loaded up on these fish and kept the plant running to capacity for several weeks, all without making inroads on the quota. In all, the plant ran at full capacity during a large part of the season.

Fishermen at Kodiak, however, did not approach taking the entire quota. The herring there were said to have been "wild" and did not frequent the waters in which they were traditionally found.[96] Total production was 8,777 tons of meal and 2.3 million gallons of herring oil.[97]

In May 1944, Edwin Dahlgren and Lawrence Kolloen, herring specialists at the Fish and Wildlife Service, issued a forecast of herring abundance in Alaska, which *Pacific Fisherman* characterized as "rosy." Of the Kodiak area, the researchers wrote:

> Future heavy contributions by year classes not yet in the fishery may permit of further increase in the catch, but even if the next year classes are of low numerical strength there will be sufficient reserves of older fish to avert a sudden decline. The influx of 1939, 1940 and 1941 has bolstered the stocks to a level considerably above that of recent years.

Of Prince William Sound, they wrote:

> Notwithstanding the favorable prospects for a high abundance of herring in the summer run, good management requires that only a moderate withdrawal be made during this period. This follows because it is the same body of fish, if allowed to mature past the fifth year, which will support the future fall fisheries . . . Good catches can be anticipated in the fall fishery. The decline of recent years has been checked and a new cycle of abundance begun. If carefully guarded, the resource should continue to yield well over a long period.

Regarding Southeast Alaska, they wrote:

There should be an abundance of herring at Cape Ommaney in 1944. This augmented abundance should permit an increased harvest, and if the year classes which make this increase possible prove to be as abundant as the data at hand indicate, then the operations in this area may soon again be expanded to approach their former importance.[98]

1945

Secretary of the Interior Harold Ickes's desire for maximum production of seafood products to further the war effort manifested itself in several regulatory changes for the 1945 herring season. The quota in the Southeast area was increased 50,000 barrels, to 250,000 barrels. In Prince William Sound, the quota for the fishing period June 15 to August 20 was 150,000 barrels. Significantly, the quota on the "fall" fishing period—August 20 to October 15—was removed entirely. The Kodiak-area quota, 300,000 barrels, was unchanged.

To encourage prospecting in waters that were currently not exploited, such as Resurrection Bay and Day Harbor, on the south shore of the Kenai Peninsula, the Fish and Wildlife Service established seasons that opened before the traditional herring-fishing areas. In these areas, there was no quota on the amount of herring that could be taken. As well, the service lifted the restriction on the size of herring purse seines. This was, as the industry pointed out, a matter not of conservation but of efficiency. By utilizing deeper nets, purse seiners—increasingly outfitted with echo sounders (fathometers) that could help locate schools of herring at depth—could access previously unavailable fish.[99]

Noteworthy in 1945 was fishing in Day Harbor, where a large body of herring appeared early in the season. The fish disappeared at the end of the first week of July, but not before seiners caught 116,000 barrels.[100]

Cured herring production remained at a nominal level—7,259 barrels. Among these were some 600 barrels of large herring that were packed at Dutch Harbor. This was the first herring production there since 1942, when due to the Japanese bombing, the Aleutian Islands had been turned into a war zone.[101]

Eleven reduction plants—five in Southeast Alaska, two in Prince William Sound, and four in the Kodiak Island area—produced 2.8 million gallons of herring oil and 11,188 tons of meal.[102]

Perhaps emulating British Columbia, where during the war canneries were packing more than a million cases (each case contained forty-eight one-pound cans) of herring annually, a total of 2,308 cases were packed in Alaska, all in Southeast Alaska.[103] It was the first time in many years that herring had been canned in Alaska, but it was not repeated.

1946

Although wartime demands no longer applied, fisheries regulators continued to support large herring harvests throughout the territory. Catch quotas were increased to 350,000 barrels in Southeast Alaska and to 400,000 barrels in the Kodiak area. The quota in Prince William Sound, 150,000 barrels, was the same as in 1945, but it applied to the entire season. The previous year's regulations had allowed unlimited catches between August 20 and October 15.

The prudent approach to the Prince William Sound fishery was warranted; fishermen there caught only a little more than one sixth of the quota. They did, however, catch significant amounts of herring outside the quota area, likely in Day Harbor. In all, some 149,000 barrels of herring were taken outside the quota areas of the Prince William Sound and Kodiak districts.

In Southeast Alaska, where three reduction plants operated, the 1946 season was considered successful. There, a dozen seiners took 300,000 barrels of herring, of which 214,000 barrels were taken in the Cape Ommaney area and the remainder from the Kuiu Island grounds. According to Ward Bower, the Fish and Wildlife Service's Alaska agent, this catch indicated "the extent to which these populations have been restored."[104] Service biologist Lawrence Kolloen proudly reported that it could now be "safely asserted that the restoration of the stocks in this area has been achieved."[105]

Production of cured herring, 8,324 barrels, was slightly above the previous year. With a small exception, Kodiak accounted for nearly all of the territory's production. The exception was 108 barrels packed at the Nesskaug saltery at Unalaska, which closed its doors at the end of the season, marking the end of herring curing at Unalaska/Dutch Harbor. *Pacific Fisherman* noted that although herring were abundant in the Kodiak area and the market for cured herring was fair, the high cost of maintaining a gibbing operation curtailed herring curing there. A maritime strike in Seattle that tied up the pack for nearly two months didn't help matters.[106]

Overall, the eleven reduction plants that operated in the territory—the same number that operated in 1945—produced 15,868 tons of meal and 3.5 million gallons of herring oil, considerably more than was produced in 1945.[107]

The industry celebrated its success at a banquet at which the trophy herring belt—originally a rusty barrel hoop studded with herring scales, but now a copper barrel hoop emblazoned with herring scales and ornamented in front with a miniature wooden barrel—was awarded to Andrew Buchan, of the Buchan & Heinen Packing Company, which operated a reduction plant at Port Armstrong, in Southeast Alaska. *Pacific Fisherman* considered the eighty-year-old Buchan to be "the dean of the Pacific herring business." He had been involved in Alaska's herring industry since at least 1921. With the award, which had been suspended during the war, came a poem:

> The Herring Belt these many years
> Has gone unclaimed and rustin',
> For herring has been full of tears
> An' packers took a bustin'.
>
> Come forty-six and this is changed
> As Fate her hand relaxes,
> With fish and price in manner nice—
> Enough to pay the taxes.
>
> The Herring Belt was fairly won,
> And fairly is awarded.
> Unto our host, who caught the most—
> A Scotsman well regarded.
>
> The Laird of Armstrong gets the belt;
> The Laird of Armstrong done it.
> 'Twill be re-built to fit a kilt—
> For Andrew Buchan won it.

As was the custom, the winner of the belt paid for the award banquet.[108]

1947

Herring quotas for 1947 were unchanged from the previous year. Spring began with contentious negotiations between fishermen and packers. The fishermen demanded thirty cents per barrel per share, with nine shares per boat, the captain receiving two shares. This was nearly twice the seventeen cents per barrel, with eight shares per boat, that had been agreed to in 1946. The packers wanted no change. Each party dug in their heels, and it was June 30 before they agreed on twenty cents per barrel, nine shares per boat. But by that time, it was too late for the seiners to reach Prince William Sound in time for the early fishing season, which in recent years had been essential to profitable operations. Because of this handicap, Prince William Sound packers deemed it too risky and chose to forego operations.

Cured herring production, except for a small amount produced at the Arentsen & Company plant, at Big Port Walter, in Southeast Alaska, all occurred in the Kodiak area. The total was 8,037 barrels, a shade more than had been produced in 1946.[109] Fourteen reduction plants, eight in Southeast Alaska and six in the Kodiak area, produced 14,232 tons of fish meal and 3.7 million gallons of herring oil.[110] Even though Prince William Sound did not contribute to the production, these quantities were not substantially less than those produced in 1947. Max Jacobs, of the Oceanic Fisheries Company, which cured and reduced herring at its plant on Raspberry Island, in the Kodiak area, was awarded the barrel-hoop herring belt, which *Pacific Fisherman* now described as being "jeweled" with herring scales.[111]

1948

The Fish and Wildlife Service initially established quotas of 400,000 barrels in Southeast Alaska, 150,000 barrels in the Prince William Sound area, and 300,000 barrels in the Kodiak area. In Prince William Sound and the Kodiak area, the quotas were subsequently increased to 180,000 barrels and 390,000 barrels, respectively. Sixteen reduction plants—seven in Southeast Alaska, four in Prince William Sound, and five in the Kodiak area—operated in the territory. Herring was cured at two plants, both in the Kodiak area: that of the Oceanic Fisheries Company, on Raspberry Island, and that of the Chatham Strait Fish Company, at Zachar Bay. Both companies also operated reduction plants.

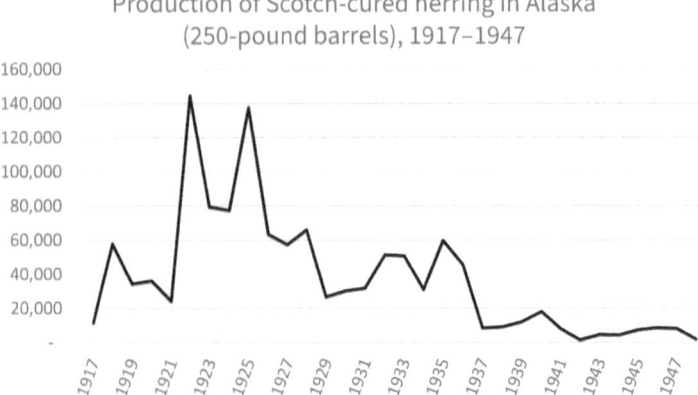

Figure 6.3. Data sourced from *Pacific Fisherman* yearbooks, 1917–1948.

Southeast Alaska's 1948 herring fishery was a failure. The usual schooling of herring in the vicinity of Cape Ommaney didn't occur, and fishermen landed only 128,869 barrels. In Prince William Sound, production was curtailed by a lack of mature fish during the second half of the season. The Kodiak area, however, had excellent fishing throughout the season, with fishermen landing some 360,000 barrels. When all was said and done, territory-wide meal and oil production in 1948 was less than the previous year: 12,683 tons of meal and 3.4 million gallons of oil.[112]

The production of cured herring was almost negligible: 1,728 barrels. *Pacific Fisherman* summed up the cured-herring industry's challenges: "Adverse market conditions, plus high costs of operation, plus the uncertainties of Alaska shipping plagued by strikes." These, the journal wrote, had combined to curtail curing to the "vanishing point."[113]

7
A CHRONICLE OF ALASKA'S HERRING INDUSTRY
1949–1966

This chapter provides a timeline of Alaska's herring industry during the years 1949–1966, a period that marked the demise of dedicated herring-reduction operations in Alaska.

1949

The lack of fish in Southeast Alaska and the lack of mature fish in Prince William Sound did not bode well for the near-term future of Alaska's herring industry. In February 1949, the Fish and Wildlife Service held its usual herring hearing in Seattle. For herring men, the news wasn't good: the agency was considering drastic cuts to quotas in Southeast Alaska and Prince William Sound.

The reason for the cuts, according to the Fish and Wildlife Service, was that sampling had shown the 1945 and 1946 year classes to be very weak, and that the 1949 fishery would have to be sustained on the dwindling remnants of the 1944 year class, which had been strong. The strength of the incoming 1947 year class was unknown.

The agency subsequently reduced Southeast Alaska's quota by more than half, to 150,000 barrels, and it reduced Prince William Sound's quota by almost half, to 100,000 barrels. In the Kodiak Island area, where fishing had been excellent in 1948, the quota was cut to 250,000 barrels. The territory's total quota, 500,000 barrels, was a little more than half that of the 940,000-barrel total quota in 1948.

Fishermen and plant operators protested. They ascribed low catches in 1948 to, among other factors, "unfavorable westerly weather" and herring that "were so wild they could not be held." One seine-boat captain said fluctuations in the herring catch at Kodiak was a matter of variations in the availability of herring rather than their abundance. Another argued that shallow-water fishing did not give a true indication of the age composition of the herring population. Citing the herring industry's conservation concerns, another seine-boat captain said limiting operations would curtail the sampling necessary for effective management of herring stocks.[1]

The industry's pushback apparently had some effect. In April, the Fish and Wildlife Service announced that the quotas for Southeast Alaska and Prince William Sound would each be 50,000 barrels more than the agency had proposed in the February hearing. *Pacific Fisherman* implied that Julius Krug, who had succeeded Harold Ickes as secretary of the interior in 1946, had interceded to authorize the increase.[2]

The entire quota exercise had almost no impact. The market for fish oils had tanked, and, at the same time, Alaska's Legislature had increased the cost to operate a fish processing plant. In its overhaul of the territory's tax code was an income tax on businesses at a rate of 10 percent of the total federal income tax obligation.[3]

The proverbial straw that broke the camel's back was the demand by fishermen for an increase in their per-barrel shares. The fishermen wanted at least twenty-three cents per barrel, while reduction plant operators—wary of the fish oil market—offered fifteen cents. Negotiations broke down, and, as a result, there were no herring reduction operations in either Prince William Sound or the Kodiak area.

In Southeast Alaska, however, negotiations between fishermen and operators, though protracted, were successful. Each fisherman was to be paid fifteen cents per barrel or the fishermen would divide among themselves 25 percent of the gross revenue derived from the products of each barrel of herring, whichever was greater. Three established reduction plants were able to operate, and they were joined by the *Pacific Rim*, a floating reduction plant that was leased by the Alaska Fishermen's Cooperative.

Because of the delay caused by the negotiations, however, the season got off to a late start. Adding to the industry's troubles, fishing was poor, and the catch fell far short of Southeast Alaska's 200,000-barrel quota. Production totaled 3,608 tons of meal and 601,186 gallons of herring oil—the smallest quantities in a quarter century. No herring were cured in the territory.[4]

In addition to levying an income tax on businesses, Alaska's Legislature created the Alaska Department of Fisheries. The new department's duties were to assist in conservation and perpetuation of the territory's fisheries resources; to promote more resident ownership, management, and control of the fisheries; and to cooperate with the federal Fish and Wildlife Service.

The department was overseen by the simultaneously created Alaska Fisheries Board, which consisted of five members (three fishermen, one processor, and one public), who were appointed by the governor and confirmed by the legislature.[5] J. Howard Wakefield, whose Wakefield Fisheries Company operated a herring plant on Raspberry Island, in the Kodiak area, was appointed to a two-year board term as the processors' representative.[6] *Pacific Fisherman* considered the creation of the fisheries board as having "launched the territory into a pre-Statehood program of developing a fishery administrative agency."[7] The journal was correct: Alaska's Legislature had specified that "on Alaska's assuming Statehood, the Board shall assume all powers now vested in the Fish and Wildlife Service relating to the regulation and conservation of the fisheries of Alaska."[8] The Alaska Fisheries Board appointed Clarence Anderson, who was no stranger to Alaska or to herring, as director of the Department of Fisheries.

1950

After the debacle that was the 1949 Alaska herring fishery, 1950 was a return to more "normal" operations—if there ever was such a thing as normal in Alaska's herring industry.

The Fish and Wildlife Service took a new approach to management of Alaska's herring fishery by instituting annual quotas in the Prince William Sound area (whose waters now extended west to Resurrection Bay) and the Kodiak area that would be maintained for three years. In doing so, the service's goal was to provide an element of stability. The annual quota for the Prince William Sound area was 180,000 barrels, while at Kodiak it was 275,000 barrels. Additionally, the quota period was shortened in each area, and herring could be caught after the end of the quota period in unlimited quantities. The Fish and Wildlife Service believed late-run fish comprised distinct populations that were healthy and did not need protection.

Because of the low abundance of herring in Southeast Alaska and the unfavorable age composition of the herring stocks there, the three-year quota program was not applied in that region. The quota for 1950 was set at 150,000 barrels, the same as it had been in 1949.[9]

Among the Alaska Fisheries Board's first recommendations to the Fish and Wildlife Service was that the 1950 herring quota in the Kodiak area be increased by 25,000 barrels, the reason being that there had been no herring fishery in the area in 1949.[10] This recommendation fell on deaf ears.

Negotiations between fishermen and operators in 1950 were completed in a timely manner. In Southeast Alaska, the arrangement from 1949 was unchanged. The arrangement in Prince William Sound and the Kodiak area was identical with that in Southeast Alaska, except that fifteen cents was deducted from the sale of the products of a barrel of herring before the 25 percent gross stock for the fishermen was computed.

Fully a dozen herring reduction plants operated in 1950, four in each of the primary production areas. The Alaska Fishermen's Cooperative's floating reduction plant *Pacific Rim*, which had operated in herring-deprived Southeast Alaska in 1949, operated in Prince William Sound in 1950.[11]

Fishing was good in both Prince William Sound and in the Kodiak area, but Southeast Alaska had another extremely disappointing season. Meal production throughout Alaska in 1950 totaled 13,694 tons; oil production totaled 2.9 million gallons. Marking the end of an era that began during World War I, the Scotch-cured herring produced in 1950—1,056 barrels—represented the last substantial production of cured herring in Alaska.[12]

1951

> Season of 1951 will go down as one of the dizziest in the vertiginous . . . history of the Alaska herring business.—*Pacific Fisherman*, October 1951[13]

Based on the poor showing of herring in Southeast Alaska in 1950, the area's quota for 1951 was reduced to 100,000 barrels, 50,000 barrels less than it had been in 1950. The Prince William Sound and the Kodiak areas entered their second year of the three-year continuing-quota program.[14]

At the end of the first week of June, operators, fishermen, and plant workers reached an agreement for the upcoming season. The fishermen increased their share from the fifteen-cents-per-barrel "floor" of their previous year's agreement to twenty cents per barrel. The agreement read: "Each man of an 8-man crew will be paid 20 cents for each barrel of fish caught by the boat during the season, or [to divide among themselves] 25 percent of the gross income realized from the products of one barrel of herring, whichever is greater." Shore workers negotiated a 15 percent across-the-board wage increase, subject to approval by the federal Wage Stabilization Board.

A dozen herring reduction plants were scheduled to operate: four in Southeast Alaska, four (including the floating plant *Pacific Rim*) in Prince William Sound, and four in the Kodiak area.[15] For most of them, the season would be disappointing, to say the least.

Herring in Prince William Sound failed to arrive until late August, by which time all the boats except those of the Port Ashton Packing

Company had pulled out. Meanwhile, the Kodiak area was a bust; by late September, only about 50,000 barrels of the area's 275,000-barrel quota had been caught. The failure at Kodiak had not been anticipated, because the age composition and volume of the catch in 1950 were favorable to a continued abundance of fish. Fishing in Southeast Alaska, too, was slow.[16]

Meal and oil production in the territory, 5,887 tons and 1.3 million gallons, respectively, was less than half the previous year's production. Production in the Kodiak area, the most reliable producer of herring in recent years, was measly: 789 tons of meal and 85,957 gallons of herring oil.[17] Because of what the US Fish and Wildlife Service's Alaska agent Seton Thompson termed "unfavorable market conditions," processors chose to forego curing herring.[18]

1952

At least for bait fishermen in Ketchikan, 1952 began auspiciously. Herring were plentiful in Tongass Narrows (Ketchikan's waterfront) early in the year, and seiners were able to fully supply the town's two cold-storage plants with fish caught almost at their front door.[19]

For operators of herring-reduction plants, however, the season's prospects were bleak. Probably first on their mind was the diminished price of herring oil. Late in 1951, it had been selling in Seattle for thirteen cents per gallon; by spring 1952, the price had fallen to about half that amount. Second on operators' minds were the demands of fishermen and plant workers. Fishermen demanded the same arrangement as in 1951 but with an expensive addition: a guarantee of $500 per month (per man) for a three-month period—even if no fish were caught. Plant workers, who received a 15 percent wage hike in 1951, wanted another 15 percent hike. *Pacific Fisherman* wrote of the "feeling of hopelessness" that pervaded the Alaska herring industry.[20]

Given the low price of herring oil, the demands of fishermen and plant workers were untenable. Fishermen ultimately settled for twenty cents per barrel or to divide among themselves 20 percent of the gross sales price of the products of a barrel of herring, whichever was greater, with no guarantee. Plant workers received no raise. Still, given the price of oil, fish costs and labor costs were still disproportionately high.

Six reduction plants elected to operate in 1952. In Southeast Alaska, there were three: Buchan & Heinen, Port Armstrong; Storfold & Grondahl, Washington Bay; and Oceanic Fisheries, Port Conclusion.

In Prince William Sound, there were two: Oceanic Fisheries, at Thumb Bay, and the Alaska Fishermen's Cooperative, which operated the floating reduction plant *Pacific Rim*. In the Kodiak area, the only herring reduction-plant operator was the Wakefield Fisheries Company, on Raspberry Island.

Fishing in Prince William Sound was very slow, but it was even worse in the Kodiak area. *Pacific Fisherman* called the herring run there a "severe failure"—the second year in a row at Kodiak.[21] Wakefield Fisheries was so desperate for herring that it sent its boats to Chignik, about 250 miles from its Raspberry Island plant, to fish for herring.[22] Only Southeast Alaska showed favorable returns, and even there fishing was good only to the degree that it was better than had been expected. The region's entire 100,000-barrel quota was caught by August 1. Nevertheless, because of the high prices for fish and labor and the low price of herring oil, 1952's herring season in Southeast Alaska was a money loser for processors.

Production in Alaska in 1952 totaled 4,590 tons of meal and 685,467 gallons of herring oil.[23] No cured herring was produced.

In the fall of 1952, the Fish and Wildlife Service amended its fishing regulations to permit winter fishing for herring in the extreme southern part of Southeast Alaska. Canadians had successfully been fishing for herring during the winter on the south shore of Dixon Entrance, and American fishermen wanted to determine the feasibility of doing so on the north shore. The waters open to winter fishing were those south of 55° north latitude, approximately the latitude of the south end of Annette Island.

The Fish and Wildlife Service's ninety-three-foot research vessel *John N. Cobb* was outfitted with seine gear and spent six weeks in November and December exploring Southeast Alaska's waters to determine the abundance and availability of herring during the winter months. Neither the *John N. Cobb* nor the four seiners that fished the region's south end in an experimental fishery experienced any measurable success. As part of the *John N. Cobb*'s operation, researchers tagged 2,000 herring in Kendrick Bay, on the Clarence Strait shore of Prince of Wales Island, and 3,400 in Tongass Narrows with nickel-plated belly tags. The researchers hoped to gain information on the migration patterns of these fish.[24]

Another development in 1952 was the signing by the United States, Canada, and Japan of the International Convention for the High Seas Fisheries of the North Pacific Ocean. The emphasis of the agreement was salmon, but Japan agreed to abstain from fishing off the coasts of North America, exclusive of the Bering Sea, for herring of North

American origin. The United States and Canada also agreed to implement necessary conservation measures for the species.²⁵ Japan had severely depleted the herring resources in its own waters and was in search of a new source of fish to exploit. It would be 1960, however, before Japanese high-seas vessels began targeting herring in the eastern Bering Sea (see below).²⁶

On a sad note, Jack Storfold, vice president of the Storfold & Grondahl Packing Company, which since 1925 had maintained a herring operation at Washington Bay, in Southeast Alaska, died in November. The cause of his death was inhaling fumes of a solvent being used to clean engine parts at the company's reduction plant. *Pacific Fisherman* described the thirty-seven-year-old Storfold as "one of the most active and best-known younger men in the Alaska herring industry."²⁷ In 1937, another family member, Olaf Storfold, captain of the company's herring seiner *Limit* perished with seven crewmen when the boat sank in the Cape Ommaney vicinity (see chapter 6).

1953

The year began somewhat ominously for Southeast Alaska herring reduction plant operators. The Alaska legislature, which had a long history of opposition to the manufacture of meal and oil from herring in Southeast Alaska, passed a memorial asking federal fisheries managers to "prohibit fishing for herring for reduction purposes in Southeastern Alaska immediately and until such time as the supply recovers its previous abundance."

The resolution noted that the catch of herring in the region had been "subnormal" since 1938, that seiners had "annihilated local populations of herring through over-intensive fishing," and that the "herring industry contributes virtually nothing to the economy of Alaska."²⁸

It is likely that salmon trollers, the fishing-gear group most adamantly opposed to herring-reduction plants, were instrumental in getting the memorial passed. Commercial fishing for salmon with troll gear is allowed only in Southeast Alaska, hence the memorial's focus on this region. By far, most trollers were Alaska residents. Like all the previous anti-herring-reduction legislative memorials, this one was ignored by federal fisheries managers.

Herring continued to be scarce in both Prince William Sound and the Kodiak area, where production was limited to a relatively small quantity of frozen bait. In Southeast Alaska, two operators, Storfold &

Grondahl Packing Company, at Washington Bay, and Buchan & Heinen Packing Company, at Port Armstrong, operated reduction plants. Five purse seiners supplied the two plants. At least early in the season, herring fishing in Southeast Alaska was said to be surprisingly good. For the season, however, production of meal and herring oil was the lowest since the early 1920s: 2,544 tons of meal and 427,555 gallons of oil.[29]

Given the scarcity of herring in the Kodiak area during recent years, one reduction plant operator suggested an alternative material to supply his reduction plant: sea lions.[30] Such was not to be, although salmon fishermen, whose catch sometimes suffered from sea lion depredations, would likely have supported it.

This was the last year Alaska herring were commercially cured for human food. Production was a miniscule 775 pounds.[31]

On a more positive note, the Fish and Wildlife Service's research vessel *John N. Cobb* was deployed to Prince William Sound in mid-October for a three-month exploration for and study of herring. The vessel employed a midwater trawl to catch the fish.[32]

1954

Aside from the bait fishery, which produced some 3.6 million pounds of fresh and frozen herring, Alaska's herring industry in 1954 was limited to the operations of three reduction plants. Two were in Southeast Alaska (which operated under a 50,000-barrel quota, except for fish to be used for bait or caught by gillnets), while the other was in Prince William Sound. Herring continued to be scarce in the Kodiak area. Total territorial production was 2,794 tons of meal and 457,550 gallons of herring oil, most of which was produced in Prince William Sound, where fishing was particularly good early in the season.[33]

1955

Early in the year, the Fish and Wildlife Service announced that the herring quota in Southeast Alaska would be doubled, to 100,000 barrels. Alaska's Legislature weighed in, reworking its 1953 herring memorial to apply to the current situation. The legislature said the Fish and Wildlife Service had "prematurely and dangerously" relaxed the quota and asked the secretary of interior and the director of the agency to "return to and maintain the 50,000 bbl. allowable annual take of herring for reduction purposes in Southeastern Alaska until such time as the supply recovers its previous and natural abundance."[34]

Similarly, the Alaska Department of Fisheries—"in view of the controversial nature of the herring problem"—recommended that the Southeast Alaska quota for reduction purposes not be increased until justified by a "complete and thorough" study of the resource.[35]

Despite the territory's concerns, the Fish and Wildlife Service proceeded with the quota increase. And perhaps for good reason: herring were abundant in Southeast Alaska in 1955, which drove a sharp increase in meal and oil production. It was a good season also in the Prince William Sound area. Total meal production in Alaska in 1955 was 4,480 tons; oil production was 1 million gallons.[36]

1956

Based on the resurgence of herring in Southeast Alaska, reduction plant operators asked the Fish and Wildlife Service for the 1956 quota to be increased to 200,000 tons, double that of the previous year. Salmon trollers, on the other hand, proposed a five-year moratorium on herring fishing in Southeast Alaska. Absent that, they wanted protection for the herring that frequented Shelikof Bay, on Kruzof Island, near Sitka. This was an important area for salmon trollers, but in recent years it had been the single most productive herring grounds in the region.[37]

As a compromise, the Fish and Wildlife Service established a Southeast Alaska quota of 140,000 barrels, with the provision that the catch in the Sitka area was limited to 50,000 barrels. Additionally, for the first time the agency established catch limits on herring to be used for bait. The limits applied to two locations: Silver Bay, near Sitka, and Fish Egg Island, near Craig. At each location, bait fishermen had quotas of 1,000 barrels of herring.[38]

Because the herring-fishing efforts in the Prince William Sound and Kodiak areas had been relatively modest in recent years, neither area was limited by a quota.

Two reduction plants operated in Southeast Alaska in 1956, and the purse seiners that supplied them caught the entire quota by July 26. Operators requested additional quota, but the Fish and Wildlife Service denied the requests. In the Prince William Sound area (which extended west to Resurrection Bay), herring fishing was decent for the third year in a row, with a harvest of about 120,000 barrels. The harvest in the Kodiak area was 108,000 barrels.

By this time, biologists believed that the primary factors in fluctuations in herring stocks were natural ones, principally those related to the success or failure of spawnings and the survival of young herring. The 1954 and 1955 year classes did not appear to have been strong, and it was expected that the 1957 herring fishery would depend almost entirely on the 1953 year class. Beyond that, prospects were not bright.[39]

Total meal production approached 8,000 tons, about 3,500 tons of which was produced in Southeast Alaska. The Prince William Sound and Kodiak areas each produced a bit over 2,000 tons. Total herring oil production in the territory was 1.7 million gallons.[40]

The year also marked the end of the publication of the annual Alaska fisheries reports that were begun by the Bureau of Fisheries in 1905 and continued by the agency's successor, the Fish and Wildlife Service. For writers of books such as this, these reports are an invaluable resource.

1957

"Relatively, the brightest spot in the Pacific fish reduction picture was Alaska, where in 1957 the herring business for the third successive year registered moderate gain," wrote *Pacific Fisherman*. The season's total catch for the reduction industry was 57,338 tons (about 115 million pounds), which was about 6,000 tons more than had been caught in 1956, and it approached the average Alaska production for the past thirty years.[41] Fishing was excellent in all three major producing areas. In Southeast Alaska, the 180,000-barrel quota was caught by July 21. The Fish and Wildlife Service then granted an additional 60,000 barrels, with the provision that the fish could be caught only in the waters south of Cape Decision, the southernmost point of Kuiu Island. Unfortunately for both fishermen and operators, the area was essentially barren of herring.[42]

Three reduction plants operated in Southeast Alaska, one in Prince William Sound, and two in the Kodiak area. Together, they produced 8,822 tons of fish meal and 1.7 million gallons of herring oil.[43]

In 1957, the Fish and Wildlife Service instituted a new management tool: aerial surveys and charting of herring spawning beaches in Southeast Alaska. That year, the agency recorded herring spawn along 132 miles of shoreline.[44]

Additionally, in its pursuit of knowledge about Alaska's herring stocks, the agency deployed its research vessel *John N. Cobb* on a seven-week exploration of the Gulf of Alaska waters between Baranof Island, in Southeast Alaska, and Prince William Sound. Vessels crossing the gulf had at times reported seeing schools of herring, but there was only fragmentary knowledge of their abundance and availability. The *John N. Cobb* operated at a minimum of fifty miles offshore and used a midwater trawl and gillnets to catch herring.[45]

In April 1957, Alaska's Legislature replaced the Alaska Fisheries Board with the Alaska Fish and Game Commission and replaced the Alaska Department of Fisheries with the Alaska Department of Fish and Game (ADF&G).[46] In 1959, by legislative action, the Alaska Fish and Game Commission became the Alaska Board of Fish and Game.[47]

1958

In 1956, herring fishery managers had predicted the 1958 herring abundance would potentially be constrained by the weakness of the 1954 and 1955 year classes. The managers were largely correct: herring fishing in the Kodiak Island area was a complete failure, and in the Prince William Sound area, in the words of *Pacific Fisherman*, "a bitter disappointment." Fishermen in Southeast Alaska, on the other hand, found very good fishing. The total herring harvest for reduction in Alaska in 1958 was 41,788 tons, of which 36,185 tons (almost 87 percent) was caught in Southeast Alaska. The catches in the Kodiak and Prince William Sound areas were the poorest on record: 1,711 and 3,892 tons, respectively. More than 80 percent of the herring catches in Southeast Alaska and the Kodiak Island area were from the 1953 year class.

Twenty-two purse seiners participated in the 1958 Alaska herring fishery, fourteen in Southeast Alaska and four each in the Prince William Sound and Kodiak areas. Territory-wide, meal production totaled 6,484 tons, while herring oil production was 1.5 million gallons. Notably, 1958 marked the end of dedicated herring reduction operations in Prince William Sound.

An attempted innovation in 1958 was the installation of flow scales—continuous-weighing machines—at several Southeast Alaska reduction plants to weigh the catches more rapidly. Fishermen, however, didn't trust the devices, and plant operators reverted to the traditional process of measurement: filling and dumping a barrel-sized hopper. This practice continued through 1966, after which the last dedicated herring reduction plant in Southeast Alaska, the Edible Herring Products Company plant, at Big Port Walter, was shuttered.[48]

Ever on the forefront of scientific technology, herring biologists explored the possibility of using radioactive internal tags to track herring.

The project required licensing by the Atomic Energy Commission, and this was granted in 1960. In a field experiment that same year, biologists inserted small pieces of radioactive cobalt wire into herring's body cavities. A radiation detector placed above the conveyor belt that carried fish from fishing boats to the fish plant sent a signal to an ejector that removed tagged fish from the conveyor belt. A full-scale test of the system was planned for the summer of 1961, and at least a few herring were tagged. During the summer of 1962, several tags were recovered from herring delivered to the Storfold & Grondahl reduction plant at Washington Bay.[49]

1959

Notable in the 1959 herring-fishing season in the Kodiak area was the preponderance of young herring, which resulted in poor yields at the two reduction plants that operated there. This, coupled with a weak market for meal and oil, rendered reduction operations unprofitable, and both plants closed down early.[50] Herring would not be reduced on a large scale at Kodiak until 1966, and even then it would continue only a few years.[51] Fishing in Southeast Alaska was good, with the main quota of 180,000 barrels filled in early July. Statewide, meal production totaled 8,640 tons; herring oil production totaled 1.7 million gallons.[52]

Of significance to the coastal Alaska herring fisheries, in 1959 Soviet vessels began an investigation of fisheries potential in the eastern Bering Sea. Of primary interest was herring. The Soviets found quantities of herring northwest of the Pribilof Islands and within two years figured out the herring's annual cycle there. In early 1961, before the herring had dispersed for spawning, Soviet midwater trawlers harvested about 10,000 metric tons. Production rapidly increased. The largest production of eastern Bering Sea herring by the Soviets was attained during 1962–1964, when the combined annual catches of 100–150 trawlers engaged in the fishery ranged from 150,000 to at least 175,000 metric tons.[53]

Alaska gained statehood on January 3, 1959, but the legislation that created the new state specified that Alaska would not obtain jurisdiction over its fish and wildlife resources until it could convince the federal government that adequate provision for the administration, management, and conservation of those resources had been made. A year later, on January 1, 1960, Congress granted Alaska full authority over its state-waters fisheries (except for the halibut fishery, which is regulated in US and Canadian waters by the International Pacific Halibut Commission).

1960

Under its new authority, Alaska quickly began the process of managing its fisheries according to the sustained-yield principle, as mandated in the state's constitution. Surprisingly, given Alaska's Legislature's long-standing animosity toward herring reduction, the state maintained what *Pacific Fisherman* called "a liberal attitude" regarding the commercial utilization of the herring resource.

In Prince William Sound and in the Kodiak area, herring fishing was basically unregulated, save that fishing for herring was not permitted during the salmon-fishing season in any waters closed throughout the year to fishing for salmon. In Southeast Alaska, the herring-fishing season opened on June 1 and continued until the end of February 1961. At the same time, ADF&G wanted to more closely manage the fishery, and it divided Southeast Alaska into districts and assigned quotas to each district.[54]

As prepared for market, herring products ranked a distant fourth in value among Alaska's fishery products, behind salmon, king crab, and halibut. The value of herring products, $1.39 million, represented slightly more than 1.4 percent of the value of all Alaska fishery products.[55]

Meal and oil production, all of which occurred in Southeast Alaska, totaled 4,126 tons and 1 million gallons, respectively. The market for these products, however, was depressed. The problem was Peru.

In the early 1950s, Peru had claimed a territorial limit 200 miles seaward from its coast and had taken control of the fisheries within this zone. Among the fish that populate these waters are anchovetas (*Engraulis ringens*), which thrive in the nutrient-rich Humboldt Current, a cold upwelling zone along the west coast of South America that supports what is perhaps the world's most productive marine ecosystem. Peruvians then quickly began building reduction plants to utilize the vast anchoveta resource. By 1960, there were about 100 such plants, and Peru was the world's leading manufacturer of fish meal, capable of producing more than a million tons annually. Unfortunately for traditional producers of fish meal, the addition of Peru's production glutted the worldwide market.[56] Moreover, fisheries scientists estimated that Peru's anchoveta resource could sustain an annual harvest of approximately 9.5 million metric tons.[57] By contrast, the maximum harvest in Alaska, in 1937, was about 126,000 metric tons. Alaska reduction plants, with relatively high operating costs, found it increasingly difficult to compete.

In another development, in 1959 the United States, Canada, and Japan had amended the 1952 International Convention for the High Seas Fisheries of the North Pacific Ocean to allow Japan to fish for herring outside Alaska territorial waters. The reason for relaxing the restriction was that the International North Pacific Fisheries Commission, the organization tasked with implementing the treaty, had determined that herring stocks in these waters were not being fully utilized.

The amendment became effective in 1960, but the Japanese did not take immediate advantage of it. Instead, they sent a mothership and its associated fishing fleet to the eastern Bering Sea, with a target of taking 8,700 tons of herring. Fishing with midwater trawls, the fleet found fishing to be very good.[58]

1961

The volume of fish meal and herring oil produced in Alaska continued to fall. Fish meal production in 1961 was 3,576 tons; herring oil production was 625,786 gallons. Three Southeast Alaska plants accounted for all of the production.[59]

Also in 1961, Clarence Anderson retired as commissioner of the Alaska Department of Fish and Game. He had been appointed commissioner of the newly created Alaska Department of Fisheries in 1949. Reflecting on his twelve years of service, Anderson wrote that the department had "been developed from an idea on a few scraps of paper to a well-functioning organization of 175 carefully chosen, well-trained, dedicated employees."[60]

1962

In 1960, *Pacific Fisherman* had praised the State of Alaska's "liberal attitude" regarding the commercial utilization of herring. That changed in 1962, when Alaska's Legislature passed a resolution urging the Department of Fish and Game and the Board of Fish and Game to "take immediate steps to eliminate the taking of herring inside Alaska's territorial waters for any purpose except for bait or for food for human consumption."[61]

According to the federal Bureau of Commercial Fisheries' *Commercial Fisheries Review*, in 1962 state law permitted the operation of only one herring reduction plant in Southeast Alaska.[62] No state law dictated so, but the journal may have been referring to a Board of Fish and Game decision. At any rate, in 1962, only the Storfold & Grondahl Packing Company operated a reduction plant, at Washington Bay. The three seiners fishing for the plant landed 14,000 tons of herring, about half of the previous year's catch.[63] *Pacific Fisherman* lamented the near demise of "an industry which once operated profitably along the entire Alaska coast from Cape Ommaney to Kodiak Island." The Storfold & Grondahl plant's 1962 production totaled 3,533 tons of meal and 647,180 gallons of herring oil, about the same quantities as three plants had produced the previous year.[64]

Probably the most foretelling event in 1962—at least so far as the development of a market for new herring products was concerned—was the mid-April seizure by Alaska agents of two "American-style" Japanese purse-seine boats in Alaska waters in Shelikof Strait. The vessels had entered the strait accompanied by the mothership-transporter *No. 31 Banshu Maru*. They were in search of roe herring—egg-laden female herring that were ready to spawn. The eggs from these fish, *kazunoko*, were a highly valued delicacy in Japan and a popular gift during the New Year's celebration. As previously noted, the Japanese had depleted the herring resource in their waters and were looking for new sources of the fish.

As they do today, state waters extended three miles from the shore, but beyond them at that time were international waters. Herring, of course, spawn along the shore, so while the *No. 31 Banshu Maru* could have stayed in international waters, the seine boats likely fished in Alaska's waters. Both boats were seized while at anchor, one within the three-mile limit, the other beyond the limit.

State officials arrested the captains of the seine boats and the mothership. About four days later, the men, who had been taken to Kodiak for arraignment, were released on bail and were allowed to return to Japan, and the seine boats were released on the condition they not conduct fishing operations in Shelikof Strait.

Complicating matters was the fact that Alaska officials claimed Shelikof Strait was an "inland water" over which the state had jurisdiction. According to Alaska governor Bill Egan, the claim was based on historical and geographical grounds. The US Department of State did not share this opinion.[65] The State of Alaska intended to extradite the three captains but chose not to do so after Japanese interests proposed to purchase herring in the Cordova area (and chum salmon in the Ketchikan area). To Alaska officials, it appeared the Japanese had changed their policy regarding fishing in Alaska-claimed waters, and, moreover, there was potential business to be had.[66]

Regarding their proposed purchase of herring, the Japanese wanted to bring a mothership into Prince William Sound, where US fishing vessels would supply the ship with about 3,000 tons of roe herring for which they would be paid $40 per ton. Japanese workers would process the herring, brine-curing the roe and freezing the remainder. Because of a lack of firm knowledge regarding the timing of herring spawning in the sound, however, the proposal—which would also have required the approval of the US Department of Labor—did not reach fruition.[67]

1963

The herring industry began 1963 with a new, draconian Alaska Department of Fish and Game regulation: the prohibition of herring reduction in Southeast Alaska. Fortunately for the industry, at least for the Storfold & Grondahl Packing Company, the regulation was rescinded in early February. The company's Washington Bay reduction plant, the only herring-reduction operation in Alaska that year, produced 2,232 tons of meal and 572,054 gallons of herring oil.[68]

1964

Possibly related to the 1962 failed Japanese proposal to process American-caught herring aboard a Japanese mothership in Prince William Sound, in February 1964, a Japanese company completed negotiations to purchase 1,200 tons of large frozen roe herring. The fish were to be shipped to Japan, where roe would be removed and the carcass processed for human consumption.[69] It is likely the contract stipulated a minimum roe content, likely 10 percent. There were no reports of this contract being fulfilled.

However, 1964 was significant because it marked the advent of Alaska's commercial roe-herring fishery. Over the next decade, the roe-herring fishery would develop into a valuable component of Alaska's fishing industry, and it is the subject of part II of this book.

Despite Peru's prodigious production of fish meal and fish oil, the market for these products was stronger in 1964 than it had been in recent years. As a result, for the first time since 1961, the Strofold &

Grondahl Packing Company had competition in the herring-reduction business: the Edible Herring Products Company, at Big Port Walter, and the Washington Meal & Reduction Company, at Zachar Bay, on Kodiak Island. In addition to the improved prices for meal and oil, the herring in Southeast Alaska that season were especially fat and yielded the highest amount of oil that had been experienced in recent years—about forty-five gallons per ton of herring. Purse seiners fishing for the plants delivered about 22,000 tons of herring. When operating at full capacity, the Storfold & Grondahl plant was capable of reducing about 260 tons of herring per twenty-four hours. Thirteen employees operated the plant, seven on the day shift and six at night.

Production of meal totaled 3,472 tons, while production of herring oil was a healthy 995,162 gallons.[70] In its report for 1963–1964, ADF&G observed that "[e]conomic and marketing problems, rather than herring abundance, were currently the limiting factors for the reduction industry."[71]

As had occurred during the past few years, in January Soviet vessels began herring fishing in the Bering Sea, northwest of the Pribilof Islands. By early February, the Soviet fleet comprised more than 150 vessels. Fishing continued until April, when the herring began to migrate to their spawning grounds, and the vessels dispersed to other fisheries.[72]

1965

As in 1964, two reduction plants operated in Southeast Alaska in 1965, the Storfold & Grondahl Packing Company plant, at Washington Bay, and the Edible Herring Products Company plant, at Big Port Walter. Excellent spawning in the spring was an encouraging sign, but fishing that summer was poor. Fishermen attributed the lack of herring to unseasonably cold water that persisted throughout the summer. Their herring catch was 74,371 barrels (9,234 tons), less than half their catch of the previous year, and it yielded just 1,501 tons of meal and 377,988 gallons of oil.

The shortage of herring, moreover, was likely to persist because the strong 1958 herring year class, which had supported Southeast Alaska's reduction industry since 1960, had run its course, and there was nothing on the horizon to replace it. When Storfold & Grondahl shuttered its plant at the end of the season, it did so permanently. The company had been a stalwart in the herring reduction business since 1926. The plant was dismantled in 1967.

Figure 7.1. Edible Herring Products Company plant at Big Port Walter, undated. (US Forest Service)

Meanwhile, there was relatively small-scale herring reduction in 1965 in the Kodiak area. There, the Washington Fish & Oyster Company operated its reduction plant at Zachar Bay in connection with a fishery that harvested 597 tons of roe herring. The operation was able to process herring offal and carcasses in its reduction plant.[73]

1966

The Edible Herring Products Company plant at Big Port Walter (figure 7.1) was the sole reduction operation in Southeast Alaska in 1966. Fishing was slow, and the herring catch, 5,073 tons—56 percent of which was caught in the vicinity of Noyes Island—was the lowest since 1941. (The catch of bait herring in Southeast Alaska totaled 2,422 tons.)[74] Moreover, production at the plant had been steadily declining in recent years, from 909 tons of herring meal in 1964, to 452 tons in 1965, to 415 tons in 1966.[75] Alaska's only other reduction plant that operated in 1966 was Zachar Bay Fisheries, in the Kodiak area. Its production totaled 732 tons of herring meal.[76]

A victim of diminished herring stocks in the traditional fishing areas, the closure of additional areas to fishing for herring for reduction purposes, and the massive quantities of relatively inexpensive meal and oil being produced in Peru, the Edible Herring Products Company plant shut its doors forever on August 30, 1966, marking the end of dedicated, large-scale herring reduction in Southeast Alaska.

A Chronicle of Alaska's Herring Industry

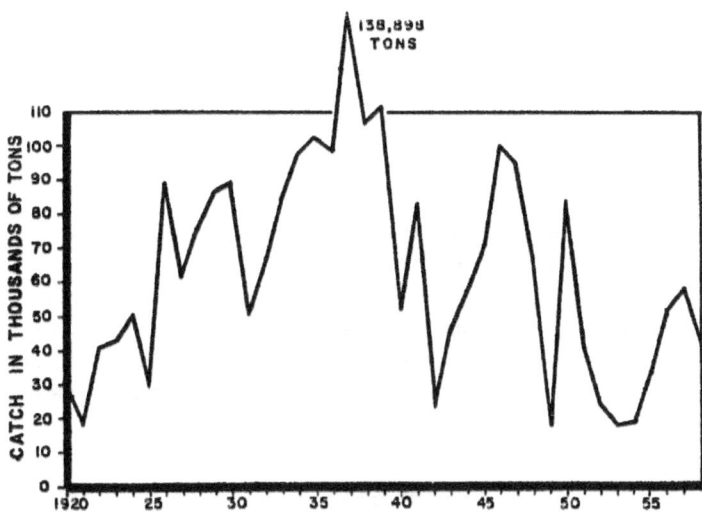

Figure 7.2. Alaska herring catch, 1920–1958. (*Pacific Fisherman*, January 25, 1959)

Figure 7.3. Data sourced from Bureau of Fisheries and ADF&G annual management reports and *Pacific Fisherman* yearbooks.

At Zachar Bay, relatively small-scale herring reduction done in conjunction with herring-roe operations would continue through 1971. The plant there, however, produced meal but no oil after 1967.[77]

8
BAIT HERRING

> Traditionally the first catch of herring bait signals the start of the commercial fisheries in the Sitka area. The whistles at the cold storage plant are sounded and anyone wanting a mess of fresh herring can take it free of charge. Usually herring bait is taken in the immediate vicinity of Sitka in the month of February.—Alaska Department of Fish and Game, 1967[1]

Herring is the favored and traditional bait of halibut fishermen but is also utilized by cod fishermen, salmon trollers, crab fishermen, and others. Halibut fishermen prefer large herring with firm flesh (because they stay on hooks better) and a high oil content.

Unlike using herring to make meal/fertilizer and oil, the use of herring as bait engendered no controversy. Early Bureau of Fisheries regulations did not restrict fishing for bait herring. Modern regulations accommodate the bait fishery, which has never been overly large.

Alaska's bait-herring fisheries in Southeast Alaska, in Prince William Sound, and in the waters around Kodiak Island typically occur in the fall and early winter, when herring are firm-fleshed and thus most suitable for bait. At Dutch Harbor, the fishery occurs in late summer, when herring congregate there after spawning at Togiak. The Norton Sound fishery occurs in late spring and early summer. The market there in recent years has been limited by the local processing company to an amount far below ADF&G's quota.[2]

SOUTHEAST ALASKA

> The business of supplying herring to the immense fleet of halibut fishing boats that gather in Southeastern Alaska in the winter has become quite an industry.—*Pacific Fisherman*, January 1911[3]

The demand for herring to be used as bait began in the late 1890s, when the Seattle halibut fleet began fishing in Southeast Alaska. The fleet fished during the fall, winter, and early spring, when it was difficult and dangerous to fish on the halibut banks at Cape Flattery, at the north entrance to Puget Sound. By 1899, the fleet had made Petersburg, a fledgling fishing community on Wrangell Narrows, its winter headquarters.

Figure 8.1. Brailing bait herring, Washington Bay, Southeast Alaska, June 1971. Purse seiner *Little Lady*, tender *Howkan*. Notice bycatch king salmon in brailer. (James Mackovjak)

The location had two advantages: it was along the main steamship route to Seattle, and icebergs from the not-too-distant LeConte Glacier could be ground up and used to ice the catch. Facilitating the development of the fishery, in 1899 the Icy Strait Salmon Company, the pioneer cannery at Petersburg, constructed a wharf and arranged with steamship companies to make regular calls for freight. Rather than make the long journey to Seattle with their catches, the halibut fishermen sold their fish at Petersburg. The halibut were packed with ice in locally made wooden boxes, each holding about 500 pounds of fish, and sent to Seattle on steamships.

By 1905, three large scows, each capable of handling 200 to 400 boxes of halibut at a time, were anchored at what became known as Scow Bay, about three miles south of Petersburg. The fishing boats found it much easier to come alongside and discharge on the scows rather than on the wharf, and it was relatively easy for the steamships to transfer the boxes from the scows to their cargo holds.

At Scow Bay, halibut fishermen could sell their catch and get ice, but they also needed bait, and herring—preferably fresh herring—was their preference. At times herring was used almost exclusively. And for good reason: herring are a favored food of halibut. During the summer of 1907, a Bureau of Fisheries agent opened a halibut measuring about forty-two inches long. Inside its stomach were twenty-two good-sized herring.[4] A bureau report that same year on the fishes of Alaska noted that herring had recently "come to be in great demand as bait in the halibut fisheries."[5]

Herring were sometimes very abundant in the Petersburg area, but at other times they could not be found for periods of a week or two. For halibut fishermen, there was no guarantee herring would be available. In 1909, John Cobb and his colleague at the Bureau of Fisheries, Millard Marsh, wrote that the supply of herring for halibut bait had not kept pace with demand. They characterized this bait shortage "the most serious problem confronting the halibut fishermen."[6]

The unreliable supply forced the halibut fleet to sometimes rely on salted herring, which were usually put up during the summer months. To salt herring that would be used for bait, about 200 pounds of whole herring were sprinkled with about fifty pounds of salt and dumped into a barrel. Herring used for halibut bait were smaller than those used for human consumption, the number of fish per barrel ranging from 900 to 1,200.[7] In 1906, Alaska salteries catering to halibut fishermen packed 4,450 barrels—890,000 pounds—of bait herring. (An additional 440,000 pounds of fresh bait herring was also provided.) That year, Scow Bay accommodated forty-one halibut vessels, and John Cobb pointed out the desirability of having a small cold storage at Petersburg that was capable of freezing bait herring.[8]

By 1910, the New England Fish Company, at Ketchikan, was freezing bait herring.[9] Nevertheless, bait herring could still be a precious commodity. As *Pacific Fisherman* reported in February 1911:

> Bait has been a very serious matter with the fleet fishing for halibut in Southeast Alaska. Two purse seines at Ketchikan, three at Scow Bay and Petersburg, and two at Juneau have been doing their best to supply the demand, but, owing to the very large fleet fishing this season, they have been finding great difficulty in doing so. The New England Fish Company has bought and frozen all the herring it could at its Ketchikan plant in order to supply its own steamers and the fleet of smaller vessels which sells its catch there.[10]

Production of frozen bait herring in Alaska for the year 1911 totaled 750,146 pounds. Additionally, 1.1 million pounds of herring were sold as fresh bait, and 2,080 barrels (about 416,000 pounds) were sold as salted bait. All bait-herring production that year occurred in Southeast Alaska.[11] Among the producers of fresh bait herring was the Halibut Bait Company, of Petersburg, which *Pacific Fisherman* said was "catching and selling bait herring on a large scale." The company's seiner was the small steamship *Agnes W*. At Juneau, the steamship *Peerless* supplied herring to the halibut fleet.[12]

> Herring, as is well known, is the only uniformly successful bait that has been found for halibut, and without a good supply for this purpose the great halibut industry which at present is growing very rapidly, would soon dwindle to almost nothing.[13]

By the 1920s, halibut stocks off the Washington coast had become depleted, and the fleet moved to Alaska waters for the entire season. This increased the demand for bait herring. Production in Southeast Alaska in 1922 was 2.1 million pounds, and in central Alaska it was 1.8 million pounds. Total bait herring production in Alaska increased to 5.2 million pounds in 1923, but it declined to 3.6 million in 1924. That year, a scarcity of bait idled practically the entire halibut fleet for three weeks in September.

Production, though, rebounded and increased, and in 1926 it totaled seven million pounds, almost all of which was caught in Southeast Alaska.[14] Annual bait-herring production remained at about the 1926 level for the next four decades.

Reflecting the lack of controversy regarding Alaska's bait-herring fishery, 1925 Bureau of Fisheries regulations that placed some limitations on the herring fisheries were designed to not interfere with the taking of herring for halibut bait.[15]

The herring fishery that supplied the halibut fleet with bait was primarily conducted during the winter and spring. Most of the herring were frozen, but those to be sold as fresh bait were often held in impoundments. Circa 1930, the New England Fish Company plant at Ketchikan was freezing about two million pounds of herring into fifty-pound blocks each winter. In 1933, seven companies in Southeast Alaska maintained pounds to provide fresh herring to the halibut fleet.[16]

Bait herring were impounded at Indian Cove, in Auke Bay, near Juneau, since at least the early 1950s.[17] Likewise, bait herring were impounded at Sitka beginning in the early 1960s. In 1983, the Alaska Department of Fish and Game authorized bait pounds at five locations in Southeast Alaska: Tee Harbor, Indian Cove, Farragut Bay, Scow Bay, and Sitka Sound. Three herring pounds were operated near Sitka during the 2001–2002 season.[18]

Herring is a preferred bait by trollers targeting king salmon.[19] Sport fishermen also use herring as bait when trolling for salmon or fishing for halibut. At Sitka in 1965, the Katlian Packing Company and

Figure 8.2. Ketchikan waterfront, 1936. Brailing herring from the net of the purse seiner *Seaketch* onto the *Pirate*, which is acting as a tender. (Image courtesy of Ketchikan Museums, Tongass Historical Society Collection, THS 8L9.5.125)

Sitka Cold Storage produced a new product: sport-pack herring. Each pack contained a dozen herring laid out individually in a tray and frozen. Together, the companies produced a total of about 200,000 trays. During each of the following two years, they produced about 300,000 trays. The herring were graded by size, and each pack held six to twelve herring.[20]

By the early 1960s, a store that catered to sport fishermen kept live herring in an impoundment at Indian Cove, in Auke Bay, near Juneau.[21] It was an honor-system arrangement: fishermen paid at the store and then, on their way to the fishing grounds in their boat, stopped at Indian Cove to dipnet from the impoundment the quantity of herring they had purchased.

There was no closed season and no quotas whatsoever on bait herring until 1956, when limits of 125 tons of herring were applied to two locations in Southeast Alaska: Silver Bay, near Sitka, and Fish Egg Island, near Craig.[22] At least since the early 1970s, bait herring for the frozen market in Southeast Alaska have been caught by purse seiners primarily targeting discrete wintering schools in major bays and inlets.[23] The fish is then frozen at cold-storage plants, typically in forty-pound blocks.

At the request of fishermen, in 1974 and 1975, ADF&G opened the Seward small-boat harbor to fishing for bait herring. The harvest was directed toward young herring that would be used in the local recreational salmon troll fishery. (The City of Seward hosts an annual Silver Salmon Derby.) The quota was unusual: 50,000 dozen herring. No harvest, however, was reported.[24]

In 1979, the Alaska Board of Fisheries authorized a "tray pack pound" fishery in Southeast Alaska to which it allocated a harvest of up to 100 tons. Only limited harvests occurred in the early 1980s, and by 2002 there was little participation in the fishery.[25]

Annual statewide bait-herring production was typically 2,000 to 3,000 tons until the early 1970s. The expansion of the state's crab fisheries during the 1970s increased the demand for herring. Since 1984, bait-herring harvests have averaged about 4,200 tons, with a peak of 9,126 tons in 2016.[26]

PRINCE WILLIAM SOUND

The first recorded substantial effort to harvest bait herring in Prince William Sound was in June 1975, when four purse seiners landed 227 tons. There was no bait-herring harvest in the sound in 1976. Beginning in 1977, ADF&G changed the bait fishery to a fall/winter fishery that began October 1 and ended February 28. But the department was flexible. It opened the 1979–1980 season in the southwestern and western regions of the sound on September 15 by emergency order after a large concentration of herring was observed. The September 15 opening and February 28 closure for the entire sound became the norm until the 1984–1985 season, which opened on September 1 and remained open until the guideline harvest level was reached or at the end of the season on January 31, whichever occurred first.

During the fall and winter, however, herring were often too deep for purse seiners to catch them. A new development during the 1977–1978 season was the use of midwater trawls to catch bait herring. Five trawlers were involved. One was a conventional trawler, the others were pair trawlers, two vessels that each towed one side of a large trawl. Together, the five trawlers landed 859 tons of herring out of a 1,400-ton guideline harvest level. An additional seventeen tons were landed with seine gear.

A guideline harvest level (GHL) represents a conservative preseason estimated allowable harvest volume that would not jeopardize the viability of herring stocks and may vary annually based on stock status. ADF&G reports use the terms *guideline harvest level* and *quota* interchangeably.

During the 1978–1979 season, seven trawlers landed 989 tons of bait herring. The catch by pair trawlers declined substantially in the following years, and their use was abandoned after the 1981 season.

Meanwhile, during the 1978–1979 season two seiners deepened their seines by adding additional netting to their bottoms and harvested

185 tons. Since then, purse seiners, especially those with deepened seines, accounted for essentially all of the bait-herring harvest in Prince William Sound.[27] The catch by seven purse seiners during the 1981–1982 season was 1,030 tons, all of which was caught in September 1981. During the next decade, the seasonal harvest averaged 1,070 tons. The maximum number of seiners involved was eight. In 1986, for the first time in several years, no bait herring were tendered—transported by boat—out of Prince William Sound for processing elsewhere. Six local processors were involved in the fishery.

Though the GHL was almost 1,700 tons, there was little interest in Prince William Sound bait herring in 1989, the year of the *Exxon Valdez* oil spill. Three seiners fished in the first half of November and landed 646 tons. Interest increased in 1990, and herring stocks had increased as well. The GHL was 3,151 tons, almost double that of the previous year. Five seiners and one trawler delivered 1,995 tons of herring to seven processors during two months of fishing in the fall.

There was a great demand for bait in the fall of 1991, and during two weeks of fishing in October, fourteen seiners harvested 4,258 tons—the all-time record. Ten processors bought the fish. Most were located in Cordova, but Icicle Seafoods (Seward) and Inlet Salmon (Kenai) purchased a good portion of the catch. The processors paid fishermen an average of $250 per ton, which put the ex-vessel value of the fishery at a bit over a million dollars.

Herring remained abundant, and for the 1992 season, ADF&G set the GHL at 3,416 tons. The number of seiners increased to seventeen, and during three weeks of fishing that fall they caught 3,900 tons.

Unfortunately, the herring population then experienced a steep decline. The guideline harvest for 1993 was 978 tons, less than a third of what it had been the previous year. Eight seiners landed 1,087 tons of herring.

The fishery was closed in 1994 and 1995, reopened in 1996, and continued through 1998. The guideline harvest during those years averaged 912 tons while the actual harvest averaged 872 tons. The fishery was closed following the 1998 season and as of 2021 has not reopened.[28]

KODIAK

From 1960, when the State of Alaska obtained jurisdiction over its fisheries, until 1973, there were no harvest quotas or closed seasons for food and bait herring in the Kodiak management area. In 1974, regulations were introduced that limited the fishing season to the period from July 1 to February 28. In 1979, GHLs were established. Seasons and GHLs have been adjusted over the years to accommodate changes in the stocks and the fishing effort. The fishery, though, is comparatively small: between 1960 and 1992, food and bait herring production ranged from zero to 381 tons, essentially all of which was used for bait. Bait herring at Kodiak were harvested almost exclusively by purse seines until 1977, when midwater trawls were introduced. (As noted above, trawls were also used to catch bait herring in Prince William Sound.)

In early 2001, the Alaska Commercial Fisheries Entry Commission (which is discussed at length in part II of this work) designated the Kodiak food and bait fishery as a limited-entry fishery and issued interim-use permits to thirteen fishermen. The permits were valid until the commission determined which fishermen were eligible for limited-entry (permanent) permits.

ADF&G that season was concerned about controlling the harvests in areas with relatively small guideline harvest levels (sixty tons in the Uganik District and forty-seven tons in the Eastside District). The department notified fishermen that it would not allow a competitive fishery to occur, and it suggested that the fishermen form a cooperative, as had happened during some openings in the Sitka Sound roe-herring fishery. The permit holders agreed to do so. Under the cooperative agreement, only one vessel was allowed on the grounds, but the ex-vessel revenues were shared by all the permit holders. The cooperative allowed ADF&G to provide more liberal fishing time and to open a broader area. The cooperative fishery worked well, and all the Kodiak food and bait fisheries that have occurred since 2001 have been cooperative efforts.

In 2002, the Commercial Fisheries Entry Commission issued limited-entry permits to nine Kodiak-area herring fishermen. Five of the permits were for purse-seine gear, and four were for trawl gear.

The average GHL from 2009 to 2018 was 300 tons, while the average harvest was 160 tons.[29]

DUTCH HARBOR

An ancillary of the Togiak roe-herring fishery (see chapter 13) is the Eastern Aleutian Islands food and bait herring fishery, which is commonly referred to as the Dutch Harbor food and bait fishery. It is, for all practical purposes, a bait fishery.

After spawning, Togiak herring migrate southward along the Alaska Peninsula and concentrate in the vicinity of Unalaska Island in the late summer. The fishery, which began in 1981 and primarily supplies bait for the crab and longline fisheries, occurs in ADF&G's Unimak, Akutan, and Unalaska Districts and that portion of the Umnak District east of Samalga Pass. Most of the harvest, however, occurs within several miles of seafood-processing plants in Unalaska and Akutan Bays.

In the past, the season typically opened in mid-July and closed in mid-September, but in some recent years it has opened in late June. In 2015, ADF&G delayed the opening for a day because of numerous humpback whales feeding on herring in the area. There was fear of collisions between fishing boats and whales.[30]

The Dutch Harbor fishery quickly grew from two purse seiners harvesting 704 tons in 1981 to six to nine purse seiners harvesting an average of about 3,200 tons from 1982 to 1987. Gillnetters, too, participated in the fishery. Their numbers peaked at thirteen in 2002 and 2003, but no gillnetters have participated since 2009.[31]

During the 1981 and 1982 seasons, there were no commercial herring harvest restrictions in the fishery. From 1983 through 1986, the Alaska Board of Fisheries implemented a harvest ceiling, and in December 1987, as part of an amendment to the Bristol Bay Herring Management Plan, the board directed ADF&G to manage herring fisheries so the overall exploitation of a herring stock did not exceed 20 percent of the spawning biomass. In the amended plan, the board included provisions that accommodated the Togiak spawn-on-kelp fishery and the Dutch Harbor food and bait fishery. The provisions stipulated that before opening the Togiak roe-herring fishery, ADF&G must set aside 1,500 tons of the available harvest for the Togiak spawn-on-kelp fishery and then allocate 7 percent of the remaining available harvest to the Dutch Harbor food and bait fishery.[32]

The allocation for the 1988 Dutch Harbor food and bait fishery was 3,100 tons. The harvest that year was 2,004 tons. From the inception of the fishery in 1981 through 2018, the average harvest has been 1,970 tons. The average ex-vessel value of the fishery during the years 2007–2016 was $611,000.[33]

Some years, more than twenty purse seiners participated in the fishery, and they could take the entire quota—and sometimes considerably more—in a matter of minutes. The fishery evolved into a cooperative in which a limited number of boats fished and shared the proceeds with the rest of the fleet. Since 2004, the maximum number of purse seiners that participated in the fishery during any given year was three.

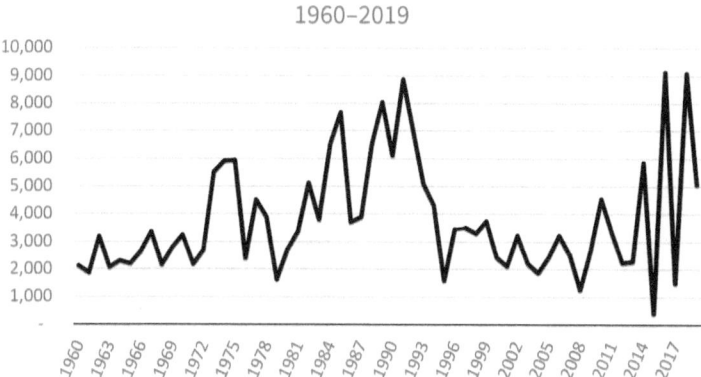

Figure 8.3. Data sourced from ADF&G management reports.[34]

In 2013, three purse seiners, each fishing for a different company and all sharing the same spotter plane, landed 1,764 tons of bait herring, one ton short of their quota. It was an efficient operation that represented a big success for fishermen and the fishery's managers alike.

In 2004, the Alaska Board of Fisheries established a food and bait herring fishery in the Adak District. Its 500-ton herring allocation was independent of the Dutch Harbor allocation. However, because of a lack of industry interest, there has never been a herring fishery in the Adak District.[35]

UPPER COOK INLET

Currently, though small in scale, the upper Cook Inlet food and bait fishery is possibly the most lucrative Alaska herring fishery today—at least in terms of price per pound. There, in an open-access fishery, ten or twenty gillnetters supply bait herring to local commercial halibut fishermen and sport fishermen. (Homer is a sportfishing center.) In 2018, the gillnetters harvested 17.6 tons, which they sold for $1.00–$1.50 per pound, or $2,000–$3,000 per ton. By contrast, roe herring harvested in Sitka Sound that year were worth about $340 per ton. The lower Cook Inlet bait-herring harvest in 2019 was 34.3 tons.[36]

Part II
Roe Herring

Figure 9.1. Kazunoko. (Yoshihiko Hirabuki)

9
GENESIS AND MANAGEMENT OF ALASKA'S ROE-HERRING FISHERY

> It is maybe the most extreme example I'm aware of how a major Alaska industry could be dependent on an extremely specialized foreign market.—Gunnar Knapp, former University of Alaska fisheries economist, 2021[1]

> Given the time frame of Kazunoko's consumption (a few weeks at the end of the year), the short duration of fishing periods (days or even hours), and the natural fluctuations in fish populations (orders of magnitude even in the absence of heavy fishing), many consider the Pacific roe herring fishery the biggest crapshoot west of Reno.—Terry Johnson, University of Alaska, 1993[2]

Salted herring roe (*kazunoko*) is a delicacy in Japan that is traditionally served or given as a gift during the new year's celebration. Mature herring roe is bright yellow, with translucent eggs and few blood vessels. It has firm consistency and a sticky feel. For retail sale, kazunoko is packaged in decorative trays and sold in department stores and other gift outlets.

Before being served, kazunoko is soaked overnight in fresh water to remove most of the salt. When ready to be eaten, kazunoko is typically marinated in spices and soy sauce. When the egg skein is bitten into, each egg pops individually, providing a mildly crunchy sensation.

Among other benefits, according to Japanese lore, eating herring roe increases fertility. Translated, *kazunoko* means "more sons and daughters."[3] The following is a description of a kazunoko marketing campaign published in *Bristol Bay Fisherman* in 1982.

> The Hokkaido Federation of Fish Processors Associations... has opened a nationwide campaign to promote the consumption of kazunoko and to stimulate the interest of the average consumer. Using the slogan "Please eat kazunoko, a Japanese taste," the campaign opened with the free distribution of 1,000 bags of kazunoko, each bag containing three pieces of the product. The bags were given away at brief promotional events in three major cities—including Tokyo and Osaka—by five "Miss Kazunokos" dressed in the distinct yellow color of the expensive delicacy.[4]

Roe herring are caught in waters on or near herring spawning grounds in the spring, just prior to spawning, when the roe is fully mature and the roe content is highest. The market favors fish that weigh from 160 to 200 grams (5.6 to 7 ounces), which puts the egg skeins at the optimum size for gift packs.[5]

Roe-herring processors typically want fish with a minimum roe content—weight of the roe in relation to the weight of the entire catch (both females and males), as determined by on-the-grounds sampling—of 10 percent, which is an indicator that the roe is mature. The price paid for fish with a 10 percent roe content is usually the base price in a fishery. Processors typically pay more for catches with a higher roe content and less for catches with a lower roe content.

The Japanese traditionally sourced roe herring in their coastal waters, but in the mid-1950s, herring stocks there began declining, and the Japanese began looking internationally for a replacement source of herring roe. Sources eventually included Alaska, California, and British Columbia, as well as eastern Canada, Russia, Korea, China, and several northern European countries, such as the United Kingdom (Scotland), Ireland, the Netherlands, and Germany. Generally, the most valuable roe came from British Columbia, followed by San Francisco Bay, and then Alaska.

Roe herring in Alaska are primarily caught using purse seines, but in some fisheries gillnets and beach seines are also employed. While seines catch both large and small fish, the mesh on gillnets can be sized to select for large fish.

In Alaska, the relatively high prices paid to fishermen caused a rapid expansion of the roe-herring fishery. Officially referred to by the Alaska Department of Fish and Game as the herring sac roe fishery, by the early 1970s it had become the largest herring fishery in Alaska. Because it occurred in the spring, before Alaska's salmon seine and gillnet fisheries opened, the roe-herring fishery provided an important economic boost to fishermen, fish-processing plant workers, and local economies during what was usually a lean time of year.

The market for kazunoko, however, has declined. By 1990, Japanese buyers were warning of falling demand. Two factors accounted for the decline: the Japanese were becoming more health conscious and eschewed the salted product, and the younger generation had acquired Western eating preferences. Nevertheless, it was a few years later, 1996, that the value to fishermen of Alaska's roe-herring fishery—a function of the volume of fish caught and the ex-vessel price—peaked. That year, the statewide ex-vessel value totaled $61 million.[6] By 2020, the market for kazunoko produced from Alaska herring was a shadow of its former self. However, kazunoko produced from British Columbia herring is considered a premium product and has retained a solid, if limited, market.[7]

Alaska's three largest roe-herring fisheries were, in chronological order based on their inception as major fisheries, in Prince William Sound, Sitka Sound, and Togiak, in Bristol Bay. Somewhat smaller roe-herring fisheries occurred at Kodiak Island, lower Cook Inlet, and Norton Sound. All six of these fisheries are chronicled in the following chapters.

Unfortunately, there is not space in this book to discuss the many small or short-lived roe-herring fisheries that have occurred along Alaska's coasts. In the following chapters, I have included redundancies to help ensure that discussions of the various fisheries can stand alone.

ROE STRIPPING

The process of extracting herring roe that was commonly employed in Alaska until the mid-1970s was crude. It was generally known as "roe stripping," but in polite circles, the process was referred to as the "decomposition method." In less-polite circles, it was "pop and dump."

In the process, herring, with several shovelfuls of salt, were loaded into totes (in the early years plywood boxes lined with plastic bags, each holding about a ton of herring). Water was added to fill the tote, and the fish was then allowed to "age"—a polite word for rot—in the brine for about four days.[8] But it was the flesh, not the eggs, that rotted; the egg skeins actually hardened a bit. Removing the skeins was simply a matter of squeezing the herring just above the visceral cavity, and the egg skeins would push through the rotted belly flesh and "pop" out. The skeins were then graded by Japanese technicians and packed in brine in five- or six-gallon plastic buckets. The filled buckets were kept under refrigeration.

Squeezing herring was low-skill, piecework employment. And it was stinky: a sign on the door of the laundromat in Seward in 1973 read "No Herring Squeezers." Little was required in the way of facilities or equipment: a table, running water (even saltwater), a scale, and some baskets were all that was necessary. Some processing facilities were makeshift, with tarps for roofs or no roof at all. Roe herring were also processed aboard vessels. Popular for this sort of operation were power scows—World War II–era wooden boats that were cheap and had considerable deck space and crew accommodations.

Figure 9.2. Squeezing roe herring aboard a power scow, *Tatitlek*, Prince William Sound, 1974. (James Mackovjak)

Roe-stripped carcasses (and male herring) were typically discarded overboard on site.

By 1970, Juneau Cold Storage, and perhaps other firms as well, was freezing whole roe herring in boxes and shipping it to Japan. By the mid-1970s, freezing whole herring had become the standard practice in Alaska. In 1977, Alaska's Legislature outlawed roe stripping, though it provided a temporary exemption for Bering Sea operations.[9]

GENESIS OF ALASKA'S ROE-HERRING FISHERY

Though there had been one-ton harvests of roe herring in Resurrection Bay and lower Cook Inlet in 1961 and 1963, the first significant Alaska roe-herring fisheries occurred in 1964 in at least three locations: Spiridon Bay, on Kodiak Island; Sitka, in Southeast Alaska; and Unalakleet, on Norton Sound. There may have also been a roe-herring operation that year at Seldovia, on the Kenai Peninsula. The company behind the operations at Spiridon Bay and Unalakleet, and the possible operation at Seldovia, was Anchorage-based Western Alaska Enterprises, a subsidiary of Taiyo Gyogyo Ltd., a large Japanese seafood company. By far the largest operation that year was at Spiridon Bay, where 23,070 pounds of roe were extracted from 235 tons of herring and salted. Sitka Cold

Storage, at Sitka, produced 9,990 pounds of herring roe, and at Unalakleet 2,520 pounds of roe were extracted from 20 tons of herring.[10]

In 1965, there was an acute shortage of herring roe in Japan, and the price of the product was at a record high. At the central wholesale market in Tokyo, top-quality herring roe was bid up to 22,500 yen per kilogram (US$28.40 per pound.)[11]

ADF&G reported that in 1965 southcentral Alaska produced 108,672 pounds of herring roe, with a wholesale value of $274,148 ($2.52 per pound). Three Kodiak Island firms, King Crab, Inc., the Kodiak Bait Company, and the Washington Fish & Oyster Company, accounted for the entire production.

Southeast Alaska produced 89,000 pounds, with a wholesale value of $32,989 ($0.37 per pound).[12] The sole herring-roe operation in Southeast Alaska that year was the Far East Trading Company, at Ketchikan.[13] The low value of the Far East Trading Company's production indicated it likely did not meet Japanese quality standards. The fish may have been processed aboard a boat or barge.

In 1966, ADF&G listed three Southeast Alaska firms—Kito Enterprises and Petersburg Cold Storage, in Petersburg, and Katlian Packing Company, in Sitka—as having produced a total of 54,475 pounds of herring roe.[14]

MANAGEMENT OF THE HERRING FISHERIES

Alaska became a state on January 3, 1959, but, pursuant to a provision in the Alaska Statehood Act, the new state did not obtain jurisdiction over its fish and wildlife resources until it convinced the federal government that adequate provision for the administration, management, and conservation of the resources had been made. One year later, in January 1960, Alaska gained control over its fish and wildlife resources.[15]

Unique among the states, Alaska's constitution states that "Fish, forests, wildlife, grasslands, and all other replenishable resources belonging to the State shall be utilized, developed, and maintained on the sustained-yield principle, subject to preferences among beneficial uses."[16] This directive is fundamental to the state's management of its fisheries resources. Overall, the state has been far more active in managing Alaska's fisheries than the federal government had been.

The Alaska Board of Fish and Game was originally responsible for developing regulations to manage Alaska's fish and game resources, but in 1975, citing the complexity of the issues involved in both fishing and hunting/trapping, Alaska's Legislature divided the board into two independent entities: the Alaska Board of Fisheries and the Alaska Board of Game. Members of each board are nominated by the governor, approved by the legislature, and serve three-year terms.

The purpose of the Board of Fisheries is the conservation and development of the state's fisheries resources, and it has broad authority to adopt regulations governing the use of fish resources in the state. The Alaska Department of Fish and Game (ADF&G) implements the board's regulations as well as statutes adopted by Alaska's Legislature. The department also researches fisheries.[17]

For management purposes, ADF&G regulates most contemporary herring fisheries by geographical management units delineated around distinct spawning aggregations, e.g., Sitka Sound and the Togiak District.[18]

Though methods and standards have evolved, the current statewide herring management strategy is to harvest 0–20 percent of a spawning population's biomass each year. The upper end of the range is applied to stocks that are in good condition; the lower end is applied to stocks exhibiting a trend of decreasing abundance and poor recruitment.[19] Since 1982, the harvest of a particular spawning stock is prohibited unless the estimated spawning biomass is above the minimum threshold level established for the fishery.[20]

ADF&G uses a variety of survey techniques to assess herring stocks, including aerial visual estimates, hydroacoustic surveys, and spawn deposition surveys. Information gained in these surveys, along with herring age, weight, and length estimates and other data, is used to establish a guideline harvest level (GHL) for each fishery. The GHL, which is often referred to as a quota, represents a conservative preseason estimated allowable harvest volume that does not jeopardize the viability of herring stocks and may vary annually, based on stock status.[21]

In-season, ADF&G manages the roe-herring fishery to maximize the value of the roe produced. The department's challenge is to work with fishermen to locate schools of ripe herring—ideally large fish—and then to allow fishing during the narrow window between when female herring become ripe and when they spawn. At the same time, the department has to limit fishing so overharvest and waste (dumped sets) do not occur. Test fishing by commercial fishermen prior to openings helps locate schools of herring and assess their value.

* * *

The state also manages its fisheries by restricting access to them. During the early 1970s, Alaska's population was growing rapidly and so was the number of commercial fishermen. Through seasonal and geographical limitations and gear restrictions, ADF&G attempted to limit catches to what it considered sustainable, but the effect of each new entrant into a fishery was to reduce the average catch—and income—of all fishermen. The additional fishing effort also presented a challenge to sound fisheries management: fishing could be so intense that it was difficult to monitor the catch, and overfishing sometimes occurred.

To rectify this situation, in 1973 Alaska's Legislature passed the Limited Entry Act to "promote the conservation and the sustained-yield management of Alaska's fishery resources and the economic health and stability of commercial fishing in Alaska."[22] It would do so by limiting access to selected state-managed fisheries.

To administer the limited-entry program, the legislature simultaneously created the Alaska Commercial Fisheries Entry Commission. Generally, the commission determined how many fishermen each fishery could support and issued limited-entry permits to those who met economic dependence and historical participation requirements for the various fisheries. Eligibility determinations often took years, and the commission issued interim-use permits to qualified fishermen whose eligibility was being determined.

Because limited-entry permits were transferable, including by sale, the commission created private wealth. The commission reports permit values based on averaging actual prices reported by fishermen transferring permits on the open market. Fishermen buying permits, of course, invest in them based on estimates of future earnings from the fishery and the potential resale value of the permit.[23]

The limited-entry program was implemented in the different fisheries over many years, though some fisheries have remained open access. The commission's first action, in 1974, limited access to the state's major salmon fisheries.

Some herring seiners, particularly those in Southeast Alaska, preferred the competitive nature of the fishery and initially opposed limiting it. The fishery, however, became extremely intensive. And short. A roe fishery opening at Sitka in April 1975 lasted only ninety minutes—barely time enough to make a couple of sets. The competition was intense, and while some fishermen caught a lot of fish, others had little to show for their effort. It was obvious that if the fleet continued to grow, fishing operations would become untenable, and those fishermen who had earlier opposed limited entry now wholly embraced the program.[24]

Likewise, herring seiners who were established in the already-overcrowded Prince William Sound and Cook Inlet fisheries favored a limited-entry program.[25] Moreover, Charles Stovall, commissioner of the Commercial Fisheries Entry Commission, noted in 1976 that the fishing effort in the sound and inlet had increased during the past three years "to the point that management of the resource has been seriously threatened and in some cases over-harvesting of the allowable take has been inevitable."[26] Case in point: in 1974, the ADF&G established a 5,000-ton herring quota for Prince William Sound. The quota was exceeded that year, and again in 1975.[27]

In 1977, the limited-entry commission began limiting herring fisheries. Specifically, it limited the roe-herring purse-seine fisheries in Southeast Alaska, Prince William Sound, and Cook Inlet. Because of the low fishing effort at Kodiak—only one purse seiner fished for roe herring during the 1976 season—the roe-herring fishery there was not limited in 1977. Participation in the Kodiak fishery grew, however, and the commission limited entry into it in 1984.[28]

Some purse seiners fished in all of Alaska's major roe-herring fisheries. They either qualified for or purchased limited-entry permits for the fisheries for which they were required. These "circuit seiners" typically started out in Sitka, then moved to Prince William Sound, then to Kodiak, then to Kamishak Bay, and then to Togiak, which was not a limited-entry fishery. Some diehards occasionally continued to Norton Sound. The vessels had experienced crews and were often equipped with state-of-the-art equipment, including side-scanning sonar, to help them locate schools of herring. The sonar enabled the fishermen to work during any time of the day or night and in weather conditions that precluded using spotter planes.[29]

For Southeast Alaska, the commission initially issued thirty-eight permanent permits, all but two of which were issued to Alaska residents. It later issued an additional fourteen permanent permits. By 1987, a Southeast Alaska roe-herring purse-seine permit was worth about $217,000. In 2011, a permit for the fishery was sold for about $519,000—an all-time record for any Alaska limited-entry permit.[30]

Roe-herring purse-seine permits for Prince William Sound and Cook Inlet were less valuable. For Prince William Sound, the commission initially issued eighty-five permanent roe-herring purse-seine permits, seventy-seven of which were issued to Alaska residents. Ten years later, a permit for this fishery was worth about $96,000. The *Exxon Valdez* oil spill in 1989 and an outbreak of disease several years later compromised the sound's herring fishery, and the value of a roe-herring purse-seine permit in 2011 was a little less than $30,000.[31]

For Cook Inlet, the commission initially issued fifty-seven resident and four nonresident permanent roe-herring purse-seine permits. In

1987, a permit was worth $111,000. Because of low herring abundance, the roe-herring fishery in Cook Inlet has been closed since 1999.[32]

Since its inception, the Alaska Commercial Fisheries Entry Commission has recognized 167 unique herring fisheries and has limited access to twenty-two of them.[33] Most of those fisheries, however, have been discontinued. Currently, the commission recognizes forty-two unique herring fisheries and limits access to seventeen.[34]

For herring seiners in Southeast Alaska, the limited-entry system had at least one unintended benefit. The region's roe-herring seine fleet, despite being limited in number, was very efficient and powerful, but the area fished (in Sitka Sound) was small. During seasons of low herring abundance, fishery managers would not have risked opening the season for even half an hour for fear that the quota would be exceeded and the herring stock damaged. But more than once, permit-holding fishermen had a solution: they unanimously agreed to fish cooperatively. A limited number of boats fished, but the entire fleet shared the revenues from their catch. This might not have been possible had the limited-entry program not clearly limited the number of and defined the stakeholders.[35]

Cooperative fishing was employed in Sitka Sound when the proposed fishing area was too confined to permit the jostling of dozens of seine boats and their skiffs or at the end of a season when just a small portion of the quota remained to be caught. The 1979 roe-herring fishery in Sitka Sound, described in chapter 10, was the first cooperative fishery. Cooperatives have also been employed in the Togiak roe-herring fishery and in the Dutch Harbor food and bait fishery, neither of which are limited-entry fisheries.

For ADF&G, cooperative fishing made in-season management of the fishery easier. Of the cooperative roe-herring fishing effort in Sitka Sound in 2015, ADF&G biologist Dave Gordon said, "If everything's going well and they're harvesting good-quality herring, there's no reason to shut it down. We can basically fish all day long and give them the flexibility to target their efforts on good fish, and take their time, and not worry about competing for schools and that sort of thing."[36]

Chapters 10–13 describe and chronicle Alaska's largest roe-herring fisheries.

10

SITKA SOUND ROE-HERRING FISHERY

> Whales and sea lions and bald eagles come to Sitka to prey on the herring. As do an elite group of fishermen who annually vie in a high-stakes, multiday competition that sometimes takes place in the harbor immediately offshore Sitka's downtown on Baranof Island in Southeast Alaska. On such occasions, stores close their doors, not because the shopkeepers have gone fishing, rather because they've gone to watch fishing. Spectators line the shore and stand shoulder to shoulder on the town's bridge to watch the frenzied action of a fishery unlike any other, a precisely timed, macho haul of massive schools of ready-to-spawn fish nowadays captured in YouTube videos with titles like "The Shoot Out," a fishery still basking in the glow of the single set that netted a lucky boat nearly a million dollars.—John Grossmann, *Gastronomica*, 2015[1]

> "The annual Sitka Sound herring sac roe fishery alternated between tedious waiting, furious free-for-alls, and orderly cooperative harvesting."—Will Swagel, *Daily Sitka Sentinel*, 1989[2]

The Sitka Sound roe-herring fishery was, with those in Prince William Sound and Togiak, one of the three largest such fisheries in Alaska. And it could be very lucrative: a *Daily Sitka Sentinel* article in 1986 referred to it as the "multi-million-dollar herring sweepstakes."[3]

Of the spawning herring's schedule in Sitka Sound, ADF&G Sitka management biologist Bill Davidson said in 2000, "The fish take their time; they always keep you guessing. Every year is different."[4] That said, herring typically begin to arrive in Sitka Sound in the latter part of March and begin spawning several days later, with the larger fish spawning first. Generally, spawning is complete about ten days after herring spawn is first sighted.

Roe herring in Sitka Sound were harvested only by purse seiners, which, as will be explained below, were limited in number beginning in 1977. The fleet represented a lot of fish-catching potential: during a fifteen-minute opening in 2001, seiners landed almost ten million pounds of herring. Because Sitka Sound's was a single-gear fishery, there wasn't competition between gear groups for allocations, and this made managing the fishery a little easier and less stressful. The last roe-herring fishery in Sitka Sound was in 2018.

The commercial roe-herring fishery in Sitka Sound commenced in 1964, when Sitka Cold Storage froze 9,990 pounds of herring roe.[5]

Sitka was home base to a large fleet of commercial salmon trollers, and they and other fishermen were initially critical of taking herring just as the fish were about to help create a new generation of fish. Likewise, herring roe was a traditional food that Natives at Sitka believed was threatened by the roe-herring fishery. Nevertheless, the fishery grew.

ADF&G data for 1965–1968, however, is sketchy, and it is impossible to determine where roe herring were harvested. The department's reports for 1965 list 89,000 pounds of herring roe being produced in Southeast Alaska but failed to list a corresponding processor(s). In 1966, there were 54,475 pounds of herring roe produced, and the Katlian Packing Company, at Sitka, was listed as one of the three roe-herring processors in the region. There was no herring roe produced in Southeast Alaska in 1967. Production in 1968 was 22,775 pounds, and the sole processor that year was Petersburg Cold Storage.[6]

ADF&G considers 1969 to be the advent of the roe-herring fishery in Sitka Sound. The harvest that spring was 575 tons, and the fishery was immediately enveloped in controversy. That fall, the Southeast Alaska Trollers Association sent a letter to the Board of Fish and Game recommending all roe-herring fishing be stopped and fishing for bait herring should be allowed only after spawning was complete.[7]

At its December 1969 meeting, the board established a 750-ton year-round quota on commercial herring in Sitka Sound. The board's goal was to provide some protection to spawning herring.[8]

The 1970 roe-herring season in Sitka Sound was basically a repeat of the 1969 season, though with a bit more volume—703 tons. Most of the fish was frozen at Pelican Cold Storage, Petersburg Cold Storage, and probably at Sitka Cold Storage, but some of it was "stripped" of its roe, probably at Sitka Sound Seafoods, and the carcasses disposed of in local waters.[9] Ever critical of the roe-herring fishery, the Southeast Alaska Trollers Association condemned the practice of disposing of carcasses at sea as "not only wanton waste but a waste of a natural resource."[10]

Roe-herring production at Sitka in 1971 was 741 tons. Sitka Sound Seafoods, owned by Bob Wyman, went big into roe stripping that year and was the only company in Southeast Alaska to do so. To "age" the volume of fish Wyman hoped to process, rather than use the wooden totes commonly used for holding fresh fish, he purchased a dozen self-standing circular swimming pools from Sears. Wyman set the pools up on the Conway Dock, where his plant was located, and filled them with herring and brine. After about four days in the pools, the flesh of the fish was sufficiently rotted, and workers stripped the herring of their roe. Sitka's mayor referred to Wyman using the "decomposition method" to process roe herring. The carcasses were loaded onto a scow that, when full, was towed to Eastern Channel, where they were dumped.

Following Wyman's lead, Petersburg Fisheries also aged herring in swimming pools.

According to ADF&G's production report, 27,894 pounds of herring roe was produced in Southeast Alaska in 1971. It was valued at $44,630.[11]

The catch in 1972 diminished a little, to 602 tons, and in 1973, eight seiners were involved in the fishery. Their catch, 597 tons, was processed at Sitka, Petersburg, Juneau, and Excursion Inlet.[12]

Propelled by an increased Japanese demand for kazunoko, 1974 was a year of expansion in Alaska's roe-herring fisheries, and the number of seiners that participated in the Sitka Sound fishery increased to twenty-five. There were eighteen tenders on the grounds, but the seiners' catch, 681 tons, was not proportionately more than those of the previous two years.[13]

The 1975 roe-herring season in Sitka Sound, according to ADF&G biologist James Parker, "got out of hand." ADF&G managers wanted to harvest about 10 percent of the estimated spawning biomass in the sound, but the twenty-eight seiners that participated in the fishery landed 1,517 tons of herring, an estimated 24 percent of the spawning biomass.

The following section, with several small digressions, is a chronicle of the Sitka Sound roe-herring fishery since 1976.

1976

In contrast to the 1975 season, ADF&G's James Parker described the 1976 fishery as "orderly." The herring were spread out enough that the seiners weren't concentrated in a small area. On April 16, during three openings that totaled about four hours of fishing, thirty-eight seiners landed 800 tons of herring, a little more than ADF&G's 780-ton GHL. Bob Wyman, whose Sitka Sound Seafoods purchased part of the catch, estimated it was worth $160,000, about $200 per ton. Wyman planned to strip roe from at least a portion of the herring he purchased and to send the carcasses to Petersburg for reduction into meal at the Petersburg Fisheries reduction plant.[14] Most likely, though, the carcasses were simply dumped in nearby waters.

1977

Limited entry, as described in chapter 9, was implemented in the Southeast Alaska roe-herring fishery in 1977. The Commercial Fisheries Entry Commission that year issued thirty-eight permanent permits, though the number would subsequently increase to fifty-two.[15]

On April 4, roe-herring seiners at Sitka were put on two-hour notice for an opening. Last-minute ADF&G surveys, however, indicated that only enough herring were present to provide for the basic spawning needs of the area, and the season was ended before it began.[16]

1978

Following the disappointing 1977 nonseason, and anticipating a relative scarcity of herring in 1978, ADF&G reduced the GHL for roe herring in Sitka Sound to 250 tons. Despite this grim projection, twenty-three seiners came to Sitka in 1978. Most of the herring that entered Sitka Sound that spring were immature, and the catch was a mere 175 tons—an average of about 7.6 tons per boat.[17]

1979

The Sitka Sound roe-herring fishery came of age in 1979. Herring abundance in Sitka Sound had vastly increased that year, and ADF&G raised the GHL to 2,000 tons, more than twice as high as it had ever been. Moreover, the ex-vessel price of herring was improved.

Forty-eight seine vessels arrived in Sitka expecting the season to last a few hours, but at the last moment the Sitka Fish and Game Advisory Committee suggested the quota could be further increased provided ADF&G could manage the fishery for a 10 percent roe recovery. Fishermen, of course, liked the idea of an increased quota, but they realized that the 10 percent roe recovery would be unattainable in a competitive, every-man-for-himself fishery. Holders of limited-entry permits (see chapter 9) then met and unanimously signed a legal document that organized themselves into a cooperative. Only four or five boats would fish, but the ex-vessel revenues would be shared equally by all.

ADF&G increased the quota to 2,300 tons and on April 12 opened a broad area to fishing. The designated seiners fished selectively. When a school of herring was located, they made a small test set to sample the fish and determine if the roe content met the required standard. Over four days of fishing, they landed 2,559 tons of roe herring, the roe content of which averaged 9.34 percent, substantially lower than had been anticipated. The average price paid for the fish was almost $2,200 per ton, and the ex-vessel value of the catch was about $5.6 million—about $117,000 per boat.[18] "Not bad for sitting on your ass for three days in the Pioneer Bar," said one fisherman.[19]

James Parker, ADF&G's fishery management biologist at Sitka, thought roe recovery would have been 6–7 percent (and far less valuable) in a competitive fishery. Parker was pleased with the management and results of the fishery. "Under the circumstances, I think it was the best possible way to conduct this fishery: best for the resources, for the fishermen, for the industry and for the general well-being of this particular kind of fishery," he said.[20]

1980

For the 1980 season, ADF&G increased the GHL to 4,000 tons. The ex-vessel base price of herring, however, had fallen dramatically from the previous year, to a little less than $500 per ton. The roe-herring fishery in Sitka Sound in 1980 was a traditional competitive fishery. Fortunately, herring were spread out and fishing was relatively slow paced, occurring during eight-hour openings on April 4 and April 5. Landings totaled 4,385 tons, and the ex-vessel value of the fishery was $2.2 million.[21]

1981–1987

The 1981–1987 roe-herring fisheries in Sitka Sound were competitive fisheries. In terms of catch and ex-vessel value, they averaged 5,178 tons and $4.9 million, respectively. The fishery during those years peaked in 1985, when the catch was 7,475 tons, and its ex-vessel value was $7.9 million.[22]

1988

The 1988 season was expected to be a banner year. Herring were abundant, and ADF&G had established the GHL at 9,200 tons. Fishermen had been paid $1,000–$1,100 per ton in 1987, and they expected about the same in 1988, which would have made the fishery worth nearly $10 million, a record. It was not to be.

While there were large schools of fish, a large proportion of the fish were immature. Hopes that the mature fish would segregate themselves from the immature fish were never realized. Moreover, the intensive spawning observed in early April meant that the schools contained a lot of spawned-out fish.

Faced with this biological reality, seiners somewhat reluctantly organized themselves into a cooperative, as they had in 1979. This time, however, fishermen insisted it was a "joint venture with processors." Each of the fifty-two permit holders' share of the 9,200-ton quota was 177 tons, and each negotiated a price with the processor they had made an agreement with prior to the opening.

Fishing began on April 4. ADF&G initially limited the number of seiners to twenty, but fishing was slow, and the department allowed more seiners to participate. The number peaked at forty-four, then began dropping as more and more of the seiners delivered their share of the quota to fourteen processors on the grounds.

The season lasted eleven days, and the catch was 9,390 tons, which had an ex-vessel value of $4.2 million. And it was encouraging to note that there were 104 nautical miles of herring spawn in Sitka Sound in 1988, the most since recordkeeping began at statehood.[23]

1989

The 1989 fishery was a duplicate of the 1988 season except that ADF&G increased the GHL 2,500 tons, to 11,700 tons. The share for each of the fifty-one permit holders who participated in the cooperative fishery was 230 tons. During eight days of fishing, the seine fleet landed 11,831 tons of roe herring. The average roe content of the fish was low, 9.4 percent, and the market was tight. Processors paid fishermen about $100 per ton, and the total ex-vessel value of the fishery was only $1.2 million—the lowest since 1978.[24]

1990

Spawning herring were far less abundant than they had been in 1989, and ADF&G reduced the GHL to 4,146 tons. The first opening of the free-for-all competitive fishery was on April 5 and lasted almost three hours; the second (and last) opening was the following day and lasted forty-five minutes. During these openings, the fleet landed 3,804 tons of fish. The roe content of the fish was a healthy 10.6 percent. Moreover, the market had improved, and processors paid fishermen an average of $520 per ton. The fishery's total ex-vessel value was nearly $2 million.[25]

1991

Herring abundance continued to decrease in Sitka Sound, and ADF&G reduced the 1991 GHL to 3,200 tons.

Complicating matters, when the herring arrived in late March, mixed in with the mature, valuable herring were a lot of immature fish that had little market value. Their presence drove the average roe content of schools of fish below acceptable levels. In a free-for-all competitive fishery, a lot of sets would be dumped because the catch did not meet roe-content standards. ADF&G considered the potential mortality from the massive dumping of sets unacceptable. The situation, however, could improve on short notice if schools of mature fish arrived or the immature fish segregated themselves from the mature fish.[26]

Optimistically, on March 29, ADF&G put fishermen on two-hour notice that an opening might be announced if test fishing showed improved prospects. It was not to be. A week later, as the *Daily Sitka Sentinel* wrote, "Today is the seventh day an armada of seine boats, tenders, spotting planes and hundreds of men and women who run them have stood by on two-hour notice for the Sitka Sound sac roe-herring season to open."[27]

But the situation did not improve, and by April 10, the average size of the fish and their roe content had diminished. As something of a last-ditch effort, ADF&G suggested that fishermen organize into a cooperative, which would enable a more selective fishing effort. If they chose not to do so, ADF&G threatened to cancel the fishery.

The fishermen immediately agreed to form a cooperative, but by this time about fifteen of them had departed, most to Prince William Sound for the roe-herring fishery there. The cooperative comprised thirty-five boats. Seven of them, each associated with one of the seven processors present, would fish.

ADF&G opened the fishery for two days. Without the time constraint characteristic of a competitive fishery, fishermen could be more careful when releasing substandard catches. ADF&G closely monitored their fishing. During the opening, seiners landed about 1,400 tons of herring. In doing so, they released about two-thirds of their sets, but ADF&G believed no significant mortality had resulted. The department then extended the fishery for two days.

On the first day of the extension, seiners landed about 400 tons. On the second morning, seiners made a lot of sets but released all the fish. Determining that the window for profitable, responsible fishing had closed, ADF&G closed the fishery at a little after noon that day. The season's total harvest was 1,838 tons—about 60 percent of the GHL. The average roe content of the fish, 8.9 percent, was the lowest in the entire history of the fishery. Processors paid an average of about $115 per ton for the fish, and the fishery had a total ex-vessel value of about $210,000, the lowest since 1978.[28]

1992

The GHL for 1992 was 3,356 tons, slightly higher than in 1991. The quality of the fish was better, and fishermen chose to fish competitively.

There were a lot of fish near Goddard Hot Springs, south of Sitka, and on April 6, during the season's single opening, which lasted eighty-three minutes, the fleet landed 5,368 tons of herring—60 percent more than the GHL. ADF&G biologists who managed the fishery said that despite the excessive catch, there had still been more than sufficient escapement to propagate the herring run in the area. The harvest was worth $1.37 million.[29]

1993

Herring were abundant in Sitka Sound in 1993, and ADF&G set the quota at 9,619 tons—the second highest ever and almost treble what it had been in 1992. At the same time, a worldwide glut of roe herring had depressed its price. For the processors that chose to participate in the fishery, the challenge was to maximize the quality of their production. Moreover, given the reduced processing capacity, the seiners' catch would have to be carefully paced to match the capacity available.[30]

The solution was an unorthodox agreement among ADF&G, permit holders, and processors that *Daily Sitka Sentinel* reporter Eben Punderson termed a "hybrid competitive-cooperative system." No contracts were involved, but each processor was allocated a share of the quota based on the number of seiners it had under contract. Each company agreed to limit its purchases to the quantity it could process efficiently and without waste. Once a processor reached its share of the quota, it would stop buying fish.[31]

Over the course of eight days, during which at times there were only a dozen seiners on the grounds because processors were catching up, the fleet landed 10,186 tons of roe herring. Fortunately, the herring in Sitka Sound were generally of good quality, and only a limited number of sets were released, minimizing stress on the fish.[32]

Assessing the season, Bruce Joyce, owner of the Seattle-based seiner *Scandia*, said, "I have nothing bad to say whatsoever about the fishery. The quality of the fish was superb, especially considering the amount of fish we took.... I think it worked out for everybody. The processors kept it spaced out and kept things from jamming up."

Harold Thompson, of Sitka Sound Seafoods, added that ADF&G "did a great job by spreading the fishery out. They did everything they could to make it work."[33] The ex-vessel value of the fishery was $3.5 million, more than twice what it had been in 1992.[34]

In the fall of 1993, the Alaska Pulp Company shuttered its pulp mill at Sitka, ending its decades-long heavy discharge of pollutants into Sitka Sound. Longtime seiners observed that after the closure of the mill, herring in the sound appeared to be larger and healthier.[35]

1994

Though the 1994 quota of 4,432 tons was less than half that of the previous year, there were a lot of six-year-old fish that had shown good growth over the past year, and the prospects for a high-quality product were good.

Fifty-one seiners and ten buyers were present, and ADF&G opened the season, which was fished competitively, on March 29. During 4.5 hours of fishing that day, seiners landed 3,332 tons of herring. Another opening would be needed to harvest the 1,100 tons of fish that remained on the quota, and seiners were eager to fish. But with the amount of fishing power on the grounds, ADF&G was concerned that even a brief opening might result in the quota being exceeded. The department's solution was to locate an area where the quality of the fish was high but there were relatively small numbers of them. Nakwasina Sound fit the bill, and ADF&G opened the area on March 31. During a little over two hours of fishing, seiners landed about 1,400 tons of herring, exceeding their quota by about 300 tons.[36] The ex-vessel value of the fishery was $3.6 million, about what it had been in 1993.[37]

1995

Along with the dominant seven-year-old class, the spawning population in 1995 was expected to include a lot of three-year-old fish. The young fish had little commercial value, but, more importantly, the future of the fishery depended to a large extent upon them. To ensure the young fish were not overly exploited, ADF&G reduced the roe-herring quota at Sitka to 2,609 tons, the lowest since the 1970s.

ADF&G opened the fishery for three hours on March 25, and seiners, fishing competitively, landed 2,050 tons of fish, leaving about 550 tons on the quota. The department scheduled a ten-minute cleanup opening on the afternoon of March 27 in Nakwasina Sound. Only eight or nine boats landed fish, totaling about 300 tons. Later that afternoon, ADF&G called a second ten-minute opening. The catch by about a half-dozen boats that connected with fish was enough to finish the season.

The total for the season was 2,908 tons, about 300 tons more than the fishery's quota.[38] And the ex-vessel value of the fishery, $3.9 million, was not greatly different than it had been the two previous years.[39]

1996

ADF&G's quota for 1996 was 8,144 tons, nearly treble that of the previous year. Volume-wise, fishing started out well, with seiners taking 4,300 tons in 3.5 hours on March 23 in a competitive fishery. But many of the fish they caught were immature and of low value. ADF&G surveys after the opening showed a prevalence of immature fish and, in some locations, a lot of spawned-out fish. The department was reluctant to reopen the fishery until the quality of herring improved.

Eight days later, on March 31, ADF&G reopened fishing in a larger-than-usual area—except along beaches where spawning had occurred—under a controlled-fishing agreement. Seiners agreed to limit the number of boats on the water to seventeen, but all the proceeds of their catch would be divided equally among the fifty-one permit holders. The fishery would remain open from 6:00 a.m. to 6:00 p.m. every day until either the remaining quota was caught or ADF&G

closed the fishery because of a lack of suitable-quality fish or if excessive sorting and handling of recruit-sized herring was observed.

Fishing was slow, and on April 6, ADF&G ended the time-of-day restriction, and on the following day it increased the number of boats that were allowed on the water at one time to twenty-eight. On April 8, an abundance of high-quality herring was located at Crescent Bay. Seiners were successful at catching the fish, and the season was closed at a little after noon on April 9. ADF&G recorded the season's total harvest as 8,144 tons—exactly the quota.

A combination of ADF&G's effective, if unconventional, management of the fishery and a strong market for kazunoko rendered the 1996 Sitka Sound roe-herring fishery a huge success: "This will be by far the most valuable Sitka Sound herring sac roe harvest in history," said Bill Davidson, ADF&G's management biologist at Sitka. The final ex-vessel prices were $1,400 per ton in the cooperative fishery and $2,000 per ton in the competitive fishery, and the total ex-vessel value of the fishery was about $14.4 million. This was almost double the previous record of $7.9 million in 1985. And it would never be exceeded.[40]

1997

The 1997 quota was 10,900 tons, the second largest ever. Herring showed up early that year and caught some seiners off-guard. Eleven weren't yet in Sitka or didn't have their boats ready when ADF&G opened the season on March 18, the earliest ever. The fishery was competitive, and during five openings over the course of the next six days, seiners landed 11,147 tons of roe herring. The quality of the fish was high, but a lagging Japanese economy and a surplus of kazunoko in Japan resulted in a lower ex-vessel price than in 1996. The ex-vessel value of the fishery was about $4.7 million, about a third of what it had been the previous year.[41]

1998

ADF&G decreased the quota to 6,900 tons for the 1998 season. The market for kazunoko remained weak, and a large catch of small herring in British Columbia had saturated that market. Before the season, Bill Davidson, ADF&G's management biologist at Sitka, emphasized the need for seiners and processors to pay careful attention to the size and quality of the fish they produced. ADF&G would do its part: in scheduling the timing and locations of openings, the department would focus on providing opportunities for the fleet to harvest large fish. At the same time, to help ensure that a high-quality product was produced, it would pace the fishery to match processors' capacity to freeze herring.[42]

Spawning herring arrived in Sitka Sound even earlier than in 1997. As had happened that year, not all the seiners were on the grounds when ADF&G opened the fishery on March 16, a new record. The season consisted of three competitive openings spread over four days, the one nonfishing day provided so processors could catch up. The fifty-one seiners landed 6,705 tons of herring, just a little short of their quota. Bill Davidson judged the season a success regarding the size and quality of herring taken.[43] Regarding its value to fishermen, however, it was a disappointment. The ex-vessel value was about $1.6 million, the lowest it had been since 1992.[44]

1999

Spectator Sport

News of the sac roe herring opening brought a huge catch of Sitkans out to witness the annual gold rush–like dash for riches.

Dozens of onlookers lined the beach side of the Sea Mart parking lot, where seiners were making sets only a few yards offshore. Some spectators sat warmly in their cars with engines running, while others shivered in the cold, windy and rainy weather, which had replaced early-morning blue skies.

Massive flocks of seagulls had gathered for a feed, and high above them circled a squadron of buzzing spotter aircraft.

Surges of black smoke from the stacks of the seine boats at full throttle contrasted with the white spawn form lining the beaches.—Troy Etulain, *Daily Sitka Sentinel*, March 23, 1999[45]

ADF&G's 1999 quota was 8,476 tons, and the department's strategy for managing the season was to have three openings and catch approximately one third of the quota during each. Fishermen would take a day off after each of the first two openings to allow processors to catch up.

The first opening, on March 22, lasted one hour, during which seiners landed 3,492 tons of herring. It was a bit more than fishery

Figure 10.1. Sitkans watch from shore as seiners jockey for position during the 2009 Sitka Sound roe-herring fishery. (James Poulson)

managers had hoped, but the size and quality of the fish was satisfactory. ADF&G management biologist Bill Davidson knew of only one set that had been released because of small fish or poor roe quality.

The second opening, on March 24, lasted but twenty minutes, during which the fleet landed an impressive 4,192 tons, with at least a couple of sets that each yielded on the order of 350 tons.

Only 792 tons remained on the quota, and at the rate the fleet had been catching fish, a third competitive fishery would almost certainly lead to a substantial overharvest.

Davidson asked for consensus among the permit holders to have the third opening fished cooperatively. If the fishermen were unwilling to unanimously agree, "we'll close the fishery," he said. The fishermen agreed, and the fishery was opened on March 26. Each of the eight processing companies present had one or two boats on the water.[46]

Fishing was slow, and at the end of the day about seventy-five tons still remained on the quota. ADF&G reopened the fishery the following day, and only one seiner, the *Winning Hand*, fished. It was successful at filling (and exceeding) the quota. The season's total catch was 9,217 tons, with an ex-vessel value of $4.9 million.

"This fishery ended with a fizzle," commented Davidson. That said, the 1999 fishery was overall a success because the herring it produced were of very high quality, in part due to the pacing of fishing.[47]

2000

The GHL for 2000 was 5,120 tons, a significant reduction from the previous year. ADF&G opened the season for forty-five minutes on March 19, and in a competitive fishery, the fifty-one-boat seine fleet landed about 3,320 tons of herring. The roe content was disappointingly low. The fishermen stood down on March 20 to allow processors to catch up; then they waited another day for suitable herring to arrive. The fish cooperated, and ADF&G opened the fishery for fifteen minutes on March 22. The seiners landed about 1,600 tons, and the season was over. Landings totaled 4,572 tons and had a total ex-vessel value of $2.7 million.[48]

2001

The GHL for 2001 was 10,597 tons, double what it had been in the previous year. Spawning herring arrived in mid-March, and fishermen were on two-hour notice for a week before the season was opened for fifteen minutes on the morning of March 22. The catch was about 2,000 tons, and ADF&G scheduled a second fifteen-minute opening that afternoon, during which seiners landed about 4,000 tons. More than half the season's quota had been landed in the day's thirty minutes of fishing.

Importantly, the fish were of some of the highest quality ever landed at Sitka. Generally, a 10 percent roe content is considered satisfactory, but the roe content of some of the fish landed that day was 13 percent. To allow processors time to process the unexpectedly large catch, ADF&G temporarily closed the fishery.

Five days later, on March 27, the fishery was reopened for fifteen minutes. It wasn't as productive as it had been during the earlier openings, but fishermen nevertheless landed 1,568 tons of herring. However, fishing was much improved during another fifteen-minute opening that afternoon, and fishermen landed 4,887 tons of herring. This unexpectedly large catch brought the season's total harvest to 12,034 tons, about 1,500 tons more than the GHL. Bill Davidson, ADF&G's management biologist at Sitka, said the overrun was not of great concern, that the stock of herring was "healthy and robust."[49] The fishery, which totaled sixty minutes of fishing, had an ex-vessel value of $5.8 million—almost $100,000 per minute of fishing time.[50]

2002

The roe-herring fishery at Sitka had always been controversial. Among the staunchest critics were the Native people who had a long tradition of harvesting herring eggs in the sound. The commercial fishery constrained their opportunities to do so.

At its January 2002 meeting, the Alaska Board of Fisheries adopted a new policy that required ADF&G to distribute the commercial harvest both geographically and temporally to provide a "reasonable opportunity" for subsistence users to take 108,000–158,000 pounds of herring eggs in traditional locations in Sitka Sound. The regulation required ADF&G to work closely with Sitka Tribe of Alaska and other users, and in November the department and the Sitka Tribe of Alaska signed a memorandum of agreement that formalized a consultation process.[51] In 2009, based on surveys of subsistence use of herring eggs between 2002 and 2008, the Board of Fisheries increased the subsistence allocation of herring eggs in Sitka Sound to 136,000–227,000 pounds.[52]

ADF&G's GHL for 2002 was 11,042 tons, a small increase over the previous year and the fourth largest in the history of the fishery.

Spawning herring showed up late at Sitka, and the first opening was on March 27. During two hours and eighteen minutes of competitive fishing that day, seiners landed about 3,900 tons of fish. To give processors time to catch up, fishermen stood down on March 28. There was a thirty-minute opening on March 29, with fishermen landing 1,995 tons. After waiting a day for processors to catch up, on March 31 another thirty-minute opening yielded 1,035 tons of fish. Overall, the quality of fish landed during the three openings was good. On April 2, a fourth opening, which lasted 2.5 hours, yielded 1,400 tons of herring that averaged smaller in size than the fish caught during the first three openings. About 2,750 tons of fish still remained on the quota, but herring were becoming increasingly scarce. Moreover, those that were present tended to be small and have a low roe content.

On April 10, the eighth day straight of no herring fishing, ADF&G met with seiners, processors, and the Sitka Tribe of Alaska to determine how—assuming herring of sufficient quality actually showed up—the remaining quota might be caught and the fishery brought to an end. A cooperative fishery was one option. Permit holders voted on but failed to provide the required unanimous support for that option.

The groups met again on April 11, and this time the permit holders voted unanimously to enter into a cooperative fishery to finish the season. ADF&G opened the fishery at 12:30 p.m. on April 12 and scheduled it to remain open from 6 a.m. until 6 p.m. each day until 1,372 tons—half of the remaining quota—was caught. Once that quantity was caught, ADF&G would evaluate the situation and decide whether another opening was possible or warranted. The herring—or lack thereof—pretty much made the department's decision for it. Fishing was slow, and on April 14, ADF&G, noting a lack of developing spawn, closed the season. The total harvest—which from the time seiners were put on two-hour notice to the end of the final opening encompassed a record twenty-one days—was 9,885 tons, more than a thousand tons less than the GHL. Its ex-vessel value was $4.4 million.[53]

2003

The quota for 2003 was just shy of 7,000 tons, which was a little more than 60 percent of the previous year's quota. The first opening was on March 22, and seiners landed 2,452 tons during an hour and forty minutes of competitive fishing. Normally, fishermen would stand down the next day to allow processors to get the catch processed and ready themselves for more fish, but the herring caught during the first opening were especially large and the roe content was high. Everyone involved agreed that it would be advantageous to fish while these fish were available, and a second opening—this one for an hour—was held on March 23. Fishermen landed 2,668 tons of fish that day, and processors spent the next day catching up.

With about 1,900 tons remaining on the quota, the challenge for ADF&G in reopening the fishery was to locate an area with high-quality fish that was large enough for the fifty-boat seine fleet to operate in, would not interfere with the subsistence gathering of herring roe, and would also offer a good likelihood of keeping the catch within the quota. The area between Halibut Point and Middle Island fit the bill, and during a ten-minute opening that day what ADF&G management biologist Bill Davidson later referred to as "50 hotshot boats" landed about 2,200 tons of herring, slightly exceeding the season's quota.[54] The total season's catch, 7,051 tons, had an ex-vessel value of $3.2 million.[55]

2004

The 2004 return of spawning herring to Sitka Sound, an estimated 54,468 tons, was the highest in the history of the fishery.[56] ADF&G established the GHL at 10,618 tons, about 3,500 tons more than in 2003.

All of the 2004 season's three openings were fished competitively. The first was on March 22, and in an hour and forty minutes of fishing, the fleet landed 5,200 tons of fish. Seiners then stood down for two days while processors caught up and ADF&G, at the request of processors, waited for the quality of herring to improve. The March 25 opening, which lasted fifteen minutes, yielded some 3,900 tons of fish, and the final opening, a fifteen-minute affair on March 27, yielded 1,700 tons. During that opening, only sixteen of the fifty-one permit holders made landings.

Testament to the state's close regulation of the fishery and the demanding pace of short openings was the citing of a fisherman for being twenty-eight seconds late closing his seine after ADF&G announced the closure of a fifteen-minute opening. His was a big set, containing 330 tons of herring, and it may have been difficult to get the seine closed on time. The fisherman was subsequently found guilty of illegally retaining the herring and forfeited the value of the catch: $170,000.

The total harvest for the season was 10,492 tons, about 100 tons short of the GHL. It had an ex-vessel value of $5.2 million.[57]

2005

The 2005 GHL was slightly increased over that of the previous year, to 11,292 tons.

The season began on March 23, with fishermen landing 3,700 tons of herring during 130 minutes of fishing. As had been the routine, seiners then stood down for a day while processors got the catch in the freezer.

On March 25, ADF&G scheduled an opening in a limited area, but by the time the opening was called the herring had moved out. Almost no fish were caught, and the department closed the fishery after fifteen minutes. A seventy-five-minute opening later that day in an expanded area, however, yielded 1,700 tons.

The next opening, on March 27, lasted two hours, during which fishermen landed 4,800 tons of fish. To land the remaining 1,000 tons, seiners fished cooperatively on March 28 and March 29.

The final tally for the season was 11,366 tons, remarkably close to the GHL and testament to how cooperative fishing facilitated effective management of the fishery. The harvest's ex-vessel value was $6.1 million.[58]

2006

The GHL for 2006 was 10,412 tons. The season opened March 24 with 125 minutes of competitive fishing that yielded 3,955 tons of herring. The seiners stood down the next day while processors got the catch in the freezer. A ninety-minute competitive opening on March 26 yielded 2,015 tons, and a thirty-minute competitive opening on March 27 yielded 3,500 tons.

With less than 1,000 tons remaining on the quota, ADF&G management biologist Dave Gordon was concerned that a competitive fishery would result in the quota being exceeded, and on March 29 he suggested to fishermen that they form a cooperative to harvest the remainder of the fish. The fishermen did so, and Gordon immediately opened the fishery. The final tally for the season was 9,967 tons, almost 500 tons short of the GHL. Of fish size and quality, Gordon said it was "good [but] not outstanding." The harvest's ex-vessel value was $2.6 million, less than half that of the 2006 season.[59]

2007

> The threshold conditions for an opening include herring with mature roe percentages of more than 10 percent in an area large enough to allow an orderly harvest while also giving the managers the ability to keep the catch within the guideline harvest level.—*Daily Sitka Sentinel*, March 2007[60]

The 2007 GHL was 11,904 tons, about 1,500 tons more than it had been in 2006. The first opening was on March 26, and during nearly three hours of fishing, seiners caught and estimated 3,300 tons, almost a third of their quota. Because ADF&G had difficulty in locating a suitable area, the next opening wasn't until March 30. During 160 minutes of fishing that day, seiners caught 3,600 tons of herring. Almost six hours of fishing two days later yielded an additional 2,000 tons. The final opening, which occurred on April 3, lasted almost two hours and netted 2,730 tons of herring. The total catch during the season's four competitive openings was 11,571 tons, about 300 tons shy of the GHL.[61]

For the season, the average roe content of the catch was more than 11 percent, which was very desirable. Moreover, the females were ripe for an extended period, which facilitated ADF&G's pacing of the fishery. Dave Gordon, the department's area management biologist, said the 2007 season was "as good as it gets." The ex-vessel value of the catch was $5.7 million, more than double what it had been the previous year.[62]

2008

The GHL for 2008 was 14,723 tons, the highest ever. Moreover, preseason sampling showed that older fish dominated the spawning stock, with 57 percent of them eight years old or older. And weights at age were higher than average, with the average fish weighing a hefty 206 grams. Roe content averaged 12.5 percent, considerably higher than the usual 10 percent threshold. The outlook was rosy.

The season opened on March 25, but bad weather, including snow squalls, hampered fishing, and a four-hour opening yielded a lackluster 950 tons of fish. There was, however, a bit of excitement: a fisherman making a test set caught a humpback whale. Fortunately, fishermen managed to quickly free it, and the whale appeared to be unscathed.[63] The presence of humpback whales, as well as of sea lions, is often an indicator of the presence of herring.

With the small catch on March 25, there was no need to take a day off to allow processors to catch up, and ADF&G opened the fishery on March 26. The weather was much improved, and herring began moving into the shallows, where they were easy to spot and catch. In one hour of fishing, seiners caught 9,700 tons of herring, a record one-day catch. Five seiners had sets that surpassed 600 tons, and one, the Homer-based *Infinite Glory*, had a set estimated to contain 1,500 tons of herring. Fortunately, the boat was sitting in only about thirty feet of water, so there was no threat that the herring might dive and capsize the boat. Nevertheless, to help stabilize the *Infinite Glory*, another seiner tied to its side opposite the pursed-up seine. It took four days for twenty tenders working in relays to pump the fish out of the net and deliver it to processors. The price of herring that year was on the order of $550 per ton, so the set was worth more than $800,000.

The local processing capacity was about 1,200 tons per day, so some of the fish were taken to Petersburg and Ketchikan to be frozen. There was no fishing for four days while the processors caught up. In the end, the catch was 14,386 tons, which had an ex-vessel value of $8.9 million.[64]

The wild and lucrative 2008 fishery gained national attention: National Geographic produced a one-hour special titled *Cowboys of the Sea: Combat Fishing* that aired in March 2009. *New York Times* writer Katheryn Shattuck described the program: "In the Sitka Sound of Alaska, 50 boats battle one another for the same huge schools of herring, often bumping to gain position, in order to make a six-figure payday in a matter of minutes."[65]

2009

The GHL for 2009 was 14,504 tons, not appreciably different from the previous year's. In consideration of local production capacities, ADF&G's Dave Gordon thought the harvest should be made in four openings, with at least one day between openings. Although it took seiners five openings—all of which were done competitively—that totaled seven hours of fishing time to complete the harvest, the season worked out pretty much as Gordon thought it should.

The first opening was on March 22, and the final one was on April 2. The 14,755 tons of fish caught were considered to be of good size and quality. Processors paid $700 per ton for fish of 11.5 percent roe content, and the ex-vessel value of the fishery was $12.7 million, the highest it had been since 1996.[66]

2010

ADF&G's GHL for 2010 was 18,263 tons, an almost 4,000-ton increase over the previous year's. The harvest, which fell just shy of the guideline, was completed in four competitive openings. The first opening was on March 24 and lasted eighty-five minutes. Fish were abundant and their roe content was high, and seiners landed 6,600 tons. The adjusted ex-vessel price ended up being $690 per ton, so the value to fishermen of the fish caught during the opening was about $4.5 million.

The opening was particularly eventful for the Chignik-based seiner *Shady Lady*. The boat sustained damage early in the opening when it collided with another seiner. Not long after, the herring in a set that was being pursed by the *Shady Lady* dove, tipping the boat so far that its keel was out of the water. Fortunately, tenders and other seiners came to its aid, and the *Shady Lady* was righted without any injuries or damage.

The fourth and final opening was on April 2 and yielded 4,065 tons. About 500 tons still remained on the quota, but ADF&G elected to not reopen the fishery. Dave Gordon, the department's management biologist at Sitka, said being within 500 tons of the GHL was "a reasonable expectation," noting that there was "much fishing power and some of [the nets] can hold 300 to 400 tons, and 49 boats out there." Another opening, especially if were conducted competitively, could have easily resulted in the GHL being substantially exceeded.[67]

The catch totaled 17,874 tons, and the ex-vessel value, bolstered substantially by an off-the-charts average roe recovery of 12.5 percent, was $12.2 million.

Despite the large catch, ADF&G mapped eighty-eight miles of herring spawn along the shoreline of Sitka Sound, the second highest since the department began keeping records.[68]

2011

ADF&G set the GHL for the 2011 season at 19,490 tons, the highest it had ever been. The department envisioned a harvest strategy of taking the season's quota in five or six openings of 3,000–4,000 tons each. The number of days between the openings would be determined in-season, based on the available processing capacity and the immediacy or progression of spawning.[69]

A banner year looked to be in the making, which was reflected in the sale of a Sitka Sound roe-herring limited-entry seine permit for more than $500,000.[70] Unfortunately, it was not to be.

On March 11, a couple of weeks before the season would begin, a magnitude 6.8 earthquake struck off Japan's Pacific coast. The earthquake generated a tsunami that flooded the Fukushima nuclear power plant's lower grounds, knocking out the emergency generators that supplied coolant water to the reactor. What followed were three nuclear meltdowns, three hydrogen explosions, and the release of radioactive materials into the sea. Japan was a mess.

Despite an uncertain market, fifty seiners showed up at Sitka to fish for roe herring, and ten processors were there to buy and process their fish. Over the course of five openings that totaled 9.5 hours of fishing, forty-eight seiners made landings. Their catch totaled 19,429 tons and represented the peak harvest in Sitka Sound's roe-herring fishery. Moreover, the roe content averaged 13.3 percent, the highest ever.

And there was considerable excitement. During a seventy-minute opening, the seiner *Infinite Grace* nearly capsized when the herring in a set estimated to contain about 400 tons suddenly sounded (figure 10.2). The crew managed to release the seine, and the vessel righted itself, but the entire catch was lost. Though shaken by the ordeal, none of the crew were injured.

The combination of the large harvest and the high roe content should have resulted in a hefty payday for fishermen. But because disaster-wracked Japan was the only market for the roe, the ex-vessel

Figure 10.2. A large haul of herring nearly capsized the seiner *Infinite Grace* at Sitka in 2011. (James Poulson, *Daily Sitka Sentinel*)

price of herring had fallen to $200 per ton. The ex-vessel value of the fishery was just shy of $4 million, not even a third of what it had been in 2010.[71]

2012

The GHL for 2012 was 28,829 tons, almost half again more than in 2011. Unfortunately, the volume of herring that entered Sitka Sound in 2012 was only a little over half of what ADF&G biologists had expected, perhaps due to a low at-sea survival rate. The season opened on March 31, and during three openings seiners landed 13,232 tons of fish, 47 percent of the GHL. Fortunately, the demand for kazunoko in Japan had rebounded. The ex-vessel value of the catch was $8.9 million, more than twice the value of the previous year's catch.[72]

2013

The 2013 GHL, 11,055 tons, was 38 percent of the 2012 level. The first opening occurred on March 27 and resulted in the harvest of 2,100 tons of herring. A three-hour opening the following day yielded 3,600 tons. With almost half the quota yet to be caught, herring suddenly became scarce, and during an opening on March 30 seiners landed only 175 tons of the fish. After several days of waiting for herring that never showed and perhaps didn't exist, ADF&G determined the season could be finished only through a cooperative effort. Fishermen agreed to do so, and ADF&G reopened the season on April 3 for a full day. Pursuant to the cooperative agreement, one seiner out of every six was permitted to fish, and the revenues would be shared by all. They found but few fish, and the season was closed later that day.

The season's catch, 5,688 tons, was a little more than half the GHL. It had an ex-vessel value of $4.4 million.[73]

2014

A lot of the herring that had been too young to harvest in 2013 were expected to be of harvestable size in 2014, and ADF&G increased the GHL accordingly, to 16,333 tons. One could say the season went mostly according to plan.

It opened on March 20, and the catches during each of the four openings averaged a healthy 4,300 tons, ranging from 3,654 tons to 4,998 tons. Fishermen stood down for two days after each of the first three openings to allow processors to catch up. During the final opening, which entailed forty-five minutes of fishing on March 29, seiners landed nearly 4,000 tons of herring, putting them about 600 tons in excess of the GHL. The season's total catch was 16,957 tons, but the demand for kazunoko was again in steep decline, and the ex-vessel value of the catch was $3.1 million.[74]

2015

The GHL for 2015 was 8,712 tons, about half of what it had been in 2014. Moreover, ex-vessel prices were anticipated to be low. To increase the efficiency of the fishery, the forty-eight permit holders who chose to participate decided prior to the season to operate cooperatively. A limited number of seiners would fish at their own pace and coordinate with processors to ensure their catch did not exceed processing capacities.

The first opening was on March 18 and lasted four hours. Each day after that until the fishery closed, fishing was open for nine hours, from 9:00 a.m. until 6:00 p.m. Fishing was reported to be slow during the final days of the fishery, with the *Daily Sitka Sentinel* reporting that seiners caught the last 150 tons of their quota on March 25. The final tally for the season, however, showed the harvest to be 9,833 tons, about 1,000 tons more than the GHL. The total ex-vessel value of the catch was only $2.2 million.[75]

2016

The GHL for 2016 was 14,941 tons, almost double what it had been in 2015, but the run of spawning herring in Sitka Sound in 2016 included a lot of young, less-marketable fish.

The fishery opened on March 17, and during 140 minutes, the fleet of forty-eight seiners landed 3,600 tons of fish that were of less-than-desirable quality. An opening two days later, however, yielded 4,500 tons of nice fish. After that, it was all downhill.

It was March 23 before ADF&G reopened the fishery, but the result of two one-hour openings on that day was 1,900 tons of mediocre-quality fish. ADF&G continued to look for marketable herring, but, finding only small fish, it closed the fishery on March 28. More than 5,000 tons of quota remained uncaught. The total catch was 9,769 tons, and it had a total ex-vessel value of $2.2 million.[76]

"I think the industry started to look at weighing the value of the fish now and the value of those fish later," ADF&G management biologist Dave Gordon later said. "They'll grow, they'll come back next year and be more marketable."[77]

2017

The 2017 GHL was 14,649 tons, about the same as it had been in 2016. The difference was that the small herring that plagued the 2016 fishery had grown and were now of marketable size. The fishery itself was largely a textbook example of what should happen if the herring follow approximately their normal routine and fishery managers and fishermen properly do their job.

The season began on March 19 with three hours and twenty minutes of fishing that resulted in a catch of 3,300 tons. On March 22, high-quality herring were well distributed on the grounds, and a fifteen-minute opening yielded a catch of 4,800 tons. Fish were not as easy to find on

March 25, but they were definitely present. A three-hour-and-twenty-minute opening that day yielded a catch of 5,400 tons of herring.

With about 1,000 tons remaining on the quota, seiners decided to finish the fishery cooperatively. ADF&G opened a wide area of the sound on March 28, and about a dozen seiners caught the remainder of the quota.

Overall, everyone was pleased with the season: the size of the fish was larger than expected, the roe content was higher than expected, and there was adequate time between openings to properly process the catch. Moreover, the fishery had an ex-vessel value of $4.3 million. It was a pittance compared with the glory years of the fishery, but it was also the biggest payday since 2013.[78]

2018

Faced with a declining roe-herring market, processors at Sitka set a high standard for herring they were willing to purchase. The fish needed to weigh a minimum of 125 grams and have a mature-roe content of at least 11 percent.

The herring caught in a competitive fishery would be unlikely to meet these standards, and at a preseason meeting on March 19, the forty-eight Sitka Sound roe-herring fishery permit holders agreed unanimously to conduct the fishery as a cooperative. The quota, 11,128 tons, would be divided equally among all the boats, which translated into roughly 230 tons per boat. Cooperative fishing would allow the boats to be more selective regarding the fish they set on, and their catch would be more likely to meet processors' standards.

ADF&G opened the fishery on March 25 for five hours, and seiners landed about 2,800 tons of herring. A 4.5-hour opening the following day, however, yielded only 400–500 tons. Fortunately, the herring caught during both openings were larger than average and had high roe content.

In Sitka Sound, as is the situation in Alaska's other herring fisheries, the larger, older herring typically spawn first, followed by the smaller, younger fish. After the March 26 opening, ADF&G was unable to find a body of suitable fish, and department surveys showed that by March 29 the larger-weight class had already spawned and the fishery might be over. The fleet stood by in hopes that larger fish might arrive, but the fish never materialized, and ADF&G closed the fishery on April 3.

The season's total harvest was 2,926 tons, 8,330 tons short of the GHL. The estimated ex-vessel value was about $1 million.[79]

2019

For the 2019 season, ADF&G established the GHL in Sitka Sound at 12,869 tons. The department expected the fish to be on the young side, with an estimated 76 percent of the fish being three to five years old. At the same time, processors were continuing to face a constrained (and falling) market for kazunoko and made it known that they would adhere to the same buying standards as in 2018: herring needed to weigh a minimum of 125 grams and have a mature-roe content of at least 11 percent. The young herring expected in Sitka Sound would not meet this threshold.

Nevertheless, more than thirty seiners—fewer than in previous years—arrived in Sitka in early March. It was for naught because test fishing later that month showed the herring were indeed too small for the market.[80] "We test the fish and if it doesn't meet market requirements, [the fishery] doesn't open," said ADF&G fishery management biologist Eric Coonradt.[81]

By the end of March most of the seiners had left, and the fishery never opened. The last time there was no roe-herring fishery in Sitka Sound was in 1977.

2020

ADF&G established the target level harvest level for 2020 at 25,824 tons, almost double what it had been in 2019. Unfortunately, the data suggested the season would again be dominated by three-year-old fish.

In late February, ADF&G contacted all processors with a known interest in roe herring to assess their interest in the upcoming season. Each responded that given the demands of the market and the small size of the herring expected, they would not participate in the fishery. Uncertainty caused by the COVID-19 pandemic was also a consideration in the companies' decisions: a significant quantity of Alaska roe herring was reprocessed in China before going to Japan. ADF&G then canceled the season.[82]

Herring arrived in their expected abundance and, unhindered by fishermen, spawned along 58.5 nautical miles of Sitka Sound's shoreline. The spawn deposition along the Kruzof Island shoreline was exceptional: the spawning area was extensive and the egg density was high.[83]

Roe Herring

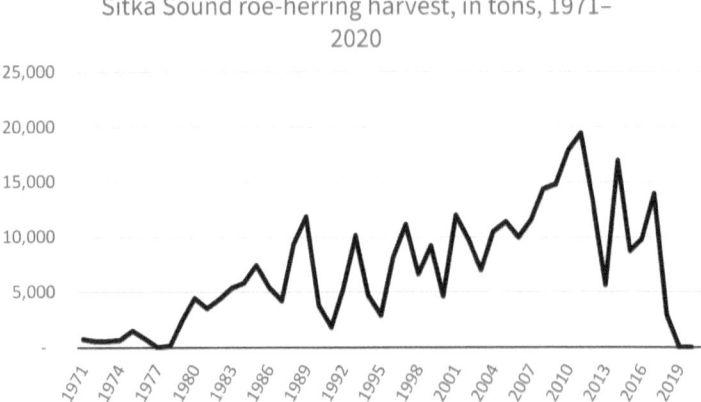

Figure 10.3. Data sourced from ADF&G management reports.[84]

LYNN CANAL

In addition to the fishery in Sitka Sound, there was a roe-herring purse-seine (and gillnet) fishery on the east shore of Lynn Canal, between Auke Bay and Berners Bay, from 1971 through 1982. The harvest averaged about 650 tons and never exceeded 975 tons. At least in the fishery's early years, the catch was frozen at Juneau Cold Storage and then shipped to Japan. ADF&G closed the fishery prior to the 1983 season because the spawning stock was depressed and did not meet the 4,000-ton minimum abundance threshold. The fishery has remained closed since that time.[85]

11
RESURRECTION BAY AND PRINCE WILLIAM SOUND ROE-HERRING FISHERIES

The roe-herring fisheries in Prince William Sound began in the spring of 1969, when one operator purchased 389 tons of herring from five seine boats, salted the fish on a barge, and transferred them to a waiting Japanese freighter.[1] As will be discussed below, the Prince William Sound roe-herring fishery would grow to become one of Alaska's largest and most important fisheries. The *Exxon Valdez* oil spill in 1989, however, marked the beginning of its end.

The year 1969 also marked the inception of the Resurrection Bay area roe-herring fishery on a substantial scale. One ton of roe herring had been harvested in Resurrection Bay in 1961, and seven tons were harvested in 1966. There had been no roe-herring operations in 1967, and Ray Anderson had a small operation at Seward in 1968.[2]

Management of the Resurrection Bay herring fisheries was almost nonexistent. It was open on an unlimited basis, and the harvest was determined by fishing effort, the availability of herring, and the market for roe.[3]

Whitney-Fidalgo Seafoods, which had salmon canneries in Port Graham (lower east shore of Cook Inlet) and Anchorage, wanted to get into the herring-roe business. The company furnished herring seines to four Homer-based purse-seine fishermen, who were experienced salmon seiners but had little or no experience seining for herring. The company guaranteed them $40 per ton for herring, from which $10 per ton would be deducted as a lease fee for the seines. Among the fishermen was Beaver Nelson. Nelson had never seined for herring, but he was smart and aggressive and over the years would help pioneer the Prince William Sound and Togiak herring fisheries.

Because salmon seiners in southcentral Alaska, particularly at Kodiak, fished mostly in shoal waters, the boats they used were smaller than the fifty-eight-foot deep-draft limit seiners (vessels of the maximum overall length allowed to seine for salmon in Alaska) commonly used in Southeast Alaska. A typical "Kodiak" seiner was about thirty-eight to forty-two feet long and drew only about three feet of water. The boats partially made up for their short length and shallow draft by being broad-beamed (wide). Nelson's boat in 1969, the *Robby*, was smaller than most. It was a thirty-three-foot plywood combination gillnetter/seiner powered by a 300-horsepower diesel engine.

Figure 11.1. Purse seiner *Nuka Point*, with bag of roe herring, Columbia Bay, Prince William Sound, April 1974. (James Mackovjak)

There were, however, very few herring in Kachemak Bay that spring. Nelson had grown up in Seward and knew that herring were often plentiful in Resurrection Bay. (Recall that in earlier years fishermen had fished in Resurrection Bay to supply herring to reduction plants in Prince William Sound.) Hoping for the best, Nelson made the twelve-hour mostly open-ocean voyage to Seward, where he found an abundance of herring. News traveled quickly, and the other seiners soon joined him.

Before long, the Homer-based seiners were joined by seiners based in Seward. Mainly, they fished in Seward's small-boat harbor, which Nelson likened to a gigantic fish trap. They delivered their catches to a Whitney-Fidalgo Seafoods tender, which took the fish to the company's cannery at Port Graham, where the roe was stripped. Unfortunately, the seines that Whitney-Fidalgo had provided were too shallow to fish effectively.[4]

In addition to the Homer- and Seward-based fishermen, three limit seiners came from Southeast Alaska. Their owners were experienced bait-herring fishermen and had large, deep seines.

The fishermen from Southeast Alaska had arranged with a Japanese company to deliver their catches to a small Japanese processing ship that anchored in the bay. Over the next several weeks, Japanese workers salted about 500 tons of herring, and the ship later delivered the product to Korea. There, the roe was stripped for the Japanese market and the carcasses saved as food for Korean troops fighting in Vietnam.

Additionally, one operator in the Cook Inlet–Resurrection Bay area, though not identified in ADF&G's 1969 reports, extracted roe from nearly fifty tons of herring. The carcasses were frozen for use as halibut and king crab bait, but fishermen who used the carcasses reported them to be unsatisfactory. The total roe-herring harvest in Resurrection Bay in 1969 was 758 tons.[5]

In 1970, Resurrection Bay and Kachemak Bay (see chapter 12) were the centers of roe-herring fishing in Alaska. Three companies operated at Seward: Whitney-Fidalgo Seafoods, Herring Northwest, and Ray Anderson's Seward Marine Services. Eleven seiners supplied the approximately 2,100 tons of herring that were landed there.

For fishermen such as Beaver Nelson, herring fishing in Resurrection Bay was much better in 1970 than it had been the year before. Importantly, all the fishermen had built their own, deeper seines. Whitney-Fidalgo Seafoods again purchased their fish, but this time the company wanted the fish delivered not to tenders but to the dock. The fish were trucked to Whitney-Fidalgo's plant in Anchorage, where the roe was stripped.

The harvesting process was a pretty straightforward affair. As they had the previous year, the fishermen mainly fished in Seward's small-boat harbor. Once they had a load of fish, fishermen towed their herring-laden seines to the dock, where the fish were brailed into totes aboard a semi-trailer for the three-hour drive to Anchorage.[6]

The second herring buyer at Seward in 1970 was Herring Northwest, established by Alaskans Bob Holmstrand and Peter Bersch. The company's operation was unique, to say the least. It involved two World War II–era power scows: the 105-foot *Diver 1*, owned by Holmstrand, and the 86-foot *Chichagof*, owned by Petersburg Fisheries. The *Diver 1*—which was in a state of disrepair—was beached at Lowell Point, on the west shore of Resurrection Bay, about two miles south of Seward. The *Chichagof* was tied to the dock at the Petersburg Fisheries (later Seward Fisheries) plant.

Early that year, Holmstrand and Bersch placed an ad in the *Seattle Times* offering work in Alaska. Having no office in the city, they interviewed applicants in the lobby of the Washington State ferry terminal. They hired about fifty women and two men. Herring Northwest would fly the workers to Anchorage and then bus them to Seward. If the workers stayed for the entire season—a minimum of six weeks—Herring Northwest would pay for their return flight to Seattle.

To accommodate the workers, Holmstrand and Bersch leased a house in Seward, and Holmstrand converted the basement of his

Figure 11.2. Herring Northwest roe-stripping operation, Seward, Alaska, 1970. (Ann Holmstrand)

house there into a makeshift dormitory. The Japanese technicians who oversaw the roe quality were housed in a trailer in Holmstrand's backyard. Workers were transported between their accommodations and the scows in a Volkswagen bus.

To feed their crew, Holmstrand and Bersch rented the kitchen and dining room at Tony's Bar, a legendary establishment in downtown Seward. Bersch's wife, Gretchen, did all the cooking, including baking bread. Her only help was her six-year-old sister, who sometimes helped wash the dishes.

Four seiners—two from Kodiak, one from Homer, and one from Seldovia—provided the herring. They towed their herring-filled seines to the scows, where a vacuum pump was used to unload them. After the herring were "aged" in brine (see chapter 9) for about four days, the roe was stripped from the partially rotted fish on the deck of the *Chichagof* (figure 11.2) and on the upland beach alongside the *Diver* 1.

Roe strippers, all of whom were women, were paid $0.35 for each pound of roe they produced, and a fast worker could produce about

Figure 11.3. Inside the Seward Marine Services' makeshift roe-stripping structure, probably 1971. (James Mackovjak)

fifty pounds in a day. To help motivate the workers, Bersch each day awarded a bonus to the woman who produced the most roe. The men, who mostly shoveled herring onto processing tables, were paid by the hour. Workdays began at 7:00 a.m. and ended at 5:00 p.m. Herring offal was pumped overboard onsite.

At one point, due to a weather-related delay, it was ten to twelve days before a mass of herring on the *Diver 1* could be stripped. By that time, maggots were feasting on the rotting fish flesh. "It was very disturbing to the workers, since most of them were women who thought they were taking a vacation, young girls who had never seen or smelled maggots, and hippies who thought life should be easier," wrote a Herring Northwest employee.

Unfortunately, Herring Northwest's colorful operation didn't pan out financially, and the company folded after the 1970 season.

The third roe-herring operation at Seward in 1970 was Ray Anderson's Seward Marine Services, which had a roe-stripping operation aboard the power scow *Chatham*. Anderson established a shore-based roe-stripping operation at Seward the following year (figure 11.5).[7]

ADF&G reported that during the 1970 season, 224,354 pounds of roe were produced in its central district, which included Resurrection Bay. The wholesale value of the roe was $373,000.

Resurrection Bay's 2,100-ton roe-herring harvest in 1970 was a record, but stocks were in decline. In 1971, the harvest was 831 tons, and in 1972 it was a mere 95 tons. ADF&G called the fishery a "dismal failure," which it attributed partly to the closure of Seward's small-boat harbor to herring seining. In fact, the most effective way herring were caught that year was during the ebb tide, when some of the herring in the small-boat harbor would move out and be caught by seiners waiting at the entrance.[8]

Stocks in Resurrection Bay continued to decline, and in 1975 ADF&G closed the fishery. It was reopened in 1985 on a very conservative basis, but there was little interest, and only 204 tons were harvested. The last roe-herring fishery in Resurrection Bay occurred in 1987. The harvest that year was 202 tons.[9]

PRINCE WILLIAM SOUND

In 1971, with herring resources in Resurrection Bay, Kodiak Island, and lower Cook Inlet in decline, the fishing effort increasingly focused on Prince William Sound.

As noted above, the first roe-herring harvest in the sound was in 1969, when five seiners landed approximately 389 tons. One vessel fished for roe herring in the sound in 1970 but made no deliveries. Another vessel that year landed ten tons of herring that were used as crab bait.[10]

The following paragraphs present a chronicle of the Prince William Sound roe-herring fishery since 1971.

1971

In 1971, fourteen purse seiners in Prince William Sound landed 939 tons of roe herring. Three processors were involved: Seward Fisheries had a roe-stripping operation at its plant at Seward, Seward Marine Services had a makeshift roe-stripping operation on Seward's waterfront, and Chatham Fisheries processed aboard its power scow *Chatham*.[11] ADF&G characterized the fishery as producing "an economic boost in the spring during a usual slack period."[12]

The roe-herring seine fleet in Prince William Sound that year comprised mostly boats from Cordova, Homer, and Kodiak. Among them was Beaver Nelson, who had helped pioneer the Resurrection Bay roe-herring fishery. Nelson was still fishing from his thirty-three-foot plywood *Robby*, but the following year would graduate to the forty-two-foot fiberglass *Nuka Point* (figure 11.1.) In addition to the more-local fishermen, some

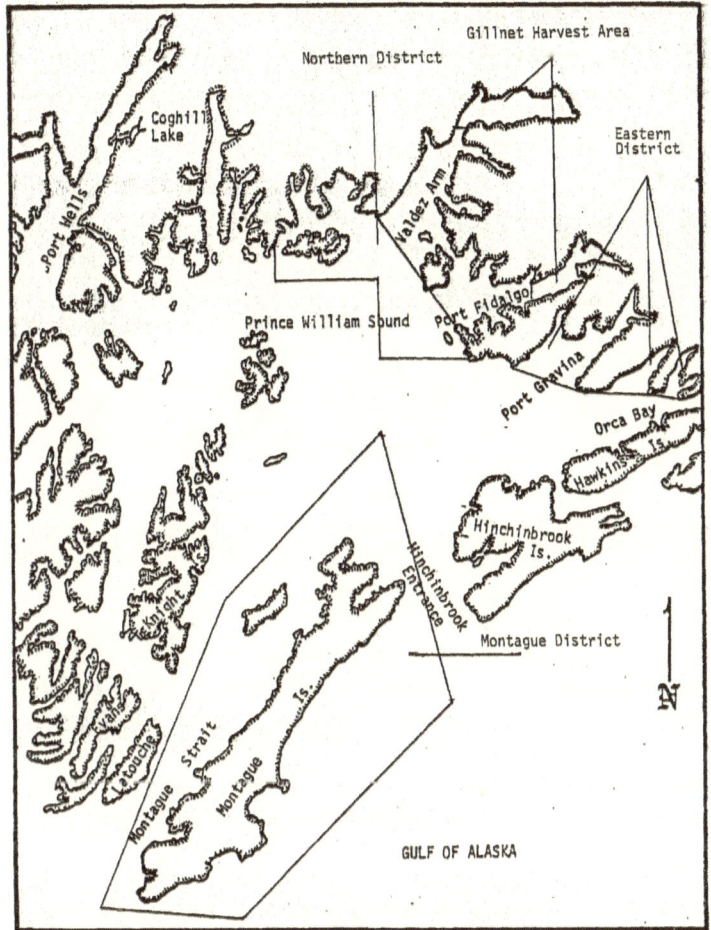

Figure 11.4. Prince William Sound roe-herring districts, 1980. (Source: Alaska Department of Fish and Game)

Southeast Alaska seiners crossed almost 400 miles of open water in the Gulf of Alaska to participate in the sound's roe-herring fishery.

1972

The Prince William Sound roe-herring catch was 1,750 tons, nearly double what it had been the previous year. The fleet size had increased by one, to fifteen. Seven companies of various sizes handled the fish.

Figure 11.5. Stripping roe herring, Seward Marine Services, Seward, Alaska, circa 1972. (James Mackovjak)

1973

In 1973, when ADF&G noted an "unlimited" market for herring roe and correspondingly high prices for roe herring, the fishing effort in Prince William Sound intensified, and the catch nearly quadrupled that of the previous year, to 6,983 tons. The harvest was promising, but it was dwarfed by the 73,450 tons that had been landed back in 1938, when eight herring-reduction plants operated in the sound.[13]

The author was in Prince William Sound during the 1973 roe-herring season, working on the Seward Fisheries tender *Frigidland*. And it was a wild affair. One of the primary areas fished was around Green Island, off the northwest shore of Montague Island. A challenge there, as Bob Holmstrand, the *Frigidland*'s skipper, pointed out, was that "there was a lot of real estate that wasn't on the charts." The 1964 Good Friday earthquake that struck Alaska—at 9.2 on the Richter scale, the second-largest earthquake ever recorded—had caused the Earth's surface in the area to be warped upward, and the marine charts the fleet used for navigation had not been updated to reflect the uplift.

Boats that year often charged around in search of herring and were not always mindful of the uncharted hazards. Several ran aground. On top of this, many of the seiners were salmon fishermen who were unaccustomed to dealing with heavy hauls of herring. In attempting to bring their catch to the surface so it could be brailed, several seiners capsized. At one point that spring, fully seven seiners had

Figure 11.6. Brailing herring onto tender *Frigidland*, Prince William Sound, 1974. The plywood box was considered to hold one ton of herring and was emptied via a quick-opening door along the bottom of one side. (James Mackovjak)

either run aground or had capsized. Fortunately, no one was injured in these mishaps.

And the fishery engendered a lot of waste. After filling the *Frigidland*'s hold from one seiner's set, there were still a lot of herring in the net and no other tenders available to take them. Unsure of what to do,

Figure 11.7. *Marine View*, rigged for seining herring, Seward, Alaska, 1973. (James Mackovjak)

Holmstrand called Seward Fisheries on the marine radio and asked if he should deckload the remaining fish, adding that seas in the Gulf of Alaska were fairly high. Seward Fisheries told him to take the fish, so we plugged the scuppers and did so. We got under way, and, once into the gulf, it took exactly four waves washing over the deck to remove any trace of the probably twenty tons of fish that had been there.

Add to this the *Marine View* fiasco. The eighty-two-foot vessel (figure 11.7) was built in 1943 as a sardine/herring seiner and at least during the 1946 season had fished for herring in Prince William Sound.[14] In more recent years, the *Marine View* had fished for king crab in the Bering Sea and tendered salmon. For the 1973 herring season, Seward Fisheries rigged the boat with a long, deep seine. The *Marine View* made its first set in Columbia Bay and netted a gigantic haul. The catch was so heavy that the big seiner couldn't bring its net to the surface to brail the fish. To retrieve the net, much of which was covered in herring spawn, crewmen were forced to cut it into pieces. That ended the *Marine View*'s career as a herring seiner.

Though spotter planes had been employed by fishermen supplying herring-reduction plants beginning in 1965, their first use to locate schools of roe herring was in 1973. The first were from Cordova, and it immediately became apparent that having a spotter plane was necessary to be competitive. During some years, on the order of sixty spotter planes participated in the Prince William Sound herring fishery.

Figure 11.8. Spotter plane camp on Nunavarchuk beach, Togiak, 2017. (Eric Rosvold)

Figure 11.9. Stripping roe herring, Seward Fisheries, Seward, Alaska, circa 1973. (James Mackovjak)

Professional spotter pilots followed the herring fleet west with the spring, beginning at Sitka, then moving on to Prince William Sound, Cook Inlet, and Togiak. Piper Super Cubs, because they could take off in short distances and stay aloft at slow speeds, were a favorite among spotter pilots. Helicopters, too, were sometimes used to spot herring. At Togiak, the base of operations for some spotter pilots were spartan: a wall tent and some fifty-five-gallon fuel drums on a flat stretch of beach (figure 11.8). Spotting herring was an inherently dangerous occupation, and fatalities—usually midair collisions—were not uncommon.[15]

Figure 11.10. Grading herring roe, Seward Fisheries, Seward, Alaska, circa 1973. (James Mackovjak)

1974

Between 1971 and 1973, the catch of roe herring in Prince William Sound had increased eightfold and gone beyond what ADF&G determined was sustainable. To gain an element of control over the fishery, the department established a 5,000-ton quota, beginning with the 1974 season.[16] The season opened March 1 and was closed June 30 or by emergency order when ADF&G determined that the quota had been reached.

The catch in 1974, however, was 6,371 tons. The inability of the fishery managers to effectively limit the catch was mostly the result of their not being able to obtain catch data in a timely manner. On Prince William Sound, seventy-two purse seiners operated in 1974, almost triple the number of the previous year. The sound is a big place, and ADF&G management biologists, stretched to the limit, were unable to continuously monitor the fishing effort everywhere. The department's aerial surveys helped, but the weather wasn't always conducive to flying. Add to this the fact that, if herring were available, seiners were capable of making huge catches within a relatively short period of time. At times, fishing was in progress for several hours or even a day before a management biologist was notified.

And because there weren't always enough tenders to handle the catch, considerable quantities of herring were wasted. Fishing had been especially good near Green Island one day during the 1974 season, and every tender there was taking herring aboard. Meanwhile, about twenty boats, their seines full of herring, were waiting their turn. Unfortunately, there wasn't enough tender capacity for all the fish, and several fishermen were forced to dump their catches. ADF&G assumed that because of the long holding period, the dumped fish were dead or dying.[17]

1975

Despite the difficulties experienced in 1974, the only management change for the 1975 season was that the fishery was opened by emergency order rather than on a set date. On April 15, ADF&G determined that sufficient fish were present in Valdez Arm to warrant an opening. Moreover, test sets by volunteer fishermen showed that the roe content of the fish was high. The department then opened a portion of Valdez Arm for two hours of fishing.

A week later, on April 22, ADF&G opened the area around Green Island for twelve hours by emergency order, after which the 1975 Prince

Figure 11.11. Early herring pump, Seward Marine Services, circa 1975. (Rhonda Hubbard)

William Sound roe-herring season was closed. In all, seventy-six seiners—the all-time-record number for the Prince William Sound roe-herring fishery—caught 5,854 tons of herring. Though the 5,000-ton quota had been substantially exceeded, ADF&G characterized the season as "quite successful," noting that the harvest was orderly. Buyers had thirty-nine tenders in the area, so no waste was apparent, and the roe quality and recovery percentage were the highest ever recorded in the sound.[18]

Hydraulically driven herring pumps developed in the mid-1970s made brailing obsolete. Pioneer roe-herring fisherman Beaver Nelson said it could take up to twenty-seven hours to brail a 100-ton set of herring.[19] According to Doug McNair,

> More than anything, the 94 million pounds of herring taken in Alaska are literally sucked from the sea. For where would this fishery be today if it weren't for the ubiquitous fish pump, the seagoing vacuum cleaners that mop out a gillnetter's hold in a matter of minutes or siphon 50 tons of fish direct from the seine in a couple of hours.[20]

Figure 11.12. Pioneer roe-herring fisherman Beaver Nelson said it could take up to 27 hours to brail a 100-ton set of herring. Hydraulically driven herring pumps developed in the mid-1970s made brailing obsolete. In this May 2009 photo taken at Togiak, the tender *Cloverleaf* is preparing to pump approximately 300 tons of herring from the net of the seiner *Theresa Morgan*. (Eric Rosvold)

Figure 11.13. Pumping herring onto deck of tender *Chichagof* at Togiak. The holds would already have been full. Note dewatering section of pipe. (Ken Moore)

1976

The Prince William Sound herring fleet grew in 1976. In early April, before fishing started, ADF&G counted 103 seiners and 53 tenders in the Valdez Arm area. At that time, the Alaska Commercial Fisheries Entry Commission (see chapter 10) was contemplating limiting entry into several herring fisheries, and some of the boats may have been in the sound mainly to establish a record of participation in the herring fishery there.

Given the large number of boats and the need to keep within the sound's 5,000-ton quota, ADF&G elected to manage the fishery by shortening the amount of time that fishing would be allowed.[21]

While they were waiting for the herring to show—and negotiating with processors over the price they would be paid for the fish—fishermen were entertained each evening by the marine VHF radio broadcasts by "Herring Rose," a play on "Tokyo Rose," of World War II infamy. Tokyo Rose had tried to demoralize American soldiers and persuade them to give up the fight; Herring Rose, with her seductive voice, tried to persuade herring fishermen to go home to their wives and girlfriends. This would leave more fish for the boat she was on.

A typical broadcast began with "Hello again, you darling fishermen out there waiting for the elusive herring to show in vain, while your girlfriends and wives are having the time of their lives partying it up with the other guys back in town." This was followed by a sigh and a deep, sexy moan. Sometimes her taunts were directed at individual fishermen: "You know who I mean, don't you, Joey. I wonder who's sleeping in your nice big bed at home tonight while you toss and turn all night out here in your tiny little rock-hard bunk."[22]

Prince William Sound fishermen were still waiting for the herring to arrive and were still negotiating a price for herring when, on the night of April 25, a huge, unexpected storm struck. Much of the seine fleet and its tenders were anchored in Gibbon Anchorage, a shallow, rock-strewn bay on the north shore of Green Island. Some seiners were tied to tenders, and others had spotter planes tied to their sterns.

Normally, Gibbon Anchorage was a fairly safe place to anchor, but the tide was exceptionally high that night and rose over the spit that protected the bay on its east side. The strong winds and large seas that suddenly developed that evening caused boats to drag anchor and crash against one another. Planes began to break loose. It was a frantic maritime traffic jam on foul grounds, as boats hauled anchor or cut loose from their tenders and sought shelter behind rockpiles and small islands. The wind blew a steady fifty knots, with gusts estimated to be twice that. And the seas were huge.

The storm blew itself out that night, and by morning the sea was like glass. But there was considerable damage: several boats and airplanes were damaged, and one boat was lost. The lost boat was a tender, the *GW King*, a World War II–era 105-foot power scow, which struck a rock pinnacle and broke into pieces. Fortunately, no one was killed or seriously injured.[23]

Fishermen and processors were still negotiating a price when marketable herring showed up in the northeast reaches of the sound a few days after the storm—and a couple of weeks later than usual. ADF&G opened the fishery for one hour on April 30, but the seiners and processors hadn't yet agreed on a price, so the seine fleet remained at anchor, as it did during a pair of successive one-hour openings over the following two days.

On May 5, after the schools of fish that might have been caught during the three nonevent openings had spawned, fishermen and processors finally settled on a price of $175 per ton. Herring showed up at Green Island two days later, and ADF&G scheduled a one-hour opening on May 9.

By then, however, much of the fleet had left, discouraged by the self-imposed strike, the late show of herring, or the need to repair damage caused by the storm. ADF&G was unsure of the number of seiners that fished at Green Island, but only sixty-six made deliveries. Others made sets but did not catch fish or had their seines torn on rocks and lost their catches. And, in some cases, gear congestion prevented boats from even making sets.

Some fishermen and processors elected to persist, and on June 6 a dozen seiners landed 417 tons of herring during a twelve-hour opening in eastern Prince William Sound. And, perhaps mercifully, the season was over. The fleet's total catch was only 2,584 tons, a little more than half the quota.[24]

Significantly, the 1976 herring year-class in Prince William Sound was a strong one and would support large catches in 1980–1982.[25]

Since 1971, Seward Fisheries (Icicle Seafoods) had been one of the main buyers of Prince William Sound herring, and in 1976 the company began operating a reduction plant. Previously, herring carcasses (and other offal from the plant) were ground up and pumped overboard near the plant. Now they were used to make fish meal and fish oil. (Icicle Seafoods' Petersburg Fisheries plant began operating a reduction plant in 1975.)[26]

1977

> Today, in terms of gross returns per season duration, the [Alaska] herring roe fishery is one of the most valuable in the world.—Howard O. Ness, economist, National Marine Fisheries Service, 1977[27]

Early in 1977, the Alaska Commercial Fisheries Entry Commission began issuing limited-entry permits to purse-seine fishermen who had met economic dependence and historical participation requirements for the Prince William Sound roe-herring fishery. One hundred and five such permits were ultimately issued.[28]

At the same time, the commission created two exclusive roe-herring districts in the sound. The Montague District encompassed the waters surrounding Montague and Green Islands, and the Northern District encompassed Port Fidalgo, Valdez Arm, Columbia Bay, Long Bay, and the waters surrounding Glacier Island and Bligh Island. The remainder of the sound was designated the General District and here fishing for herring to be used as food or bait was not restricted by the limited-entry program.

In response to complaints of the waste engendered in roe-stripping operations, in 1977 Alaska's Legislature made it illegal to waste commercially taken herring. The legislature defined waste as "the failure to use the flesh of commercially taken herring for reduction to meal, production of fish food, human consumption, food for domestic animals, scientific or educational purposes, or round herring bait." However, because there were no reduction plants in the Bering Sea, the law did not become effective there until 1979. The legislature assumed that by that time Bering Sea processors would be capable of fully utilizing herring.[29]

Fifty-four purse seiners participated in the 1977 roe-herring season in Prince William Sound, a decline of twelve from the previous year. During thirty-eight hours of fishing on April 9 and April 10, the vessels landed 2,266 tons of roe herring, less than half the quota.[30]

1978

ADF&G characterized the 1978 roe-herring fishery in Prince William Sound as "unusual" for two reasons. First, juvenile and adult (spawning) stocks were mixed, something that had never before been observed. Second, aerial observations found schools of herring throughout Valdez Arm, which should have portended good fishing. Most of the fish, however, suddenly disappeared, and no one was able to determine where they went.

The roe-herring catch for the season, 1,330 tons, was the lowest it had been since 1971, when the fishery was in its infancy. Fully seventy purse seiners participated. Untypically, they were augmented by thirty-eight gillnetters, salmon fishermen who, because of poor fishing in the Copper River delta, fished for herring during emergency salmon-fishing closures. The gillnetters landed sixty-two tons of roe herring. Beginning in 1980, gillnetters participated in the fishery each year.

ADF&G estimated the total ex-vessel value of the 1978 Prince William Sound roe-herring fishery to be about $960,000.[31]

1979

ADF&G implemented a new management strategy for the 1979 season: the season would open on April 7 and be closed by emergency order when the 5,000-ton GHL was attained. The plan would allow the fishing fleet to search for incoming schools of herring and reduced the chance of herring migrating undetected through open fishing areas. The dispersal of the fishing fleet, it was thought, would facilitate a more orderly harvest.

Initially, the plan worked well, but later in the season there were few herring within the Northern District but very large quantities at Port Gravina, just outside the district. In an effort to salvage the season for fishermen, ADF&G asked the Commercial Fisheries Entry Commission to expand the Northern District to include Port Gravina, Sheep Bay, and Simpson Bay. The commission issued an emergency order to do so, but the process took two days. The department then promptly opened fishing at Port Gravina for one hour. Fishing, however, was poor. During the two days that were required to obtain the emergency order, the herring had left Port Gravina.

That same evening, large schools of herring were located in Sheep and Simpson Bays. A one-hour opening early the following morning yielded almost 3,000 tons of herring. After that opening, the 1979 Prince William Sound roe-herring season was over.

Eighty-nine seiners caught a combined total of 4,138 tons, about triple the previous year's catch but well short of the 5,000-ton GHL. The catches in Southeast Alaska and Lower Cook Inlet were also low that year, and, in their effort to fulfill contracts in Japan, buyers aggressively bid up the price in all the state's herring fisheries. The price processors paid for Prince William Sound herring, $1,500 per ton, was the highest ever. ADF&G estimated the total ex-vessel value of the sound's fishery to be about $5.2 million, a more than fivefold increase over that of the previous year.[32]

1980

To ensure that the 1979 Port Gravina emergency-opening debacle did not repeat itself, prior to the 1980 season the Commercial Fisheries Entry Commission established a new roe-herring fishing district in Prince William Sound. The Eastern District encompassed Port Gravina, Sheep Bay, and Simpson Bay. At the same time, in the Northern District, ADF&G established gillnet-only harvest areas within Valdez Arm and Port Fidalgo.

Because of the impact on the world economy of the 1979 oil crisis (caused by a decrease in oil production in the wake of the Iranian Revolution), some processors and fishermen were hesitant to fully mobilize for the 1980 roe-herring fishery. On top of this, marketable herring showed up earlier than usual in Prince William Sound. The result was that there were not yet enough tenders on the grounds to handle the catch. Some fishermen were forced to hold their catches for long periods of time, and when the fish began dying, they dumped them. ADF&G's aerial surveys recorded five "piles" of dead herring on the seafloor, and the department estimated they represented 400 to 500 tons of wasted fish.

The seventy-four seiners that ultimately participated in the fishery—fifteen fewer than in 1979—landed 6,042 tons of roe herring, more than 5,000 tons of which was taken in the Montague District. Sixteen gillnetters landed 264 tons. The 1980 total catch, 6,306 tons, substantially exceeded the 5,000-ton GHL for the sound. The price buyers paid for herring had, however, plummeted to $320 per ton, which put the total ex-vessel value of the fishery at about $2 million, about 40 percent of what it had been in 1979.[33]

1981

ADF&G's 1981 Prince William Sound management report contained an important observation: "Because the three sac roe districts are so large and the fleet is so effective . . . when the herring become available the catch can develop so rapidly it is difficult to monitor and manage within established guideline harvest levels over such a broad area."[34] The fact that the seine fleet had increased by 25 percent, to 101 vessels, was an important factor. Probably of equal or greater importance, however, was that 1981 marked the peak in the cycle of herring abundance in Prince William Sound. There were a lot of fish to be caught.

During the season, the seiners landed 13,768 tons of roe herring, more than doubling their catch of the previous year, while the eighteen gillnetters landed 235 tons. The 14,003-ton total catch was almost three times the 5,000-ton GHL. It was also a record that would not be eclipsed for another decade. The estimated total ex-vessel value of the fishery was $5.6 million.[35]

1982

Having failed to keep the fleet within its GHL for two consecutive years, ADF&G modified its management strategy. Instead of opening all three of the sound's districts simultaneously, it opened only the Northern and Eastern Districts. The Montague District, where herring normally spawned later than in the other districts, would be opened by emergency order.

Early fishing in the Northern and Eastern Districts, however, produced only immature, unmarketable fish, and the season was closed. On April 23, ADF&G scheduled a second opening of the Northern District as well as an area around Naked Island, in the General District. The opening lasted two hours and produced 7,148 tons of herring. Following the seine fishery, ADF&G scheduled a fifty-four-hour opening for gillnet fishermen, also in the Northern District. Twenty vessels landed 394 tons of roe herring.[36] Though no fishing occurred in the Montague District, the total roe-herring catch of 7,542 tons exceeded the 5,000-ton GHL for Prince William Sound. It was the third year in a row the fishermen had exceeded the sound's GHL. The estimated total ex-vessel value of the fishery was $3 million.[37]

1983

As ADF&G had expected, the strong 1976 year-class had run its course and would not bolster the 1983 harvest. Nevertheless, the department again modified its management strategy. Districts were opened on a field-announcement basis with the emergency-order period beginning on April 1. For seiners, the entire season's fishery consisted of a single one-hour opening on April 13 in specific areas of the Montague and General Districts. A record 103 seiners participated in the fishery, but only seventy-two made deliveries. The seine catch was 2,729 tons, with the vast majority coming from the waters around Naked Island, in the General District.

Because a seiner did not make a delivery did not mean that the vessel didn't profit from the fishery. Over the years, most of the exceptionally large sets in Prince William Sound—on the order of 400 tons—were co-op sets that involved several boats. One boat was assigned to make the set while its partner boats blocked other fishermen's access to the fish. (One Prince William Sound herring seiner once hired boats with no herring permits to block for him. This caused an uproar among the fleet, and ADF&G quickly outlawed the practice.)

Gillnetters were allowed a single twenty-four-hour opening on April 21–22 in the vicinity of Galena Bay and Valdez Arm. Both locations were within the Northern District, the only Prince William Sound district in which gillnetting for roe herring was permitted. Twenty-two gillnetters landed 105 tons of herring, considerably less than in previous years. Fortunately for them, the market became highly competitive toward the end of the season—perhaps because buyers had expected a larger catch by seiners and needed to fulfill contracts—and buyers paid the gillnetters $950–1,000 per ton for herring with a 10 percent roe recovery. For fish with the same roe-recovery, seiners had been paid considerably less: $500–600 per ton. The total ex-vessel value of the fishery was $1.7 million.[38]

For the first time since 1980, ADF&G had kept roe-herring fishermen in Prince William Sound within their 5,000-ton GHL. The fact that fish were less abundant than in previous years probably helped.

1984

The only management change in 1984 was the lifting of the prohibition on gillnetters fishing outside the Northern District. For seiners, the roe-herring season comprised two openings on April 14 in the Montague District, one for two hours and one for one hour. Landings by the 105 participating seiners totaled 5,946 tons.

Twenty-four gillnetters fishing near Storey Island (north of Naked Island, in the General District) and in the Montague District landed approximately 343 tons of roe herring during fifty-nine hours of fishing spread over four openings.

The total ex-vessel value of the 1984 Prince William Sound roe-herring fishery was $4.7 million.[39]

1985

In 1985, spawning herring arrived in Prince William Sound about two weeks later than usual. The first seine fishery opening was on April 28, at Unakwik Inlet and Wells Bay, in the General District. The weather and sea conditions were perfect during the three-hour opening, and ninety seiners landed 4,817 tons of herring—very close to the 5,000-ton GHL in effect. Nevertheless, the next day ADF&G allowed a second opening, this one in the Montague District. Fishing during the one-hour opening was restricted to areas of Stockdale Harbor and Port Chalmers, and sixty-one seiners landed 2,107 tons. The total harvest by seiners had an ex-vessel value of about $5.2 million, based on a base price of $680 per ton for herring with a 10 percent roe recovery.

Gillnetters had a thirty-four-hour opening in the Montague District, during which they landed 413 tons of roe herring. The average roe recovery for the gillnet-caught herring was 11–12 percent, and the fish commanded an average price of $930 per ton. The total ex-vessel value to the twenty-one gillnetters was about $384,000—a per-vessel average of about $18,000.[40]

1986

The trend in herring abundance in Prince William Sound during 1984–1985 was up, and ADF&G anticipated the roe-herring harvest in 1986 would be average to above average. Accordingly, the department established a GHL that ranged from 5,000 to 7,000 tons.[41]

In its 1986 Prince William Sound management report, ADF&G noted that the sound's roe-herring purse-seine fleet had "explosive harvest potential."[42] It wasn't an understatement.

In mid-April, substantial schools of herring moved into Unakwik Inlet and Wells Bay, on the sound's north shore in the General District. ADF&G scheduled a three-hour opening there on April 17. The weather was almost ideal, and—every fisherman's dream—more fish than were anticipated flooded in. In three hours of fishing, 105 purse seiners landed 9,828 tons of herring, greatly exceeding the upper limit of the GHL.

In addition to the great abundance of herring, the fish were much larger than normal. As well, processors rated the overall quality of the fish, which had an average roe recovery of 10–11 percent, as above average. The base price was $750 per ton for herring with a 10 percent roe recovery rate. ADF&G estimated the ex-vessel value of the fishery to be $8.1 million, which would stand as the all-time record for the fishery.

Department biologists summed up the purse-seine fishery: "The exploitation was higher than planned, but considering the current status of the stocks, the quality of the product produced, and the safe conclusion, we don't feel the resource has been compromised in any way."[43]

The gillnet fishery was uneventful. During three openings that totaled ninety hours of fishing, twenty-five fishermen landed 449 tons of roe herring, which had an estimated ex-vessel value of $412,000.[44]

1987

In large part because of the unexpected large harvest in 1986, ADF&G anticipated that fewer fish would be available during the 1987 season, and it reduced the GHL for purse seiners to 3,000–5,000 tons.[45]

Herring arrived early in Prince William Sound, and some fishermen and one processor were not yet on the grounds on April 8, when ADF&G opened Wells, Cedar, Granite, and Fairmount Bays and the waters surrounding Fairmount Island for one hour. The next day, ADF&G opened the same waters, except Fairmount Bay, for thirty minutes. During the two openings, the ninety-five seiners that had made it to the fishing grounds in time landed 4,982 tons of herring. The fish were above average in quality and had an average roe content of 11 percent. ADF&G assumed the ex-vessel average price to be $1,100 per ton.

The gillnetters' season was a single twenty-four-hour opening that began on April 10. The catch, 533 tons, was the largest on record and, in ADF&G's words, "grossly" exceeded the upper end of the preseason harvest range of 200–300 tons. The unexpectedly abundant harvest was in large part due to the fact that a significant portion of the twenty-five-vessel gillnet fleet started using thirty-foot aluminum herring skiffs equipped with hydraulically operated shakers. The skiffs were capable of hauling four to six times the tonnage as the traditional Copper River bowpickers.

The total ex-vessel value of the 1987 Prince William Sound roe-herring fishery was $5.8 million.[46]

1988

Give or take a half-million fish, ADF&G was precise—at least on paper, if not in fact—regarding the number of herring that migrated to Prince William Sound to spawn in the spring of 1988: 524 million, or 58,621 tons.[47]

The catch during two one-hour openings on April 21 and 22 by the 105 seine vessels that participated was 7,896 tons, about 15 percent of the spawning biomass and greatly in excess of the 4,000- to 5,000-ton GHL. The catch had an ex-vessel value of $6.6 million.

Twenty-four gillnet vessels fished for 5.5 hours on April 23 and landed 358 tons, which was considerably in excess of their 275-ton GHL. Its ex-vessel value was $537,000.[48]

1989

A large portion of the roe-herring catch in 1988 consisted of fish from the exceptionally strong 1984 brood year, the strongest single brood year that had been observed in Prince William Sound since 1976. The strength of that year-class boded well for the 1989 season.[49]

The 1989 season was just weeks away when, just after midnight on March 24, the US Coast Guard received a terse message: "We fetched up on hard ground, north of Goose Island, off Bligh Reef, and evidently leaking some oil and we're gonna be here for a while."[50] The message was from Joseph Hazelwood, captain of the 978-foot *Exxon Valdez*, which had the evening before finished loading 1.26 million barrels (53.1 million gallons) of crude oil at the Valdez oil terminal and was heading to California. An unlicensed third mate was at the ship's helm when the grounding occurred. About 10.8 million gallons of oil spilled into Prince William Sound that day, oiling some 200 miles of coastline. A nightmare had arrived.

Immediately following the spill, ADF&G biologists began inventorying affected beach areas, including beaches in the Naked Island group, on Green Island, and on the north shore of Montague Island—all important herring-spawning areas.

There was no evidence that suggested spawning herring might avoid areas contaminated by oil. Oil contamination was thought to pose an appreciable likelihood of injuring or killing herring. Moreover, the scientific literature suggested that exposure of developing herring eggs and larva to even relatively low concentrations of Alaska North Slope crude oil would cause mortality, gross morphological abnormalities, and genetic and cellular defects. Because herring were thought to spend three years outside the sound before returning to spawn, biologists wouldn't begin to understand the effects of the oil spill on herring eggs and larva until 1992.[51]

Based on these factors, on April 3 ADF&G announced that the spring herring fisheries in Prince William Sound would not open. For herring fishermen, the impact was almost immediate. "Fishermen are always broke in the spring," said a Cordova-based spotter pilot. "A lot of them will do herring so they can get their salmon gear in tiptop shape," he said.[52] Not in 1989.

A massive effort to contain and clean up the oil spill soon began. The results, though, were dubious because even today—decades later—oil spilled from the *Exxon Valdez* can still be found along Prince William Sound's shoreline.

The sound's food and bait fishery, however, was opened by emergency order on November 1 in the Knowles Head area, on the sound's eastern shore. Three seiners landed 646 tons of herring, less than half of the 1,694-ton GHL. Buyers paid $250 per ton for the fish, which was used for bait.[53]

1990

The state of the sound's herring fishery was considered a measure of the cleanup's effectiveness, and ADF&G sent a positive message when it reopened the fishery in 1990.

Essentially, managers acted as if the spill had never occurred. Ninety-seven seiners were on the grounds that April, ADF&G noting in its management report that year that the fleet had "shown a dramatic increase in efficiency over recent years." The department's strategy for keeping the fleet within its 6,000-ton GHL while providing reasonable opportunities for fishermen was to have a short opening over a large area. The 1990 season's single opening, on April 12, lasted twenty minutes—the shortest ever—and encompassed Galena Bay, Tatitlek Narrows, Boulder Bay, and Landlocked Bay.

The department's strategy fell short: seiners landed 8,362 tons of herring, substantially exceeding their GHL. Some exceptionally large sets were made in the shallows in Tatitlek Narrows, and during a postfishery survey, hundreds of tons of herring were seen spilling over corklines. The catch per unit of effort was impressive: 290.35 tons of herring per boat per hour, by far the largest ever recorded.

Not accounting for postseason price adjustments—*retros*, in the parlance of fishermen—fishermen were paid $650 per ton for fish that yielded a 10 percent roe recovery. The ex-vessel value of the fishery was $5.4 million.

But there were implications for the future. The fish landed during the opening were relatively old, with more than 77 percent being over six years old.

Gillnetters, too, had a single opening. It was on April 13 and lasted four hours, during which twenty-four boats landed 505 tons of herring. The GHL for the gillnetters was 353 tons. As with the purse seiners, ADF&G noted that the efficiency of gillnetters had increased drastically in recent years. The ex-vessel value of the gillnet fishery was $323,000.[54]

1991

The 1991 fishery was better than business as usual. ADF&G attributed the abundance of herring to the high rate of survival of the 1984 year-class and its offspring. Based on its experience in 1990, the department recognized that processors' daily herring-processing capacity was limited and could be overwhelmed by a large harvest by purse seiners. To eliminate or reduce potential waste and to ensure that a high-quality product was produced, the department's strategy was to try to divide the harvest into several openings. It was successful at doing so. The season for purse seiners in the sound began on April 9 and ended on April 19, and it comprised three openings: two thirty-minute openings and one twenty-minute opening. The 104 participating seiners landed 11,923 tons of herring, slightly more than their 11,232-ton anticipated catch. The seiners' catch had an ex-vessel value of $7.2 million, the second highest in the fishery's history.

Twenty-four gillnetters fished for 10.5 hours on April 18 and landed 742 tons of herring, a bit more than their 657-ton anticipated catch. The gillnetters' catch had an ex-vessel value of $445,000.[55]

Marring the season was the death on April 9 of veteran spotter-plane pilot Tom Parker. Parker's plane crashed into a mountainside after a midair collision with another spotter plane above Boulder Bay, where thirty to thirty-five planes were spotting herring just prior to the opening of the season. The other pilot crash-landed his plane on the Tatitlek village runway and survived.[56]

1992

ADF&G's forecasted 1992 herring spawning biomass in Prince William Sound—121,342 tons—was the highest projection on record.[57] The forecast proved accurate. Prince William Sound seiners landed 16,592 tons of herring (18 percent more than their 14,100-ton allocation) during an unprecedented four openings between April 13 and April 20 that totaled two hours of fishing time. The areas fished were at Bligh Island, Naked Island, and Rocky Bay (Montague Island). The catch had an ex-vessel value of $6.7 million.

For gillnetters, eleven hours of fishing during openings on April 23–24 along the north shore of Montague Island yielded 940 tons of roe herring, somewhat more than the 825-ton GHL. Its ex-vessel value was $752,000.[58]

The total roe-herring catch for 1992, 17,532 tons, was a record for Prince William Sound, and it contributed toward 1992 being a record for statewide roe-herring production.

1993

The fishery seemed on track for a prosperous season in 1993. ADF&G projected the spawning biomass in the sound would be 134,133 tons, which exceeded the record established the previous year. Unfortunately, only about a quarter of the expected herring returned, and those that returned were smaller than expected. Among the possible causes, according to ADF&G, were high overwinter mortality and unusually low spawning recruitment.[59] The 1989 year-class, the fish spawned the year of the *Exxon Valdez* oil spill, would have been four years old in 1993 and likely a component of the spawning population.

Moreover, many of the fish were lethargic, exhibited odd schooling behavior, and had surface hemorrhages. Pathologists later determined that the fish were infected with viral hemorrhagic septicemia (VHS), a sometimes-fatal disease.[60]

As noted above, statewide roe-herring production in 1992 had set a record. In 1993, with roe from the previous year still in inventory, the market for low-quality roe herring was very tight. ADF&G did what it could to facilitate a successful fishery, such as schedule openings that would comport with cooperative-style fishing, but processors expressed little interest in the fish, and no purse-seine roe-herring fishery occurred in the sound that year.

However, because gillnets are selective for larger herring, gillnetters were able to fish. During thirty-six hours of fishing spread over four openings between April 15 and April 19, twenty-four gillnetters landed 1,030 tons of herring, a bit over their 912-ton GHL—which would probably have been a lot more liberal had the fishery's managers known that there would not be a seine fishery. Except for 1989, when there was no roe-herring season, 1993's harvest was the lowest since 1972.

ADF&G's summary: "The 1993 season can be characterized by an earlier than average spawning period, a biomass significantly less than the forecast, a lower than expected percentage of 5-year fish, low annual growth for all age classes, a tight market, and low-quality herring."[61]

1994

ADF&G's projected 1994 spawning biomass for Prince William Sound was 29,786 tons—less than a fourth of what it had been for 1993. The department anticipated that seiners and gillnetters combined would harvest 2,768 tons of roe herring.[62] Such was not to be.

An early-season assessment determined that the spawning biomass of herring was below the 22,000-ton threshold established by the Alaska Board of Fisheries that spring, and ADF&G canceled all herring fisheries. The 14.6 miles of shoreline spawn in Prince William Sound in 1994 was the lowest ever recorded.

Though there were no commercial herring fisheries, ADF&G test sets took 151 tons of roe herring. The age structure of the fish indicated that the stock was becoming senescent. Moreover, VHS-caused skin hemorrhages were observed on about 10 percent of the herring sampled, and about 30 percent of the fish were infected by the fungus *Ichthyophonus hoferi*.[63]

1995–1996

Because the spawning biomass of herring in Prince William Sound did not reach the Board of Fisheries' 22,000-ton threshold in either 1995 or 1996, there were no herring fisheries in Prince William Sound during those years. Samples in 1996 indicated that more than half the biomass was age-eight or older, though there was an encouraging showing of age-three and age-four recruits.[64]

1997

ADF&G population modeling suggested that there would be a sufficient biomass of spawning herring to support at least a modest roe-herring fishery in 1997. For seiners, the department established a GHL of 2,965 tons. For gillnetters, the GHL was 175 tons.

Seiners, with dozens of spotter planes, began gathering in the sound in early April. In contrast to earlier years, when GHLs were considerably higher and more than a hundred seiners participated in the fishery, only seventy-one chose to do so in 1997. Part of the reason may have been the anticipated low price that processors would pay for roe herring.

Tragedy struck on April 9, when two spotter planes collided in midair over Galena Bay. One plane, piloted by Ron Gribble, a former president of the Alaska Fish Spotters Association, crashed, killing Gribble and his passenger, Mike Paoli. No one was injured on the other plane, which was able to land safely.

The same day the spotter planes collided, ADF&G opened Tatitlek Narrows to gillnetters. The opening was intended to last four hours, but because the quality of the herring was low, it was closed after 2.5 hours. Twenty-two gillnetters landed 176 tons of herring, almost exactly their 175-ton GHL. For their catch, the gillnetters were paid $80 per ton, by far the lowest since the inception of the gillnet roe-herring fishery in the sound in 1981. The total ex-vessel value was only $14,080.

Herring showed in the Montague Island area a few days later, and during a twenty-minute opening on April 13 and a ninety-minute opening on April 15, seiners landed 4,700 tons of herring—far more than their GHL. At $200 per ton, the value of seine-caught roe herring was higher than the fish caught by gillnetters, but it was still the lowest since at least 1978. The seine fishery in 1997 had an ex-vessel value of $941,000.[65]

1998

In 1998, the number of herring seiners declined again, to forty-five, likely as a result of the modest 3,367-ton GHL and an anticipated low ex-vessel price. The spring of 1998 was warmer than usual, and the herring arrived early. On April 6, during a thirty-minute opening near Gilmour Point, on the north shore of Montague Island, seiners landed 3,491 tons of herring, ending their season. Processors paid $300 per ton, which was very low by historical standards but still an improvement over the previous year, and the fishery's total ex-vessel value was just shy of a million dollars.

ADF&G opened the gillnet fishery in approximately the same area on April 11. The opening was for three hours but didn't begin until 8:45 p.m. None of the twenty gillnetters present fished, however. Fishermen had misgivings about the price buyers were offering, and they complained that it would be dark before the opening would be over. They suggested it might be a good idea to just wait and see if roe recovery improved.

The next day, April 12, ADF&G opened the area for three hours. Fishing was decent, and the department, wanting to ensure that the entire 197-ton GHL was taken in this one opening, extended the fishery a half hour. The department noted that there was enough tender capacity available to handle 400 tons of fish and reasoned that there would be no waste if the gillnetters exceeded their quota. When all was said and done, the gillnetters had landed 356 tons of herring. They received an ex-vessel price of $375 per ton, and the ex-vessel value totaled $156,000.

The section on herring in ADF&G's 1998 Prince William Sound annual management report ended on a positive note: the rebuilding phase of the sound's herring stock was ongoing, and the department expected sufficient fish to allow the sound's herring fisheries to happen in 1999.[66]

1999

ADF&G surveys in early 1999 found few herring in the fish's usual haunts around Green Island and at other locations. The only location in which significant herring biomass was found was in Zaikof Bay, on the eastern end of Montague Island. The department found the widespread lack of herring to be "both unusual and troubling."[67]

Further surveys found few fish, and on April 20, ADF&G canceled the purse-seine and gillnet roe-herring fisheries as well as the food and bait fishery in Prince William Sound. (The pound fishery, described in part III, was already under way, and closed-pound fishermen were given until April 25 to release impounded herring and harvest their kelp. Likewise, open-pound fishermen were permitted to keep their kelp in the water until April 25.)

ADF&G sampling in 1999 indicated that the Prince William Sound herring population had suffered a decline of approximately 40 percent, which was centered around age-three and age-four fish. The department's year 2000 forecast predicted an extremely low contribution of age-four

Figure 11.14. Data sourced from ADF&G Prince William Sound Management Area, 1999 Annual Finfish Management Report.

herring, and in late September 1999 it canceled all the year-2000 spring herring fisheries as well as the 1999 food and bait fishery.[68]

The scarcity of herring in Prince William Sound in 1999 and 2000 wasn't an anomaly. Herring have been scarce there ever since, and commercial herring fishing in the sound—for a century a center of Alaska herring production, including cured herring, meal and oil, roe herring, spawn-on-kelp, and bait—has ended.

The definitive cause(s) of the decline in the herring population in Prince William Sound and the failed recovery remain undetermined. The leading hypothesis, however, is that it is the result of chronic and epizootic (temporarily prevalent and widespread) mortality caused by infectious and/or parasitic diseases. In 2014, the *Exxon Valdez* Oil Spill Trustee Council, the organization that oversees restoration of the Prince William Sound ecosystem through a $900 million civil settlement, classified the sound's herring population as an "injured resource" that was "not recovering."[69] There have, however, recently been some strong recruitment events, but it remains to be seen if a commercial herring fishery in Prince William Sound will be reestablished.

12

LOWER COOK INLET AND KODIAK AREA ROE-HERRING FISHERIES

LOWER COOK INLET

The earliest roe-herring harvests in lower Cook Inlet were token efforts in 1963 and 1965, when one ton and two tons, respectively, were harvested in the inlet's southeast reaches. There was no harvest in 1966 or 1967, but in 1968 the harvest was twenty tons. Interest in roe herring was growing, though, and in 1969 fishermen harvested 551 tons in Kachemak Bay, where the city of Homer is located. Kachemak Bay is part of ADF&G's lower Cook Inlet area's Southern District. The single processor reported by ADF&G to have operated in Kachemak Bay that year was the appropriately named vessel *Kazunoko*, operated by Seattle-based International Seafoods.

The harvest in 1970 increased almost fivefold, to 2,709 tons, with eighteen seiners making landings. But Cook Inlet herring stocks were in decline, and ADF&G suspected the cause was overfishing, which was facilitated by a lack of regulation by the department. Despite that suspicion, herring fishing in lower Cook Inlet remained open on an unlimited basis until 1973, the harvest determined by the availability of herring, the fishing effort, and product marketability.

The 1971 harvest was only thirteen tons, and it would never again exceed 300 tons. After 1980, there was only one year, 1989, in which roe herring were harvested in the Southern District.

However, not long after the roe-herring fishery waned in Kachemak Bay, it developed in Kamishak Bay, on the west side of Cook Inlet. Spotter pilots and fishermen first located spawning herring there in 1973. That year, fourteen seiners caught 243 tons, but the fishery grew rapidly, and the catch increased to 2,114 tons in 1974.

The harvest rose to 4,842 tons in 1976, when sixty-six seiners participated. However, due to two weak year-classes entering the fishery, ADF&G anticipated reduced catches during the next two or three seasons. Indeed, the harvest declined to 2,908 tons in 1977.

Though ADF&G established an annual GHL of 4,000 tons for Kamishak Bay in 1974, management of the fishery from 1973 through 1977 was basically "open season until closed." In 1978, herring abundance in Kamishak Bay was still declining, and management was changed to "closed season until opened by emergency order."[1] Also in 1978, ADF&G

replaced its 4,000-ton GHL with an unofficial GHL that limited the harvest rate on the herring stock to 10 to 20 percent of the estimated biomass.[2]

Another change was the restriction of the fleet size. In 1977, the Alaska Commercial Fisheries Entry Commission (see chapter 10) issued sixty-one permanent limited-entry and twenty interim-entry permits for the lower Cook Inlet fishery. (Interim-entry permits were issued to qualified applicants who were awaiting the final determination of their eligibility for a permanent limited-entry permit and allowed them to fish in the now-limited fishery.) It took a number of years, but the commission subsequently issued thirteen additional limited-entry permits, bringing the total to seventy-four.[3]

Meanwhile, herring stocks at Kamishak Bay continued to decline. The 1978 season lasted four days, but the catch was only 402 tons. Three days of fishing in 1979 yielded a similar amount, and ADF&G closed the fishery in 1980.

Fortunately, the stocks rebuilt, and the fishery was reopened in 1985. From 1985 through 1997, the apex of the lower Cook Inlet fishery, harvests averaged about 3,100 tons, peaking at 6,132 tons in 1987.

The average fleet size during those years was fifty-nine boats. The peak fishing effort was in 1988–1990, when seventy-five seiners made landings. Some of them were so-called "circuit seiners" (see chapter 9) that typically started the season with the roe-herring fishery in Southeast Alaska, then moved to Prince William Sound, then to Kodiak, then to Kamishak Bay, and then to Togiak.

The fleet represented a lot of fishing power. Accordingly, to help keep the harvests within the GHL, ADF&G restricted the areas open to fishing and the amount of fishing time. From 1991 through 1996, the most liberal seasons allowed ninety minutes of fishing. And the entire 1992 and 1996 seasons each comprised one thirty-minute opening. But the fishery could be lucrative. The average total ex-vessel value of the Kamishak roe-herring fishery from 1985 to 1997 was $3.3 million. The high year was 1988, when seventy-five seiners shared $9.3 million.

Unfortunately, herring stocks at Kamishak Bay began declining in 1996. The harvest in 1998 was only 331 tons, and the fishery was closed after that season. It has not reopened.[4]

KODIAK AREA

The Kodiak-area roe-herring fishery is unique because it is based on relatively small stocks of herring that spawn in dozens of bays on

Kodiak Island, Afognak Island, and the adjacent mainland during a six-week period that usually begins about May 1. Because of competition among shore-based processors and the fact that relatively small quantities of herring were harvested over a comparatively long period of time—which allowed for careful handling of the fish—the Kodiak roe-herring fishery has yielded some of the highest ex-vessel prices in the state.[5]

As noted at the beginning of part II, the roe-herring fishery in the Kodiak area began in 1964 at Spiridon Bay, on the northwest shore of Kodiak Island, where Anchorage-based Western Alaska Enterprises (a Japanese-owned corporation that also had a roe-herring operation that year at Unalakleet and possibly an operation at Seldovia) extracted the roe from 235 tons of herring. The fish were caught by five purse seiners, and the 23,070 pounds of roe produced was valued at $23,460, about a dollar a pound.[6]

There were no fish-processing plants at Spiridon Bay, so the roe was likely extracted aboard a vessel. As noted previously, popular for this sort of operation were power scows—World War II–era wooden vessels that were cheap and had considerable deck space and crew accommodations. The roe-stripped carcasses (and male herring) were almost certainly discarded overboard onsite.[7]

The roe-herring harvest at Kodiak expanded in 1965, but Western Alaska Enterprises was not a part of it. The sole processor of roe herring was the Washington Fish & Oyster Company, which also operated a reduction plant at Zachar Bay.

Fishing in Zachar Bay, Spiridon Bay, Larsen Bay, and Uyak Bay, eight seiners harvested 597 tons of roe herring, more than twice the previous year's harvest. Washington Fish & Oyster stripped the fish of their roe and processed the resulting offal and carcasses in its reduction plant. Roe production totaled 108,672 pounds, which was valued at $274,148 ($2.52 per pound).[8]

Interest in the fishery continued to grow. Eleven seiners fished in 1966, and their catch of 2,517 tons represented a fourfold increase over 1965's production. Four companies processed the fish. Herring, however, were getting scarce at Kodiak, and spawning populations in the four or five bays on the western shore of Kodiak Island where the fishermen concentrated their effort were generally depressed. The catch declined to 1,511 tons in 1967, but then rose a bit to 1,819 tons in 1968.

There was a lot of interest in Kodiak's roe herring in 1969, with nine processors and twenty-three purse seiners involved in the fishery. But spawning populations in the traditional fishing areas continued to be generally depressed, and production fell to 1,007 tons and then plummeted to 311 tons in 1970. Production continued to fall, bottoming out at 150 tons in 1972 before climbing back to 756 tons in 1973.[9]

During this time, the fishery was unregulated regarding harvest quotas, gear types, seasons, and fishing periods. However, to ensure overfishing did not occur when the stocks rebuilt, in 1974 the Board of Fish and Game established a 3,400-ton GHL for the Kodiak area roe-herring fishery. Still, the fishery remained essentially unregulated: the 1974–1978 seasons ran from March 1 through June 30, and there were no restrictions regarding gear types or fishing periods. The harvest in 1974 was 789 tons, but it fell to only seven tons in 1975 and then to five tons in 1976.[10]

Coincidentally, at this time the Alaska Commercial Fisheries Entry Commission was considering limiting the number of vessels that could participate in Alaska's herring (and other) fisheries (see chapter 10). Given that only two vessels participated in Kodiak's roe-herring fishery in 1975 and only one in 1976, the commission had no reason at that time to limit participation in it.

In 1977, a strong market for herring roe kindled renewed interest in Kodiak's roe herring, and eleven seiners, assisted by spotter planes, landed 308 tons of herring, which had an ex-vessel value of $108,000. Interest in the fishery continued to grow, and in 1978 the fleet of seiners increased to twenty-seven vessels, and they were joined for the first time by a half-dozen gillnetters. The catch totaled 564 tons and had an ex-vessel value of $472,000. However, spawning populations in traditional fishing areas were depressed, and ADF&G reduced the GHL for the 1979 season to 2,400 tons. The department also shortened the season to May 1 through June 30.

Fortunately, herring stocks at Kodiak were rebuilding, and the harvest in 1979 was 1,578 tons. At the same time, the ex-vessel price of herring was high, $1,500 per ton, a price that would not be seen again until 1996. Fifty-eight seiners and ninety-five gillnetters participated in the fishery, which had an ex-vessel value of $2.6 million. The large payday attracted more effort, and in 1980 ninety-two seiners and 109 gillnetters fished for roe herring at Kodiak. Though their catch, 2,166 tons, was substantially more than the previous year, it was worth considerably less—$1.6 million—because the ex-vessel price had fallen to $680 per ton. Seiners accounted for 84 percent of the harvest.[11]

Roe-herring harvests remained fairly robust from 1980 to 1991, averaging 2,130 tons, as did the total ex-vessel value, which averaged $1.8 million.[12] The average number of seiners that participated in the fishery during that period was forty-three; the average number of gillnetters was seventy-seven. The 1990 season was a two-month-long, twenty-four-hours-on-twenty-four-hours-off affair. The fleet—twenty-seven seiners and sixty-three gillnetters—harvested 2,347 tons of herring, worth $2 million.

The increased participation in the fishery during the late 1970s and early 1980s had caused the Commercial Fisheries Entry Commission to reconsider limiting entry into it, and in 1984 the commission began issuing limited-entry permits to qualified purse seiners and gillnetters. The commission issued interim-use permits to fishermen whose eligibility was still being determined. It took until 2009 for the commission to complete its eligibility determinations, and by that year it had issued 81 purse-seine permits and 119 gillnet permits.[13]

The year-class from the 1988 brood year at Kodiak was exceptionally strong and resulted in dramatically increased stocks and large harvests from 1992 through 1995. This increase, coupled with the closures of the Prince William Sound and Kamishak Bay roe-herring fisheries in the 1990s, spurred an increase in the seine effort in the Kodiak fishery. Part of the increase was due to the purchase of inactive Kodiak purse-seine limited-entry permits by "circuit seiners"—fishermen who participated in all of Alaska's major herring fisheries, from Sitka Sound to Togiak. The circuit seiners had well-equipped boats and experienced crews, and they added a lot of fishing power to the fleet.

The harvest in 1992 was a record 4,283 tons, worth $2.1 million to fishermen. The catch and payday were similar in 1993. The harvest increased to 5,893 tons in 1994, setting an all-time record. Eighty-four percent of the harvest was landed by sixty-six seiners. The ex-vessel price of herring was $800 per ton, which put the value of the fishery at $4.7 million.

Unfortunately, herring stocks began declining after 1994, and the harvest in 1995 was 4,604 tons. But fortunately for fishermen, the demand for kazunoko was high, and the price of roe herring was increasing. The ex-vessel price in 1995 was $1,282 per ton, and the value of the fishery was $5.9 million. The catch again declined in 1996, to 3,386 tons, but the ex-vessel price hit an all-time high of $2,000 per ton, pushing the value of the harvest by the fifty-seven seiners and seventy-four gillnetters to $6.8 million, an all-time record.

But the herring catch at Kodiak continued to decline, and, adding to fishermen's woes, a lagging Japanese economy and a surplus of kazunoko in Japan caused the ex-vessel value of roe herring to fall precipitously, to $500 per ton in 1997. The 3,335-ton catch that year had an ex-vessel value of $1.6 million—less than a fourth of its value in 1996. Interest in the fishery, especially among gillnetters, fell, and in 1998 the fleet comprised thirty-five seiners and only seven gillnetters.

The catch was 2,057 tons, but it was lower during the next five years (1999–2003), averaging 1,677 tons. And it would be 2004 before it once again surpassed 2,000 tons.[14] Ex-vessel prices, too, remained relatively low, averaging $573 per ton during those years.

The harvest in 2004, 3,167 tons, was a substantial increase over the previous years, and herring would remain relatively abundant over the next decade. The average GHL from 2005 to 2015 was 4,748 tons. However, reflecting the declining market for kazunoko due to the Japanese becoming more health conscious and eschewing the salted product, and the younger generation acquiring Western eating preferences, the ex-vessel price of herring was in a general decline, falling to $100 per ton in 2014. The ex-vessel value of the 2,463-ton catch that year was $246,300, the lowest since 1977.

Since 2014, there has been little interest in the Kodiak roe-herring fishery. The seine fleet mostly comprised three vessels, gillnetters were mostly absent, and the harvest has remained below 400 tons.[15]

13

TOGIAK ROE-HERRING FISHERY

But you never really know if the herring are even there. Then, one day, the ice goes out and there they are, millions of them. It's magic.—Bart Eaton, Togiak tenderman, 1983[1]

Man has fished for herring in the eastern Bering Sea for over 2,000 years, but until very recently, there has been little or no effort to manage the stocks. Stocks have historically demonstrated great fluctuations in abundance, but there is little biological knowledge concerning the cause of these year-class failures or of the total biomass.—Jeffrey Skrade, Alaska Department of Fish and Game, 1980[2]

[A conclusion] which can be drawn from the sad history of the North Sea herring [is]: Lack of scientific information and incomplete understanding of obvious events indicating a declining trend beyond the range of normal fluctuations of this size of the stock should not serve to prevent or delay efficient management action.—Albrecht Schumacher, Federal Research Centre for Fisheries, Federal Republic of Germany, 1980[3]

The Togiak area of Bristol Bay supports the largest spawning population of Pacific herring in the eastern Bering Sea. An ADF&G aerial survey in 1968 observed a school of herring that was two hundred yards wide and two miles long, and Kulukak Bay, which is approximately seven miles long and five miles wide, was turned completely milky white by spawning herring.[4]

Jeffrey Skrade, an ADF&G fisheries biologist in Bristol Bay from 1975 to 1994, recalled walking the beach with another biologist at Ungalikthluk Bay (just east of Togiak Bay) in the early 1980s after a big storm dislodged huge quantities of herring eggs from the bay's eelgrass beds. "The band of eggs along the shore was 20–30 yards wide and was well above our ankles in most places. We estimated that band of eggs to be 3–4 miles in length." Skrade added, "One can hardly imagine how many individual female herring it took to produce that many eggs."[5] The event exemplified how an environmental event can have a major impact on a region's herring biomass.

ADF&G's annual spawning biomass estimates for Togiak ranged from 69,000 tons observed in 1980 to 239,000 tons in 1979. The average from 1979 through 2017 was about 170,000 tons.[6]

https://doi.org/10.5876/9781646423446.c013

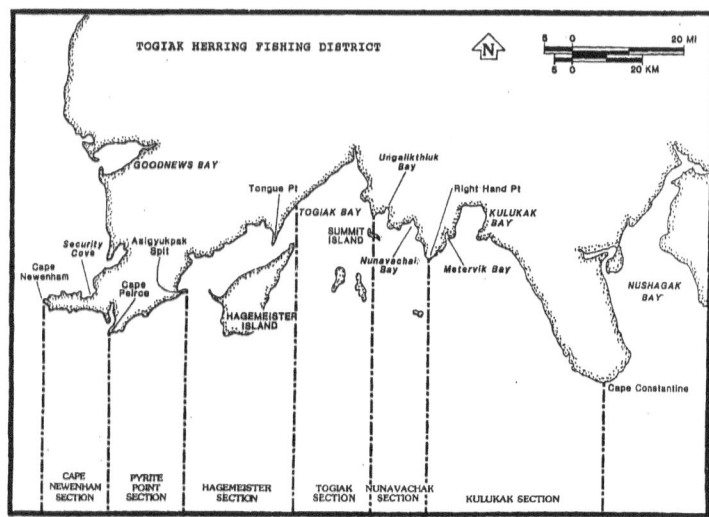

Figure 13.1. Togiak herring fishing district. (Source: Alaska Department of Fish and Game)

The roe-herring fishery that developed at Togiak, which began in earnest in 1977, became the largest herring fishery in Alaska. From 1979 through 2018, annual roe-herring harvests averaged approximately 20,000 tons.[7]

Unlike most Alaska roe-herring fisheries, the Togiak fishery is not a limited-entry fishery. The fishing effort varied considerably, with market conditions being a leading factor. Maximum effort was in 1996, when 268 purse seiners and 461 gillnetters participated in the fishery. And fishing could be very competitive—some characterized it as "combat fishing"—as well as intense: during a twenty-minute opening in 1992, purse seiners landed more than forty million pounds of herring.

Given the fleet's ability to catch huge quantities of fish in short periods of time, the challenge for ADF&G was to keep the fleet within its harvest limits. The department didn't have much infrastructure to work with. An indicator of Togiak's remoteness and the lack of facilities there was that ADF&G often held public meetings regarding the roe-herring fishery on various beaches in the area. The meetings were dubbed "beach parties." And one of the department's "offices" was a wall tent on Summit Island (figure 13.5).

For many department employees, outboard motor–powered open skiffs were the main means of getting around the Togiak area. Weather conditions, however, often limited skiff travel.

GENESIS OF TOGIAK'S ROE-HERRING FISHERY

In 1967, what ADF&G labeled an "experimental" roe-herring fishery was conducted at Togiak. During the fishery, which began on May 14 and lasted two weeks, nineteen Togiak gillnet fishermen delivered 269,000 pounds of herring to a single processor. The fishermen sorted their catch by sex, discarding the males and selling the females. The processor extracted the roe from the females, then discarded the carcasses. The product was very well received.[8]

The encouraging reception of Togiak herring roe might have been in part because the herring at Togiak are both larger and longer than other comparable stocks along the Pacific Coast. Most Togiak herring roe skeins are graded extra, extra large and are unique in the world market. Togiak herring then were also relatively old—94 percent of the herring sampled there in 1968 was age six through eleven, which was characteristic of an unexploited population.

Two processors participated in the 1968 season, but although the number of fishermen increased and two small purse seiners were employed, the catch declined to 182,000 pounds. As would be the case until 1973, all the fish sold were females. Males were discarded unweighed. Fishermen were paid $0.04 per pound ($80 per ton) for the herring.

At least one of the processors that year operated out of a herring camp set up at Kulukak Bay that consisted of several tents and a two-room plywood shelter. The routine was simple: freshly caught herring were deposited on shore and allowed to age for three or four days, after which the crew stripped the roe from the female fish and processed it for shipment. All male fish and female carcasses were discarded.

ADF&G characterized the fishery that year as "not overly profitable" for the processors, with the estimated first wholesale value of their roe production being $13,000. The department attributed the low economic return primarily to a low market price, low percentage of roe recovery (6.2 percent), and the high cost of operating in the area.[9]

And the situation would get worse before getting better. Production in 1969—all females—was 94,000 pounds, a little more than half that of the previous year. The developing fishery was, according to ADF&G, "plagued by operational difficulties, fluctuating seasonal abundance of spawning herring stocks, poor weather conditions during the fishery, an inability to forecast run timing to the extent necessary to conduct a feasible operation, and a very unstable market." A further discouraging factor, according to the department, was "the increased exploitation of these and other Bering Sea herring stocks by the Japanese high-seas herring fleet."[10]

The roe-stripping operation at Togiak in 1969 was aboard the 109-foot World War II–era power scow *Diver 1*, which carried a crew of about twenty, including four Japanese technicians who finished processing roe once it had been removed from the fish. The poor weather ADF&G referred to was a severe storm that drove much of the gillnet fleet ashore, where fishermen sheltered beneath overturned skiffs. It was a while before the *Diver 1* could even get food to the fishermen, and a while longer before the fishermen could get off the beach and resume fishing.[11]

The female component of the catch in 1970 was an even more discouraging 55,000 pounds, and in 1971, due to a lack of buyers, there was no roe-herring fishery at Togiak. The fishery resumed in 1972, with the female component of the catch being 162,000 pounds. Beginning in 1973, the Togiak herring catch record included both males and females. In 1973, it was 102,000 pounds; in 1974 it was 246,000 pounds.[12]

The roots of the dramatic expansion in the Togiak herring fishery, which began in 1977, can be traced to the failure of Japan's territorial herring fishery in the mid-1950s and the failure of the Soviet territorial herring fishery about fifteen years later. Both countries responded by sending their fleets to the central Bering Sea, particularly to an area northwest of the Pribilof Islands, where vast schools of herring wintered. The high-seas fishing effort was intense.

The Japanese/Soviet catch peaked in 1970 at 145,579 metric tons and then declined because of overfishing and poor recruitment.[13] In 1974, a National Marine Fisheries Service agent said he had "counted as many as 85 ships within a three-mile radius fishing for herring at the edge of the ice between the Pribilof Islands and St. Matthew Island."[14] And during the spring in some years, Japanese vessels were seen fishing for herring just beyond the twelve-mile limit off Togiak.[15]

The passage of the Magnuson-Stevens Fishery Conservation and Management Act in 1976 changed everything. With this legislation, the United States unilaterally established a 200-mile-wide fishery conservation zone along its entire coast, thereby creating the most extensive maritime domain of any country in the world.[16] The Bering Sea herring grounds long exploited by the Soviets and Japanese were now off-limits to foreign fishermen.

Enter Shigeyoshi Kitano, "Captain K," as he was later called (figure 13.2). In 1950, when Kitano was eighteen years old, he started his own fish company on Hokkaido, Japan's northernmost main island. His company went bankrupt, and Kitano turned to fishing for cod and pollock in the Bering Sea. After a number of years as a fisherman, Kitano became a partner in a processing company. According to Kitano, the company tried to grow too fast, and in 1967 it went bankrupt.

Figure 13.2. Shigeyosi Kitano (second from left) with Native corporation representatives at Kotzebue in the mid-1980s. (Terry Reeve)

In the late 1960s, Kitano came to Alaska as a fish buyer/technician for the Marubeni Corporation. Kitano, however, also had his own company, Kyokko Suisan Alaska, and soon transitioned to purchasing fish on his own account. He was primarily interested in salmon and herring, and his main supplier was Icicle Seafoods, an ambitious company that had fish-processing plants in Petersburg and Seward. (Kitano was part owner of Icicle Seafoods' first floating freezer-processor, the *Alaska Star*.)

In 1982, Bob Brophy, president of Icicle Seafoods, praised Kitano's understanding of Alaska's herring fisheries. "Wherever a new herring fishery has been pioneered in Alaska, Kitano was there," said Brophy.[17]

A combination of factors in 1976 fostered the development of the Togiak herring fishery:

- Japan's economy was booming, and there was high demand for seafood.
- The value of the US dollar was low relative to the Japanese yen, which made US exports especially attractive in Japan.
- There was a worldwide shortage of herring, thought to be the result of overexploitation in traditional areas and recruitment failure or adverse environmental changes.
- The termination of Soviet and Japanese herring fishing in what were now US waters in the Bering Sea made it likely that the herring biomass there would increase.[18]

A profusion of Bering Sea herring spawned along Alaska's west coast, especially at Togiak. Kitano saw opportunity in developing a roe-herring fishery there, but there was a problem: the waters at Togiak that herring favored for spawning were shallow and rocky, and purse seining—the best method for obtaining the large quantities of herring necessary for the operation he envisioned—might not be feasible.

Indeed, in 1975 and again in 1976, individual herring seiners had tried to make a go of it at Togiak. And both failed. The 1976 effort was by Seldovia-based Tuggle Entout. Herring were plentiful that year, but the tender contracted to deliver Entout's fish to the Togiak processing plant drew twelve feet of water, too much to even get close to the plant. Unable to fish commercially, Entout made a set and invited the local people, for whom herring were an important subsistence food, to come out in their skiffs and take all they wanted.[19]

To answer his feasibility question, Kitano turned to Beaver Nelson, the ambitious Homer-based purse seiner and herring processor who had helped pioneer the Cook Inlet, Resurrection Bay, and Prince William Sound roe-herring fisheries. In June 1976, Kitano chartered a small floatplane to take him and Nelson on a tour of the Togiak area.

The waters there were indeed shallow and rocky, but Nelson assured Kitano that with the right boats and gear, they were fishable. Kitano and Nelson decided to work together to establish a herring operation at Togiak in 1977.[20] Regulatory hurdles then were minimal. ADF&G's Bristol Bay herring regulations that year comprised one sentence: "There is no closed season on herring."[21]

Spawning herring typically arrived at Togiak about May 10. It would be tight, but Nelson could participate in the Prince William Sound fishery, which was typically over by May 1, then make haste for Togiak.

Figure 13.3. Beaver Nelson's "jitney" herring seiner, Togiak, Alaska, 1977. (Ken Moore)

It usually took about a week to run a boat from Homer to Togiak, with almost half the time spent waiting for suitable traveling weather. Indeed, in 1979, a severe storm during the last week of April delayed the arrival of most of the herring fleet until after peak spawning had occurred.

Catching herring is one thing, but the fish would also need to be processed. In 1975, Nelson had purchased a 160-foot military-surplus landing craft and converted it into a liveaboard floating herring-processing plant. The vessel, christened the *Maren 1*, was unpowered and had been last used to carry canned salmon from Alaska to Seattle. Operating as Nuka Point Fisheries, Nelson roe-stripped herring in Prince William Sound and Lower Cook Inlet (Kamishak Bay) during the 1976 herring season. As part of a joint venture with Kitano and Icicle Seafoods, the *Maren 1* would process herring at Togiak in 1977.

The vessel *Totem* towed the *Maren 1* to Togiak. On the *Maren 1*'s deck were two Homer-built twenty-seven-foot aluminum "jitney" seiners powered by 350-horsepower inboard gasoline engines (figure 13.3). Each boat had a small cabin forward, and in it were a marine toilet and a bench for the crew to huddle on when trying to warm up around the engine.

Nelson operated one of the seiners; another Homer fisherman, Ken Moore, operated the other. Icicle Seafoods contributed to the effort two tenders, the eighty-six-foot *Chichagof* and the ninety-one-foot *Viking Queen*, and five small gillnetters.

The *Maren 1* arrived at Togiak on May 15, a little late for the best fishing.[22] But it was a good year to start out: ADF&G aerial surveys that spring revealed "some of the largest and most numerous concentrations of herring schools ever observed" since 1967, when the department initiated aerial herring surveys. Moreover, good weather prevailed.[23]

Together in their small boats, Nelson and Moore caught an impressive 1,300 tons of herring. But it wasn't easy.

There were no bathymetric charts of the area, only maps showing the headlands, and the men spent most of their time finding out the hard way where they could make sets. According to Nelson, the most beautiful beaches had the worst snags and, complicating matters, strong tidal currents to contend with. The only way for the men to find out if a particular location was fishable was to lay the seine out around a school of fish and see if they could get it back without a catastrophe. Time and time again, the seine came back shredded, with only a tiny amount of herring. Not a day went by during which the crews didn't repair huge rips and tears in the net or broken purse lines or leadlines.

But every now and then, they managed a clean set around a big school of fish, with multiple skiffs helping keep the corkline from being pulled under by the weight of the fish, and a tender standing by. Everyone relished such moments.

About fifteen years later, after considerable prodding by ADF&G, the National Atmospheric and Oceanic Administration surveyed the Togiak area and published detailed marine charts. What Nelson termed his "hardknocks knowledge" of the area immediately lost much of its value.[24]

The tenders delivered Nelson and Moore's fish to the *Maren 1*, where they were pumped into a concrete mixer–like tumbler—Nelson referred to it as a "salting trommel"—that mixed herring and salt at a ratio of one ton of herring to fifty pounds of salt (figure 13.4). The salted fish were then transferred to a waiting Korean freighter, where they would supposedly cure while en route to Asia.[25]

The *Maren 1* did not long have the Togiak fishery to itself. About a week later, the 350-foot floating processor *All Alaskan* arrived with four Kodiak-based seiners. The *All Alaskan* froze herring in blocks.

Total production by the six seiners and forty-three gillnetters that fished herring at Togiak in 1977 was 2,795 tons, a fortyfold increase over the average annual catch of the previous years. Seiners accounted for fully 83 percent of the harvest.[26] The *Maren 1* continued to operate in the herring fishery at Togiak through the 1985 season, though the tumbler operation was abandoned after the 1979 season.

Two methods were used to process the fish: salting and freezing. Salting was the least expensive, but there was potential for damage to the roe from insufficient salt penetration, and high salt content limited the use of the carcasses after the roe was stripped.

Freezing the herring in the round was more expensive but was preferred, in large part because the carcasses could be better utilized. The frozen fish were exported to Japan, Korea, and China for processing.[27]

Figure 13.4. Salting trommel aboard *Maren 1*. (Beaver Nelson)

The Togiak roe-herring fishery would continue to grow and would become the largest herring fishery in the Pacific Ocean.[28] Its growth was bolstered by the strong 1977 herring year class at Togiak. Adding to this good fortune, the 1978 year-class was also strong. These year-classes entered the fishery in 1982 and 1983 and represented a substantial component of the herring biomass at Togiak until the early 1990s. Beaver Nelson would be there until 2007.[29]

The following paragraphs present a chronicle of the Togiak roe-herring fishery since 1978.

1978

In 1978, the seine fleet increased to twenty-five vessels, while the gillnet fleet, numbering forty vessels, remained about the same. Together, they harvested 7,734 tons of roe herring, nearly trebling the previous year's harvest. With its increased number, the seine fleet accounted for 91 percent of the harvest. Fully sixteen operators purchased herring. They primarily froze the fish, after which they were transferred to eleven foreign freighters anchored in Kulukak Bay.

All of the resident Bristol Bay fishermen were gillnetters, and they lobbied hard for a share of the developing bonanza. There was, however, no local knowledge regarding how to harvest herring with good roe content, so the local people brought in an "expert" who had experience gillnetting for herring at Kah Shakes, in Southeast Alaska.

He advised the fishermen to purchase gillnets with the same mesh size as were used at Kah Shakes. It was an expensive mistake. The herring at Kah Shakes are much smaller than those at Togiak, and the nets proved especially effective at catching immature and spawned-out herring.

Because the gillnet fleet tended to deliver poor-quality fish, processors gravitated toward purchasing seine-caught fish, which could be tested and rejected with minimal dead loss. The gillnet fleet, however, successfully lobbied the Alaska Board of Fisheries for a provision that "whenever possible," openings for gillnet fishermen should precede those for seine fishermen.

For ADF&G, the rapid growth of the fishery combined with its scant knowledge of Bering Sea herring stocks warranted a conservative management approach. Based upon aerial surveys, the department estimated that the 7,734-ton harvest constituted 2.2 percent to 4.0 percent of the spawning biomass at Togiak, well below the rates of exploitation permitted in most other Pacific herring fisheries.[30] Since 1988, ADF&G harvest policies applicable to herring in Alaska set the maximum exploitation rate at 20 percent of the exploitable or mature biomass.[31]

Prior to 1978, regulation of the fishery had been essentially nonexistent. Given the substantial increase in fishing effort in 1977 and the potential for even more effort in the future, in 1978 ADF&G began limiting fishing by scheduling fishing seasons, specifying the types and quantity of gear allowed (purse seines and gillnets, each no longer than 150 fathoms), restricting fishing to specific areas, and setting guideline harvest levels.[32]

ADF&G's center of operations at Togiak in 1978 was a tent camp at Metervik Bay. Department employees used outboard motor–powered Boston Whaler–brand fiberglass skiffs to get around.[33]

The 1978 roe-herring season at Togiak was not without tragedy. On May 30, a midair collision of two spotter planes above Kulukak Bay killed three men. One was Ray Anderson, whose Seward Marine Services had been engaged in processing roe herring since the late 1960s.[34]

1979

> The domestic herring fishery in eastern Bering Sea has developed in response to favorable market conditions and prices created by a worldwide herring shortage... The high prices paid for sac roe herring in 1979 has made herring fishing one of the most lucrative fisheries in the world.—Alaska Department of Fish and Game, 1979[35]

Figure 13.5. ADF&G Summit Island tent camp, circa 1990. Standing outside of tent, left to right: unidentified person, Chuck Meacham, Ken Florey, Norman Cohen, Dick Russell, unidentified person, Jeffrey Skrade, Bud Hodson. Sitting: Kathy Rowell. (Bob King)

The 1979 season saw a vast increase in effort. ADF&G estimated that the fishing fleet comprised 175 seiners and 350 gillnetters, and 33 processors were on the grounds. This number of processors was never exceeded, though it was equaled in 1982. The fleet's record catch of 11,126 tons was not nearly proportional to the increased effort. Nor was it an indicator of the quantity of herring present: ADF&G estimated the biomass of the spawning herring at Togiak in 1979 to be 239,000 tons, a record that still stands today.

Three factors were at play in limiting the catch. First, the herring run was early, and a severe storm during the last week of April delayed the arrival of most of the fleet until after peak spawning had occurred. Second, frequent storms during the opening hampered purse seining operations. Third, the great number of boats on the fishing grounds broke herring schools into smaller units, making them less accessible to purse seiners. Gillnetters accounted for fully 40 percent of the season's catch, as compared to 8 percent the previous year.

Although the price paid for fish varied widely, the average price per ton was $637. Thus, the ex-vessel value of the Togiak roe-herring fishery in 1979 was more than $6.5 million.[36]

1980

In February 1980, the State of Alaska, the North Pacific Fishery Management Council, and the National Marine Fisheries Service sponsored a three-day Alaska herring symposium in Anchorage. Scientists and fishery managers from Russia, Germany, Norway, Canada, and the United States presented papers on herring research and management practices. The objective was to review the research and management of Alaska herring in relation to herring research and management in other regions of the world and to learn from other countries' successes and failures. Of particular interest was how these might relate to management of the Bering Sea herring fisheries, which Jim Edenso, the state's Bottomfish Coordinator, characterized as being in their "embryonic growth stage."[37]

In a letter thanking Alaska governor Jay Hammond for his support of the symposium, Jim Branson, executive director of the North Pacific Fishery Management Council, noted that a main point of consensus among the participants was "the variability of the behavior of the herring and the need for a cautious and flexible management strategy."[38]

For the 1980 season, ADF&G established a GHL of 20,000–40,000 metric tons for the Togiak area. The fishery opened on April 15, as scheduled, but it was April 25 before the first commercial delivery was made. The season progressed well until it was closed by emergency order on May 10. The number of participating boats—140 seiners and 363 gillnetters—was similar to the previous year, but the catch increased almost 70 percent, to 18,886 tons. Purse seiners accounted for fully 84 percent of the catch.

Despite the large increase in the catch, the value of herring roe had plummeted. Fishermen were paid on a sliding scale, based on $200 per ton for fish that yielded a 10 percent roe recovery. This on-the-grounds price was typically augmented by a postseason adjustment—a *retro*, in the parlance of fishermen—based on finalization of roe prices in Japan.

Roe recovery ranged from 8 to 11 percent. The total ex-vessel value of the catch, a little more than $3 million, was less than half that of the previous year.

And there was a lot of waste. ADF&G estimated during the season that 2,500 metric tons had been wasted, but a postseason questionnaire sent to fishermen indicated that over 5,200 metric tons—more than eleven million pounds—were lost. The department attributed the waste to "(1) unsalable fish due to poor roe recovery; (2) a lack of adequate markets to handle the capacity of the fishing fleet; (3) a lack of shallow-draft tenders; (4) weather-related problems; (5) inexperience on the part of fishermen, processors, and spotters."[39]

Some inexperienced fishermen were also greedy. They made huge sets in shallow waters, and the sets went dry while the fishermen waited for a tender. If the fish proved to be immature or spawned out, however, the fishermen pulled out, leaving the fish dead on the bottom. Aerial surveyors referred to the large circular piles of dead herring lying on the seafloor as "silver diamonds." This waste did not go unnoticed by the villagers at Togiak, who claimed that the dead herring had fouled their bay and would prevent salmon from returning the following month. For ADF&G, it was a public-relations nightmare, and the department later closed the heads of some bays to fishing to prevent such waste.[40]

Waste was part of the largest set—about 700 tons—that Beaver Nelson ever saw at Togiak. It was made during a falling tide in shallow water at Tongue Point. When the tide went out, the water was only two feet deep, and the top of the bag of herring was four feet above the surface of the water. The quality of the fish suffered in big sets such as this because the weight of the mass of fish squeezed the roe out of the females.[41]

Another factor causing the waste of herring was processors' preference for seine-caught fish. It was more efficient for them to deal with the relatively few but large volumes of fish caught by individual seiners than the numerous small volumes gillnetters delivered. During the 1979 and 1980 seasons, gillnetters were sometimes forced to dump their catches because processors were too busy with seine-caught fish. An attorney for the gillnetters later told Congress that the quantity wasted by dumping in 1980 was more than 5,000 metric tons.[42]

Icicle Seafoods had a major problem one year in the early 1980s, but, fortunately, no fish were wasted. The company had its floating processor *Bering Star* on site. The vessel could freeze about 300 tons of herring per day. At one point, however, the company's tenders had over 3,500 tons onboard. Some of the tenders took their loads to Seward for processing, but there were still more fish than the *Bering Star* could efficiently handle. It all worked out, but it wasn't pretty, and the quality of the fish suffered. On the last few tenders to be unloaded, the fish had compressed into a solid mass, and getting them off the boats required the use of picks and shovels.[43]

Limiting entry into the fishery would likely have at least lessened some of the problems at Togiak, but the Alaska Commercial Fisheries Entry Commission, which had in 1977 limited entry into the roe-herring purse-seine fisheries in Southeast Alaska, Prince William Sound, and lower Cook Inlet, had chosen not to limit the Togiak fishery, which was then in its infancy. The fishery, as explained above, had grown quickly—from six purse seiners in 1977 to 175 in 1979—and would have been a good candidate for inclusion in the limited-entry program. Though some seiners over the next four decades periodically lobbied the commission to do so, for various reasons that are beyond the scope of this book, the commission elected not to limit the fishery, and to this day it remains open access.[44]

During its 1980 session, Alaska's Legislature, which has always worked to provide opportunities for Alaskans in the commercial fishing industry, passed what some called the "in-state herring processing bill." The legislation made it unlawful for a person to remove herring from the state before it had been frozen or otherwise processed for shipment. The intent was to end the practice of salting herring, then loading it aboard foreign freighters, as the *Maren 1* had been doing. The law was scheduled to become effective September 1, 1981, but, as explained below, it was overruled in federal court.[45]

1981

The big management change in 1981 was the regulation of fishing time by emergency order. Fishing periods opened once ADF&G aerial and marine surveys and test fishing by volunteer fishermen showed adequate amounts of marketable fish. Marketability was determined by a gonad index that rated egg maturity. ADF&G closed the fishing period when the amount of herring present diminished. For the 1981 season, May 2 through May 12, ADF&G allowed six openings, with a total fishing time of 101 hours.

Because fishing was channeled into discrete periods, a more normal onshore migration of herring occurred, which enabled more extensive undisturbed spawning and enhanced ADF&G's ability to assess the stocks. The department's employment of a chartered helicopter facilitated its ability to locate schools of herring and estimate their biomass, to collect samples to evaluate roe maturity, and to monitor the fishery.

There was a marked decrease in the number of vessels that participated in the fishery: only 83 purse seiners and 106 gillnetters, decreases of 70 percent and 40 percent, respectively, from 1980.[46]

As noted above, roe-herring processors at Togiak were reluctant to purchase fish from gillnetters. The Japanese, however, stepped in and offered to bring freezer ships to Togiak to purchase and process the gillnetters' herring. The processing of US-caught fish aboard foreign ships was legal under a provision of the Magnuson-Stevens Fishery Conservation and Management Act, which gave Alaska's governor authority to permit a foreign vessel to engage in fish processing within state waters under certain circumstances.[47] Alaska statute also authorized the governor to do so under certain circumstances.[48] At the same time, however, the Alaska statute prohibited the export of herring from the state before it has been "frozen or otherwise processed for shipment."[49]

In February 1981, the Bristol Bay Herring Marketing Cooperative and the Western Alaska Cooperative Marketing Association challenged the state law in federal court as being in violation of the Foreign Commerce Clause of the US Constitution.[50] The court agreed, and a US-Japanese joint venture, the Alaska Herring Corporation, was soon established. Principals in the venture were the Bristol Bay Herring Marketing Cooperative and the Japanese North Pacific Longline Association.

The association sent seven Japanese high-seas longliners to Togiak to freeze herring that spring. They were referred to as the "pumpkin boats" because they were all painted orange.[51] In 1982, a dozen Japanese longliners returned to Togiak to freeze herring. The following year, the number increased to fourteen.

The Alaska Herring Corporation joint venture, utilizing varying numbers of longliners, would continue through the 1987 season. By that time, gillnetters had gained more experience and were able to target the female fish that followed the males to the spawning grounds. As well, some had begun using nets with a larger mesh, which allowed small herring to pass through but caught the large, more valuable fish.[52]

By the season's end, fishermen had landed 12,542 tons of herring, a small fraction of which was used for food and bait. Wastage of herring was estimated to be less than thirty metric tons. The average on-the-grounds price paid to fishermen was $350 per ton for herring with a 10 percent roe recovery.[53]

1982

The fishing effort increased in 1982, with an estimated 135 seiners and more than 200 gillnetters participating. ADF&G spread fishing over three openings, one of which was extended. For seiners, the openings totaled thirty-six hours; for gillnetters, sixty hours.

The fleet set a record, catching 21,489 tons of herring, 69 percent of which was taken by seiners. The catch represented about 22 percent of estimated total biomass, somewhat above the maximum 20 percent exploitation rate that the Board of Fisheries would establish in 1987.

The average on-the-grounds price paid to fishermen was the same as the previous year: $350 per ton for herring with a 10 percent roe recovery.[54]

1983

> In late April and early May of 1983, a new floating city will appear in western Alaska or more precisely, immediately offshore of the northern portion of Bristol Bay near Togiak. This maritime community, probably numbering in excess of four thousand hearty individuals, springs anew each year, lured by the annual return of roe-laden herring to near-shore spawning areas. A month later, both community and fish have departed.—Charles Meachan, ADF&G, April 1983[55]

The Togiak fishery and the harvest continued to grow. The only thing that went down was the amount of fishing time allowed. For the 150 seiners present, fishing time was limited to fourteen hours, spread over four openings. For the estimated 250 gillnetters, fishing time was limited to forty-two hours spread over four openings. The total catch was 26,287 tons, a record that would hold until 1994. In addition to the reported harvest, an estimated 544 metric tons was lost, mainly due to accidents in the fishery and to abandoned gear.

Seiners were responsible for approximately 81 percent of the harvest. The average on-the-grounds price paid to fishermen for herring with a 10 percent roe recovery increased a bit, to $400 per ton. The total ex-vessel value of the fishery was $10.5 million.[56]

1984

In 1984, ADF&G estimated that the number of seiners and gillnetters in the Togiak roe-herring fishery had grown to 196 and 300, respectively, and the amount of time each group was allowed to fish declined to eleven hours and thirty-seven hours, respectively. The catch, 75 percent of which was made by seiners, declined significantly, to 19,300 tons. It had a total ex-vessel value of $7.2 million. Waste, too, declined and was less than 136 metric tons. Of this, about fifty metric tons was aboard a tender that sank.

Two environmental issues that had for years marred the Togiak roe-herring fishery were numerous small oil spills and trash—including lost or abandoned gillnets. Though personnel from the Alaska Department of Environmental Conservation and the US Coast Guard were

Figure 13.6. "Combat" herring seining, Togiak. (Ken Moore)

some years stationed on the fishing grounds, their presence did little to deter vessels from pumping oily bilge water overboard or dumping trash. In 1984, ADF&G expressed the need for an aggressive program to keep the waters and beaches clean before there was a "serious negative impact on the local environment." A voluntary trash cleanup on Nunavachak Bay beaches resulted in the removal of over thirty cubic yards of material, which was transported by tender to Dillingham, where it was landfilled. Additionally, ADF&G enforcement officers collected sixteen lost or abandoned gillnets.

On a sad note, two spotter pilots were killed when their planes collided south of Osviak Point just minutes before the season was scheduled to open. At the time of the collision, about thirty spotter planes were in the air.[57]

The approximately 600 people who live in Togiak were generally not welcoming of the big fishery that had developed at their doorstep. The usually quiet location was, at least during the brief roe-herring season, abuzz with activity akin to an invasion. In addition to the about 500 fishing vessels on nearby waters, there were perhaps 100 tenders, two dozen floating processors, and maybe ten foreign tramp freighters. Perhaps 150 spotter planes were at times in the air. An ADF&G employee estimated that at one point some 6,000 people (relatively few of whom were Bristol Bay–area residents) were involved in the Togiak roe-herring fishery. Were the fishery a city, it would have been Alaska's seventh largest. And it came at a price. Besides the aforementioned oil spills and trash, there were tundra fires, trespass issues, grave robbing, and disturbance of archaeological sites.[58]

Liquor was also a problem. Togiak was (and still is) a dry community in which the possession of liquor is prohibited. Nevertheless, fishermen and processors often brought liquor with them, and some processors reportedly gave beer and liquor as a bonus to fishermen who sold them their catch. In 1987, three people were killed in a drunken gunfight at a fish camp.[59]

1985

The seine and gillnet fleets were about the same size as the previous year. What was different was the duration of the openings. In 1985, seiners were limited to two openings that together totaled just three hours of fishing. Seiners caught 83 percent of the total harvest of 25,616 tons, almost half of the total Alaska roe-herring catch in 1985.

Gillnetters likewise had two openings, but theirs totaled eleven hours of fishing. The average on-the-grounds price paid to fishermen for herring with a 10 percent roe recovery was $571 per ton. ADF&G estimated the total ex-vessel value of the catch to be a record $13.8 million. The high price reflected a shortage of British Columbia product.

Of the twenty-three operators that purchased herring that year, all but two were "floaters" that froze the fish aboard or tendered them to shore plants as far away as Seward and Akutan. The two shore-based plants at Togiak were Kemp & Paulucci Seafoods and Togiak Fisheries.[60]

1986

About 200 seiners fished at Togiak in 1986, the same number as in the two previous years. The number of gillnetters, however, declined from approximately 300 in 1985 to 200 in 1986.

In its 1986 Bristol Bay management report, ADF&G noted that the Togiak herring fleet's efficiency had "increased tremendously." The overall seiner catch per vessel per hour of fishing time in 1984 and 1985 had averaged forty-one metric tons and six metric tons, respectively, while in 1986 the average was fifty-six metric tons. For gillnetters, the overall catch per vessel per hour of fishing time in 1984 and 1985 had averaged 1.2 metric tons and 0.5 metric tons, respectively. In 1986, the average was 1.5 metric tons.

The department attributed the increased efficiency to several factors, including a few days of good weather, fishermen's increased experience, and large volumes of herring in nearshore areas. On one day, ADF&G aerial surveys recorded 56.2 linear miles of herring spawn in the Togiak District.

Seiners, in particular, had to be efficient. Their entire season at Togiak in 1986 amounted to a pair of thirty-minute openings. During their hour of fishing, the seiners landed 12,845 tons of herring, 79 percent of the 16,260-ton total roe-herring harvest. Gillnetters, who were granted two ten-hour openings, landed 3,415 tons.

The estimated total ex-vessel value of the fishery was $8.6 million.[61]

1987

The herring run in 1987 was more than a week earlier than expected, and only thirty-three seiners and forty-six gillnetters were on the grounds when the first openings occurred on April 27. Their catch totaled less than a thousand tons. Because some fishermen were unable to make it to Togiak on time, at the peak of the season there were an estimated 111 seiners and 148 gillnetters on the grounds, fewer than the previous year. And there were fish to be had. Aerial surveys documented herring spawn along a record 75.8 miles of shoreline.

For the season, seiners had six openings that totaled 6.5 hours, while gillnetters had five openings that totaled thirty-five hours. Their combined catch, 15,204 tons, however, was less than expected and was probably attributable to fishermen being caught unprepared for the early run. Seiners were responsible for 83 percent of the catch.

The estimated ex-vessel value of the fishery was $8.6 million, the same as in 1986.

As noted above, 1987 was the last year of operation for the Alaska Herring Corporation Japanese American joint venture. Herring caught by the venture's gillnetters were frozen aboard eight Japanese longliners.[62]

In the fall of 1987, the Board of Fisheries limited the legal size of purse seines in the Togiak roe-herring fishery to 100 fathoms (600 feet) in length and sixteen fathoms (96 feet) in depth. Similarly, the board reduced gillnet length limit to 100 fathoms, but it also gave ADF&G the authority to reduce length to fifty fathoms in-season by emergency order if necessary.[63]

In 1995, to further control harvesting capacity, the board reduced the purse seine depth to 625 meshes, of which 600 could be no larger than 1.5 inches. Six hundred 1.5-inch meshes equals seventy-five feet. This restriction was moot because the waters primarily fished at Togiak were shallow, and the seines preferred by fishermen were about sixty feet deep.[64]

1988

The 1988 season was a hurried affair. On the afternoon of May 16, a large body of herring moved ashore and began to spawn. Early the following morning, ADF&G's aerial survey showed 22.75 miles of spawn in the area from Kulukak Bay to Anchor Point. Any delay in opening the fishery would reduce roe recovery, so at 8:00 a.m. the department announced a four-hour gillnet opening in the Togiak District beginning at 10:30 a.m. The department had planned to open the seine fishery later that day. Doing so would facilitate fishery enforcement and sampling efforts and would help the industry deploy its tenders more efficiently. The weather—specifically fog—squelched the plan.

A large fog bank was moving in from the east and threatened to envelop the area. The same pattern had occurred two days earlier, and the fog then persisted for almost twenty-four hours. Fog would ground spotter planes and make for hazardous conditions on the waters. If the fishery was not opened until the fog dissipated, however, a large volume of the herring present would likely spawn, rendering them unmarketable.

ADF&G managers made a quick decision and announced a thirty-minute purse-seine opening beginning at 12:30 p.m. The fishermen went to work, and within a half hour of when the opening ended, Togiak Bay was entirely enveloped in fog. But fishing was good: in their brief fishing frenzy, seiners caught 10,614 tons of herring. During their four hours of fishing, gillnetters landed 3,474 tons.

And the 1988 season was over. It was a record one-day catch at Togiak, and processors struggled to get it all processed before the fish started degrading.

Dean Anderson, a seiner from Seward, did especially well. A set he made off Tongue Point caught 400 tons of herring, which at the prevailing price was worth about $360,000. It took two days to pump the fish—which were sold to five companies—out of the net.

ADF&G estimated that the 14,088-ton total harvest—the smallest since 1981—had an ex-vessel value of $14.7 million, the highest ever.[65] The high value of the catch was attributable to a strong Japanese market, which propelled the average price paid to fishermen to $1,030 per ton, double what it had been the previous year.[66]

1989

The high price of roe herring in 1988 attracted additional fishermen. In 1989, fully 630 vessels—310 seiners and 320 gillnetters—engaged in the fishery, almost a hundred more than in 1988. The 310 seiners represented a peak. Another factor in the increase in the effort was that the roe-herring fisheries in Prince William Sound and lower Cook Inlet had been canceled after the *Exxon Valdez* oil spill (see chapter 11), and some of the fishermen from these areas attempted to salvage their season by going to Togiak. As noted above, entry into the Togiak roe-herring fishery was—and is—not subject to the state's limited-entry program.[67]

In contrast to the increased number of fishing vessels, there was a pronounced shortage of tenders at Togiak because many had been chartered by the Exxon Corporation to assist in the Prince William Sound oil cleanup effort. One processor said he had only about half of the approximately twenty tenders he normally employed.[68]

An element of uncertainty pervaded roe-herring fisheries that year. Emperor Hirohito of Japan had died in early January, and the traditional Japanese year-end gift-giving and celebration were reduced. How long demand for items such as kazunoko would remain stagnant and the impact on the price of roe herring were unknown. On top of this, the Japanese culture was continuing to change: much of the younger generation increasingly eschewed traditional customs and favored Western food over fare such as kazunoko.[69]

At about the same time, the Board of Fisheries issued a directive that allocated 75 percent of the "harvestable surplus" (guideline harvest level, quota) of roe herring at Togiak to the purse seine fleet and 25 percent to the gillnet fleet—almost exactly the same distribution as had occurred during the 1988 season. To achieve this ratio, ADF&G adjusted fishing time and areas for each gear type.[70]

During the 1989 season, gillnetters had two openings that totaled five hours of fishing. Seiners had two openings with a total fishing time of three hours. Together, the gear groups landed 12,258 tons of roe herring, of which 9,414 tons (76.5 percent) was caught by purse seiners and 2,844 tons (23.5 percent) was caught by gillnetters.

Unfortunately, the price of roe herring had fallen to $425 per ton, less than half of what it had been the previous year but more in line with historical prices. ADF&G estimated the total ex-value of the catch to be about $5 million—about a third of what it was in 1988.[71]

1990

The 1990 season was pretty much business as usual, though the effort, 277 purse seiners and 221 gillnetters, was slightly above average. The seiners' season consisted of a single three-hour opening on May 9, while gillnetters had six openings between May 8 and May 20 that totaled sixty-six hours. The total harvest of 12,253 tons was almost identical to that of 1989. Seiners landed 9,240 tons (75.4 percent), while gillnetters landed 3,013 tons (24.6 percent).

The ex-vessel value of Togiak herring increased somewhat, to approximately $530 per ton, and ADF&G estimated the total ex-vessel value of the fishery to be $6.5 million.[72]

A two-day beach cleanup during the season produced 313 large plastic bags of trash, which was deposited at the Togiak landfill. About a hundred fishermen participated in the cleanup.[73]

An important development in 1990 was the North Pacific Fishery Management Council's decision to limit bycatch of herring by US trawlers fishing in the Bering Sea. The Magnuson-Stevens Act defines bycatch as "fish which are harvested in a fishery, but which are not sold or kept for personal use, and includes economic discards and regulatory discards."[74]

Trawlers, especially those targeting pollock (*Gadus chalcogrammus*) with midwater trawls, often caught herring as well, and prior to the council's decision, there was no limit on the quantity of herring they could take. The National Marine Fisheries Service estimated that herring bycatch by trawlers in 1989 was 4,521 to 5,301 metric tons, but it believed this estimate was low because of unrecorded at-sea discards.

At its September 1990 meeting, the council adopted an amendment that triggered potential trawl closures in three herring savings areas for the remainder of the year or season when 1 percent of the estimated eastern Bering Sea herring biomass was taken as bycatch. Two of the areas—the summer herring savings areas—are along the Aleutian Islands and Alaska Peninsula and total approximately 20,700 square nautical miles. The third area, the winter herring savings area, is northwest of the Pribilof Islands and totals approximately 11,100 square nautical miles. The 1 percent bycatch limit for herring, in metric tons, is specified annually, based on abundance and spawning biomass as determined by ADF&G.

The limit for 1991, the first year the rule was in effect, was 834 metric tons. Pollock trawlers exceeded that limit, and the winter herring saving area was closed to trawling for pollock from September 21, 1991, until March 1, 1992.[75]

1991

Though participation in the 1991 fishery, especially for gillnetters, was considerably less than in 1990, the fleet caught considerably more fish.

Targeting herring were 170 gillnetters—more than 100 fewer than in 1990—and 200 purse seiners. Together, they landed 14,970 tons of roe herring.

Gillnetters had openings on May 10–11 and fished for a total of fourteen hours, landing 3,182 tons (21.3 percent of the total roe-herring harvest). Purse seiners had a one-hour opening on May 10 and a two-hour opening on May 12, landing 11,788 tons (78.7 percent of the total roe-herring harvest). Waste among gillnetters was minimal, but two purse-seine sets made in shallow water were unrecoverable and resulted in a loss of approximately 100 tons of herring. ADF&G estimated the value of the roe-herring fishery to be $6.2 million.[76]

1992

Spawning herring were abundant at Togiak in 1992. ADF&G documented 97 linear miles of milt, the most extensive spawn observed since 1978, when the spawn-survey program began. The department's final estimate of the spawning biomass was 156,995 tons—the highest since 1981 and more than double the preseason forecast.

The number of fishing vessels at Togiak was substantially greater than in the previous year: 301 purse seiners and 274 gillnetters. The number of gillnetters matched the ten-year average, but the number of purse seiners was 50 percent greater than that average. Fish-processing capacity, however, did not increase commensurately. There were eighteen processors, two more than in 1991.

Gillnetters had 25.5 hours of fishing spread over seven openings and landed 5,030 tons of herring, a bit short of their 5,662-ton GHL and 19 percent of the total roe-herring harvest at Togiak that year.

The entire season for purse seiners was a single twenty-minute opening on May 20. The weather, with clear skies and calm waters, was excellent for fishing, and they landed 20,778 tons of roe herring, substantially exceeding their 16,985-ton GHL. The Togiak harvest was a major factor in 1992's post-reduction-plant-era record for Alaska herring production—71,608 tons.

It was also far larger than processors at Togiak could immediately accommodate, and numerous fishermen—likely those who had come to Togiak without a prearranged market for their fish—were forced to hold their catch in their seines for up to a week. The result was large-scale deterioration of the flesh and roe, and product quality suffered. Moreover, a lot of dead herring were dumped. An ADF&G

aerial survey over several days observed more than a dozen piles of herring, and the department's "conservative" estimate was that 1,371 tons of seine-caught herring were wasted. As well as being a crime against the environment, the waste had the potential to mar Alaska's reputation for responsible fishery management. Almost no waste was observed in the gillnet fishery.

The estimated ex-vessel value of the catch was $8.8 million.

1993

A worldwide glut of herring roe in 1993 created uncertainty regarding whether roe-herring processors would operate at Togiak.[77] Despite the uncertainty, a dozen processors, six fewer than in 1992, chose to do so. Fishing effort, too, was greatly reduced: 140 purse seiners and 75 gillnetters participated in the fishery.

ADF&G's first priority for the Togiak roe-herring fishery in 1993 was to not repeat the wastage that had occurred in 1992.[78] Complicating matters was the fact that the 1993 herring run was projected to be large, about the same as in 1992. The reduced processing capacity presented a challenge in managing the fishery in a manner that both minimized waste and fostered the production of the highest-quality product possible.

The department's primary method of managing the fishery in-season was area closures, the first ever in the Togiak District. The areas opened varied from several square miles to nearly the entire district. Only areas that contained suitable-quality herring, based on test-fishing results, were opened, and the department attempted to limit the harvest during individual openings to an amount that could be processed within a reasonable period of time. The result was a larger number of openings over a longer time period. It also heightened competition within the purse-seine fleet.

ADF&G also warned fishermen that it would issue citations to anyone observed wasting herring.

The herring arrived early that year at Togiak. The first purse-seine opening was on April 27 in an area near Anchor Point, in Togiak Bay, and lasted fifteen minutes. Throughout the season, there were seventeen purse-seine openings, with two extensions, for a total of almost thirty-four hours of fishing time. The shortest opening lasted ten minutes; the longest lasted five hours. The total catch by purse seiners was 14,361 tons, 80 percent of the total roe-herring harvest at Togiak.

Gillnetters had a dozen openings and fourteen extensions, for a total of 144.5 hours of fishing. During that time, they caught 3,564 tons of herring, 20 percent of the total roe-herring harvest at Togiak. The average percentage of mature roe in the catch, 10.1 percent, was the highest ever recorded for gillnetters at Togiak. The increase was attributed to gillnetters uniformly employing larger-mesh gear.

Prior to 1993, gillnets with mesh sizes smaller than three inches (stretched measure) were common at Togiak. That changed in about 1993, when gillnetters began using three-inch mesh and larger, and this quickly became the standard gear. Other factors may have been at play, but this shift appears to have increased the percentage of female herring caught by herring gillnets substantially, from 44 percent in 1982–1992, to 57 percent in 1993–1996.

For purse seiners, the average percentage of mature roe was 9.6 percent.

Roe-herring landings at Togiak in 1993 totaled 17,925 tons, well short of ADF&G's projected harvest of 26,279 tons. The catch's estimated value was $5.2 million. Importantly, waste in the purse-seine fishery was minimal, and no waste was observed in the gillnet fishery.[79]

One benefit of the unexpectedly low harvest at Togiak was that combined with the complete bust in Prince William Sound, it helped reduce the worldwide glut of herring roe and propped up ex-vessel prices, at least temporarily. The base price for herring was $300 per ton, about $100 per ton less than the final price in 1992 but higher than fishermen had anticipated earlier in the year.[80]

1994

The 1994 fishery was characterized by a strong preseason herring-abundance forecast, poor market conditions, limited processing capacity (sixteen processors), an average number of seiners (240), and a below-average number of gillnetters (116). The number of gillnetters was lower because some processors refused to buy gillnet-caught fish because of past quality problems. This left some gillnet fishermen without a prospective market, causing them to forgo fishing.

ADF&G's GHL for purse seiners was 18,832 tons, and for gillnetters, it was 6,277 tons. Early in the season, however, an unexpected large volume of herring arrived at Togiak, and ADF&G increased the GHLs to 22,073 tons and 7,358 tons, respectively.

In response to the problems encountered in 1992, ADF&G's goal for the 1994 season was to enhance product quality and value by managing the fishery to limit the quantity of herring caught during an opening to not more than could be processed within three days. Based on company registration statistics, processors had the capacity to freeze 3,300 tons of roe herring per day. During the season, however, things didn't go as planned.

The first seine opening, held on May 11 east of the mouth of the Quigmy River, in Togiak Bay, lasted fifteen minutes. There were a lot of herring to be had, conditions for aerial spotting were excellent, and the catch was considerably larger than expected: 15,660 tons. Given the processing capacity, it would theoretically take almost five days to freeze the fish. The reality, however, was that some herring were held for up to seven days before being processed, and product quality suffered accordingly.

Overall, purse seiners in 1994 had five openings, for a total of four hours and thirty-five minutes of fishing time. The durations of the openings ranged from fifteen minutes to two hours. The total harvest by seiners was 22,853 tons, a little more than the GHL and 75 percent of the total catch at Togiak. During the course of the harvest, seiners dumped an estimated 350 tons of herring, an amount that ADF&G included in the harvest total.

Given the small areas that were opened to gillnetting, ADF&G reduced the allowable length of gillnets to fifty fathoms. For the 1994 season, gillnetters had five openings that totaled seventy-six hours. They landed 7,463 tons of herring, slightly more than their GHL and 25 percent of the total 30,316-ton catch at Togiak, a record.

The roe quality of the gillnet harvest—12.1 percent mature roe—was the highest in the history of the Togiak fishery for the second consecutive year. Roe quality in the purse seine harvests averaged 9.5 percent.

Despite the quality issue with herring caught during the first seine opening, the base price of roe herring was about the same as in 1993, approximately $300 per ton for herring whose body weight was 10 percent mature roe, with an adjustment of $30 per ton for each percentage point above or below 10 percent. The total value of the fishery was $9.1 million, the highest since 1988, and the fourth largest in the history of the fishery.[81]

The herring fishery at Togiak in 1994 wasn't entirely about roe. As an experiment, Westward Seafoods transported 750 metric tons of Togiak herring to its plant at Unalaska and made surimi from it. In Alaska, surimi is typically made from pollock. Ideally, it is tasteless and odorless, and it is used to make products such as imitation crab meat. Unfortunately, the surimi produced from herring at Westward

Seafoods smelled like herring.[82] (In 1997, Western Alaska Fisheries, at Kodiak, made surimi from male herring caught in a roe-herring fishery. The pale gray product was tinged with green and had a fishy odor.[83])

Marring an otherwise mostly successful fishery was a spate of looting of Native artifacts near Togiak. Looting, including grave robbing, had been a problem for years, but was especially prevalent that May, while the fishing and processing fleet were waiting for herring to arrive. Alaska State Troopers cited two men for grave robbing.[84]

Also in 1994, Bering Sea pollock trawlers exceeded their herring bycatch limit, and the winter herring savings area was closed to pollock trawling for three and a half months.[85]

1995

In January 1995, in response to the holding of herring in seines at Togiak in 1992 and 1994 for up to a week before they were processed, the Alaska Board of Fisheries adopted a regulation that limited the amount of time that herring could be held in purse seines to thirty-six hours after the closure of a fishing period.

In 1995, 254 seiners participated, a slight increase, while the number of gillnetters more than doubled, to 250. Given the increased size of the fleet, it was important that the number of processors also increased, from sixteen in 1994 to twenty-two in 1995. Together, processors had the capacity to freeze 4,350 tons daily, compared with 3,300 tons in 1994. Herring were abundant, and the additional processing capacity was a factor in making the season a success.

Seiners had twelve openings that ranged from ten minutes to four hours and totaled 13.3 hours. During this time, they landed 19,737 tons of herring, 74 percent of the total roe-herring catch at Togiak. Gillnetters had four openings that totaled 33.5 hours of fishing. They landed 6,996 tons of herring, 26 percent of the total catch.

The combined catch of 26,732 tons was the second largest in the history of the Togiak fishery, behind only 1994. For the first time since 1991, problems resulting from processing bottlenecks were minimal. This no doubt contributed to the fact that overall roe quality in 1995 was higher than in any previous year. Moreover, no waste was observed.

And there was more good news: initial prices paid to fishermen had doubled over the previous year, reaching $600 per ton for 10 percent mature roe, with an adjustment of $60 per ton for each percentage point difference above or below 10 percent. The large catch combined with the price increase added up to a very lucrative season for fishermen. The ex-vessel value of the fishery was $16.7 million, a record that as of this writing has not been matched.[86]

1996

Fishermen are always attempting to catch yesterday's fish, and it was no surprise that the lucrative 1995 season attracted additional fishermen to Togiak in 1996. The number of seiners increased modestly, to 268, but the number of gillnetters rose to 461, almost double that of the previous year and a new record. The number of vessels began a general decline after the 1996 season, however. The number of processors in 1996 declined somewhat from the previous year, to eighteen, but their combined processing capacity, 4,850 tons per day, represented a small increase and set a record that as of this writing has not been broken.

Herring arrived at Togiak in early May. Gillnetters had six openings that totaled eighteen hours of fishing. During that time, they landed 6,894 tons of herring, which was 27 percent of the total catch of 24,702 tons, the highest percentage since 1982. The gillnetters' catch was also about 16 percent over their GHL.

Seiners had five openings that ranged from ten minutes to one hour and totaled 145 minutes. Their catch, 17,808 tons, matched almost exactly their GHL and was 73 percent of the total harvest.

The base price paid for roe herring was $600 per ton, the same as in 1995, and the total estimated ex-vessel value of the fishery was $14.4 million. Statewide, with ex-vessel earnings of $61 million, 1996 represented the monetary apex of Alaska's roe-herring fishery.[87]

1997

A high inventory of kazunoko in Japan and a weak yen cast a shadow over the 1997 fishery at Togiak and elsewhere in Alaska. Nevertheless, fishermen are an optimistic bunch, and, though fewer than the previous season, 231 purse seiners and 336 gillnetters showed up to fish. The eighteen processors had a daily processing capacity of 4,200 tons, and ADF&G managed the purse-seine fishery closely to ensure its catches during individual openings did not exceed processing capacity.

The herring arrived early. The first opening was on May 2, and the season ended on May 6. Purse seiners had seven openings that totaled almost six and a half hours of fishing, and they harvested 18,649 tons of herring, which included an estimated 350 tons that was wasted, exceeding their GHL by 14 percent. The harvest by seiners represented 78 percent of the total catch.

Gillnetters had six openings that totaled twenty-four hours of fishing. Their harvest was 5,165 tons, which represented 22 percent of the catch and was about 5 percent below their GHL. The total harvest was 23,813 tons, a little less than in 1996.

Though the total catch was comparable to the previous year, the ex-vessel price had plummeted. The base price averaged about $170 per ton, less than a third of the price paid in 1995 and 1996. The total ex-vessel value of the fishery was $4.3 million.[88]

1998

The low ex-vessel revenue in 1997 and a weak market early in 1998 considerably diminished interest in the 1998 Togiak roe-herring fishery. One hundred and twenty-three purse seiners participated, about a hundred fewer than in 1997. And the number would continue to decline. Similarly, there were 152 gillnetters, fewer than half the 1997 number. The number of gillnetters, however, would increase over the next couple of years, after which their numbers began a long decline. The number of processors also declined, to fifteen, and they, too, would continue to decline. Their combined processing capacity in 1998 was 2,475 tons per day, a decline of about 40 percent from the previous year.

Herring arrived early, and fishing for both purse seiners and gillnetters began on April 29 and ended on May 10. The total catch was 21,120 tons. Purse seiners had twelve openings that totaled 13.5 hours of fishing. Their catch, including an estimated 400 tons of waste, was 15,840 tons, which represented 74 percent of the total. Gillnetters had thirteen openings that totaled forty-six hours of fishing. Their catch was 5,280 tons, 26 percent of the total.

The base price was $175 per ton, and the total ex-vessel value of the fishery was $4 million, the lowest since 1980.[89]

1999

A dozen companies froze herring at Togiak in 1999, and their capacity, 2,400 tons per day, was the lowest since 1990, when ADF&G began monitoring processing capacity. The fishing fleet comprised 96 purse seiners and 171 gillnetters.

Spring came late to Bristol Bay in 1999, and there was still shorefast ice in some protected locations on May 15. The roe-herring season started on May 18 and was compressed, with ice causing some problems for the fleet. Because of limited processing capacity and the restriction on holding herring for more than thirty-six hours after the closure of an opening, ADF&G had to close the season for twenty-four hours several times to allow processors to catch up. Fishing ended on May 26, with the fleet well short of reaching its GHL.

The purse seiners' 1999 roe-herring season at Togiak amounted to nine openings that totaled 4.6 hours of fishing time. Their catch, including 221 tons of waste, was 15,020 tons, almost 6,000 tons short of their GHL. Gillnetters had five openings that totaled twenty-eight hours of fishing time. Their catch was 4,858 tons, about 2,000 tons short of their GHL.

Fortunately, the base price had increased substantially, to $316 per ton, and the ex-vessel value of the fishery was $6.2 million, more than half again as much as in 1998.[90]

2000

The year 2000 marked the beginning of the end of the derby-style purse-seine fishery at Togiak. Ninety purse seiners participated, a half dozen fewer than in 1999. As had been the situation in 1999, twelve companies processed roe herring, but their 2,100-tons-per-day capacity was less than in 1999. Twice during the season most processors were plugged with fish and could not handle additional herring.

ADF&G attempted to accommodate the processors that could handle additional fish by scheduling "limited capacity" openings for purse seiners who still had markets available. In all, purse seiners had thirteen openings that totaled 15.75 hours of fishing time. Nevertheless, the reduced processing capacity curtailed the catch: the 14,632 tons of herring landed by purse seiners was about 15 percent below their GHL. On the plus side, their herring averaged 10.3 percent mature roe, one of the highest in the past ten years.

The 227 gillnetters represented an increase of seventy-five over the previous year. They had seven openings, with a total of sixty-seven hours of fishing, and landed 5,442 tons of roe herring, about 300 tons less than their GHL but about 27 percent of the total roe-herring harvest at Togiak. Mature roe content was 10.5 percent.

The base price for roe herring at Togiak ranged from $100 to $200 per ton. The total value of the fishery was $4 million.[91]

2001

In January 2001, ADF&G amended its Bristol Bay herring management plan so that 70 percent of the roe-herring catch at Togiak could be taken by purse seiners and 30 percent by gillnetters. The prior allocation was 75 percent for purse seiners and 25 percent for gillnetters, so the change represented an increase in the gillnetters' allocation at the expense of the seiners. The GHLs for seiners in 2001 was 14,624 tons; for gillnetters, it was 6,268 tons.

The concept of limited capacity purse-seine openings, such as were held in 2000, had appeal, particularly if it could be applied more broadly. After considerable preseason discussion with company representatives and purse seiners, ADF&G modified its management to allow purse-seine cooperatives. Each cooperative consisted of several permit holders tied to a particular processor. The department would schedule longer-duration openings and open larger areas, and the cooperative members would fish at a rate that fully utilized processing capacity, essentially self-regulating within ADF&G's time and area constraints. At the end of an opening, the ex-vessel proceeds generated within a cooperative were divided equally among the member fishermen.

Ideally, the cooperative approach would enhance the quality of the product and minimize waste. No one knew if the new management strategy, which ADF&G called a "leap of faith," would work. It probably helped that the purse-seine fleet was the smallest since 1978, just sixty-four vessels.

Eleven processors, with a total processing capacity of 2,255 tons per day—up slightly from the previous year—operated in 2001. Five were floating plants and six tendered fish to shore-based plants in Bristol Bay.

During the 2001 season, purse seiners had thirteen openings that totaled twenty-six hours of fishing. The cooperative method worked for the first eleven openings, but during the last two several fishermen were heard on the radio "fishing" for a buyer for fish they already had in their seines. ADF&G reported 219 tons of waste in 2001 and later alluded to partial pumping of seine sets. Tenders would pump the amount of fish from a seine that they could carry or that the processor they were associated with wanted, and released the remainder of the fish—mostly, if not all, dead.

Some processors also wasted herring. In Alaska's roe-herring fisheries, male fish are generally an encumbrance rather than a source of profit. Processing them is subsidized by the value of roe-laden females. In 2001's tight market, three processors at Togiak dumped male herring rather than freeze them. Alaska state law required full utilization of herring, and one company's records revealed that it had ground up and then dumped about a half-million pounds of male herring.

Throughout the 2001 season, purse seiners landed 15,320 tons of herring, a little above their GHL and slightly more than 70 percent of the total 21,801-ton roe-herring harvest at Togiak.

The gillnet fishery was a traditional, derby-style event. Ninety-six gillnetters had eight openings that totaled eighty-four hours of fishing, and they landed 6,481 tons of herring, just shy of 30 percent of the total harvest and slightly more than their GHL.

Unfortunately for fishermen, the ex-vessel prices of roe herring in 2001 remained low. Base prices ranged from $100 to $135 per ton, and the ex-vessel value of the fishery was $2.6 million.[92]

2002

> The kazunoko custom is fading away with war-era generations. Young Japanese have new, fast-food tastes. And the market for Alaska herring eggs has cracked.—*Anchorage Daily News*, 2002[93]

Reflecting the poor market for roe herring, participation declined again in 2002, with thirty-seven purse seiners, eighty-two gillnetters, and eight processors participating. Combined daily processing capacity totaled 1,920 tons.

As had been done the previous year and would be the norm in the future, the purse-seine fishery was done cooperatively. Constrained by limited markets, processors, to maximize their efficiency, chose the makeup of their purse-seine fleets. This gave them the power, in collusion with fishermen members of the cooperative, to exclude entrants into the fishery. To prevent a recurrence of the massive dumping of herring that had occurred in 2001, both ADF&G and state troopers warned processors and fishermen that anyone responsible for releasing dead herring would be cited for wanton waste.

The combined catch by seiners and gillnetters in 2002 was 17,049 tons, about 5,000 tons less than in 2001. Its ex-vessel value was proportionately less: $1.9 million. Estimated waste in 2002 was forty tons.[94]

2003

The winter of 2002–2003 was especially warm in Bristol Bay, and herring arrived earlier than usual at Togiak. Fishing began on April 26 and ended on May 7. Seven processors participated, one fewer than in 2002. Their combined daily processing capacity, however, was 1,920 tons, the same as it had been in 2002.

The cooperative method of fishing, as had been practiced since 2001, had definitely changed the pace and intensity of the fishery, as did the reduced size (thirty-five vessels) of the purse-seine fleet. Because fishing was less aggressive, ADF&G at one point opened purse-seine fishing for a twelve-hour period—mostly from 8:30 a.m. until 8:30 p.m.—for eight days in a row.

These extended openings allowed fishermen to take the time to search for desirable herring rather than making a set on the first fish they encountered. And some fishermen, rather than making a set to sample a school of fish, would jig with sportfishing poles. If the fish they caught met quality and size standards, the fishermen would then make a set. And in some cases, several processors shared sample information. Cooperation such as this likely reduced the number of sets made and the number of sets released. Processors also limited the size of sets their fleets could make and limited harvests for their individual fleets, based on their processing capacity. The traditional "race for fish" was slowed substantially.

Overall, the 2003 purse-seine herring fishery at Togiak comprised nine twelve-hour openings, two one-hour openings, and a final ten-minute opening. (The ten-minute opening yielded 1,088 tons of herring from twenty-five deliveries.) The 110 hours and 10 minutes total fishing time offered a stark contrast to the single twenty-minute opening that comprised the area's season in 1992.

Overall, the liberalized system increased product quality by allowing fishermen to be more selective and to make smaller sets, the latter of which reduced damage to roe. Also, it minimized another impediment to quality: the time herring were held in seines. Given the declining market for kazunoko, quality was a paramount consideration. The purse seiners' total catch in 2003 was 15,158 tons, 98 percent of their 15,467-ton quota.

For the seventy-five gillnetters, it was business as usual. During 142 hours of fishing spread over thirteen openings, they landed 6,505 tons of herring, 98 percent of their 6,624-ton quota.

The total ex-vessel value of the 2003 Togiak roe-herring fishery was approximately $3.2 million.[95]

2004

> The seine fleet is now divided into processor-controlled cooperative fleets that harvest just enough herring to keep the processing lines full from day to day.—Alaska Department of Fish and Game, 2004[96]

In December 2003, the Board of Fisheries slightly modified the Togiak herring fishery management plan to allow for the in-season 70/30 allocation of the total GHL between seiners and gillnetters to be uncoupled after each gear type had harvested 80 percent of its allocation.

Another change was that the gillnet fleet began organizing into cooperatives. Six processors were present at Togiak in 2004, one fewer than the previous year, but their combined processing capacity of 2,150 tons per day was an increase over 2003.

The run at Togiak was not as strong as expected, and ADF&G closed the fishery early. During seventy-eight hours of fishing, thirty-one purse seiners landed 13,888 tons of herring, 79 percent of their GHL. The fifty-two gillnetters had 162 hours of fishing time and landed 4,980 tons of herring, 66 percent of their GHL.

The base price for herring was $138 per ton, and the fishery's total ex-vessel value was $2.5 million.[97]

2005–2019

> The seine fishery has turned into a relatively self-regulating fishery. Processing companies manage their cooperative fleets such that they harvest enough fish to keep the processing lines running at full capacity after the daily gillnet harvest has been accounted for.—Alaska Department of Fish and Game, 2005[98]

The years 2005–2019 were the final "normal" years of the Togiak roe-herring fishery. Herring were generally abundant, and the combined catch by gillnetters and purse seiners averaged about 21,000 tons, ranging from 14,879 tons in 2016 to 27,610 tons in 2013. In 2019, twenty seiners set an all-time record, harvesting 23,060 tons of roe herring.

Moreover, though the seiners wasted an estimated 1,660 tons, waste was generally not as large a problem as it had been prior to the establishment of the cooperatives. For four years during this period, ADF&G reported zero waste.

Due to the declining market for kazunoko, however, the value of the catch was lower than it had been during the earlier years of the fishery. The low was in 2015, when ADF&G estimated the ex-vessel base price of herring to be $50 per ton. In other years, though, it reached $150 per ton. Still, this was far less than the $600-per-ton base price in the mid-1990s. The average ex-vessel value of the fishery during the years 2005–2018 was $2.6 million, ranging from $1.0 million in 2015 to $4.2 million in 2013.

The reduced value attracted fewer fishermen, especially gillnetters. The number of gillnetters declined from fifty-six in 2005 to one in 2018, and the number of seiners declined from thirty-three in 2005 to nineteen in 2019. The number of processors halved, from eight in 2005 to four in 2018. Processing capacity, nevertheless, remained fairly steady, at about 2,200 tons per day.[99]

Excessive herring bycatch by pollock trawlers was an issue only in 2012, when pollock trawlers exceeded their limit, and the winter herring savings area was closed to pollock trawling for five months.[100]

2020

Even before the COVID-19 pandemic, the market for roe herring looked especially grim. During the 2019 fishery, four companies sent floating processors to Togiak, but only two committed to doing so for the 2020 season. Complications related to the pandemic made it impossible for one of the companies to follow through, but Icicle Seafoods sent its floating processor *Gordon Jensen*. The fishing fleet comprised two purse seiners and one gillnetter.

Spawning was well under way when they arrived, and their harvest totaled less than 3,000 tons, the exact amount being unavailable to the public because of ADF&G's confidentiality policy. The fishermen were paid $100 per ton for the fish.[101]

For the 2020 pollock season, the herring bycatch limit was 2,532 metric tons. Bycatch in that year was especially high, and by May 24 pollock trawlers had caught 2,763 metric tons, exceeding their 1 percent threshold. On June 4, the National Marine Fisheries Service announced the closure of all three herring savings areas to pollock fishing for periods

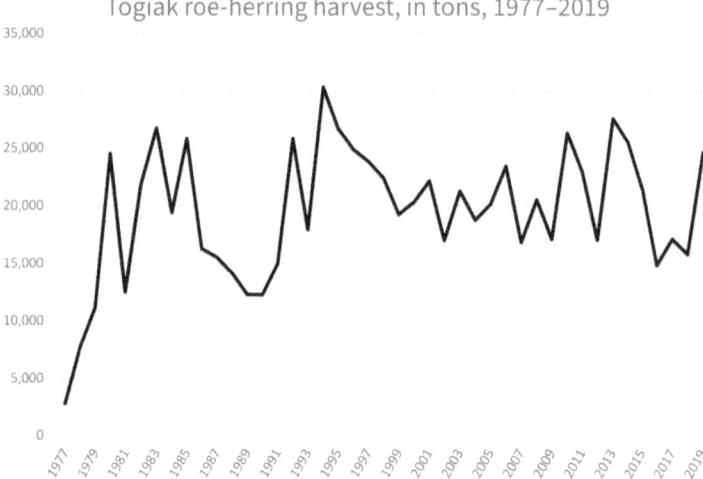

Figure 13.7. Data sourced from ADF&G Bristol Bay management reports, 1990, 1991, and 2018.

ranging from fifteen days to six months.[102] Two weeks later, the service rescinded the closure of one of the summer herring savings areas "to prevent the underharvest of the 2020 pollock total allowable catch."[103]

The Togiak herring population, however, appears to be healthy. ADF&G's 2021 mature herring biomass forecast was 236,742 tons, the highest since the department's age-structured assessment model was first employed, which was for its 1993 forecast. Based on a 20 percent exploitation rate, ADF&G allocated 42,639 tons to the 2021 roe-herring fishery.[104] However, given the continued decline in the market for kazunoko, interest in the fishery was minimal: only ten seiners, three gillnetters, and two processors (Silver Bay Seafoods and Icicle Seafoods) chose to participate. The base ex-vessel price was reported to be $75 per ton, but because of low fishery participation, the harvest volume is confidential.[105]

ADF&G's 2021 forecast for roe herring at Togiak may have set a record, but it was dwarfed by the 2022 forecast: 357,536 tons, a whopping 51 percent increase over 2021.[106] All indications are that the herring population that spawns at Togiak is flourishing. Nevertheless, given the aforementioned greatly declined market for kazunoko and the resultant low prices to roe-herring fishermen, the vast majority of the herring that arrive at Togiak to spawn in 2022 will almost certainly be able to complete their business unhindered. ADF&G Bristol Bay fisheries biologist Tim Sands put the current Togiak roe-herring situation in perspective: "There's a certain amount of herring that can be sustainably harvested. But if you look at a broader ecosystem perspective, not harvesting herring isn't necessarily a bad thing either, especially if there's no money in it."[107]

14

NORTON SOUND ROE-HERRING FISHERIES

As noted in chapter 3, Norton Sound was the location of the northernmost commercial herring fishery in Alaska. The herring that spawn in Norton Sound tend to be large and fat and arrive in late May or early June. They favor the eastern reaches of the sound, where the extensive kelp beds are a preferred substrate habitat for spawning.[1]

In 1964, Unalakleet, on Norton Sound's east shore, was the location of one of three or four fisheries that year that marked the inception of the roe-herring fishery in Alaska. The others were at Sitka, in Southeast Alaska, and Spiridon Bay, on Kodiak Island, and possibly at Seldovia, on the Kenai Peninsula.

At Unalakleet that June, Anchorage-based Western Alaska Enterprises (the same Japanese-owned firm that had a roe-herring operation at Spiridon Bay) employed local fishermen who used mostly beach seines but also gillnets to harvest 40,000 pounds of herring. The males were discarded, and 2,520 pounds of roe was stripped from the females. The product was reported to be of low quality due to poor handling methods.

ADF&G knew little of the herring that spawned in Norton Sound, and one benefit of this fishery was that the department was able to sample 350 fish to obtain age, length, weight, and sex-composition data. A high proportion of the herring sampled were in the older age groups, which was characteristic of an unexploited population. Also, the average weight of the fish (166.4 grams) was considerably larger than that of the general North American herring population.[2]

Japanese fishermen also became interested in Norton Sound herring in the 1960s—perhaps too interested. A considerable portion of the sound at that time was international waters, beyond the twelve-mile-wide contiguous zone claimed by the United States.

In 1968, Japanese high-seas gillnetters began fishing in the sound and that year caught about 130 tons of herring. The Japanese effort in Norton Sound peaked the following year, when forty gillnetters, accompanied by two factory ships and two freighters, landed some 1,400 tons of herring.

Many of the Japanese gillnetters in 1969 routinely and brazenly fished in US waters. On June 7, the US Coast Guard cutter *Storis* steamed into Norton Sound and apprehended and seized two of them, the *Zenpu Maru 8* and the *Coei Maru 11*, near St. Michael. Other Japanese vessels avoided apprehension by cutting loose of their gillnets and fleeing into international waters.

The *Storis* escorted the two seized vessels to Nome. Because of bad weather and other reasons, US officials were unable to retrieve the nineteen gillnets that had been abandoned, and the Japanese, under bond, were allowed to retrieve them.

The captain of the *Zenpu Maru 8*, who had attempted to elude the *Storis*, was subsequently fined $5,000, while the captain of the *Coei Maru 11*, who had surrendered peacefully, was fined $3,500.

At Nome, an ADF&G biologist attempted to sample the more than twenty tons of herring aboard the Japanese vessels. The fish had been segregated by sex and then salted, but they were too dehydrated for measurements to be taken.

The Japanese continued to fish for herring in international waters in Norton Sound, but at a much reduced intensity, until the implementation in 1977 of the Magnuson-Stevens Fishery Conservation and Management Act (1976), which created an exclusive economic zone that extended 200 miles out to sea along the entire US coast.[3]

Stepping back a few years, there was no domestic commercial roe-herring catch in Norton Sound in 1965, and the effort would be sporadic and relatively small—the annual harvest averaging about 27,000 pounds—until 1979. The demand for herring roe was strong that year, and sixty-three fishermen and seven buyers participated in the Norton Sound fishery. The fishing fleet comprised thirteen purse seiners and fifty gillnetters, and together they harvested 1,292 tons of roe herring. Purse seiners—only a few of which, if any, were Norton Sound residents—took 70 percent of the catch, which had an estimated ex-vessel value of $762,000.[4]

One of the seven buyers of the catch in 1979 was Seward Fisheries, a subsidiary of Icicle Seafoods, which is now OBI Seafoods. It partnered with the Norton Sound Fishermen's Co-op and purchased 309 tons of herring at Unalakleet. The fish were loaded into totes and flown in chartered cargo planes to Anchorage, where they were frozen.[5]

For ADF&G, managing this new, remote fishery was a challenge. There wasn't a method to reliably forecast herring returns to Norton Sound, and the department relied on in-season assessments of biomass for management of the fishery. Such was the situation until 2013, the final year of the roe-herring fishery in Norton Sound.[6]

The herring-fishing effort in Norton Sound greatly increased in 1980, when 294 fishermen participated. Sixty-six percent of the fishermen were Norton Sound residents, 26 percent were nonlocal Alaska residents, and 8 percent were from out of state.

The season opened May 21 and closed June 4. Fishermen harvested 2,451 tons of roe herring, almost double the 1979 catch. In contrast to 1979, fully 98.5 percent of the 1980 catch was taken with gillnets. Local residents accounted for 55 percent of the catch.

Disappointingly, the ex-vessel value of the catch was approximately $500,000, about $250,000 less than fishermen received in 1979, when they caught only half as much herring. A depressed market was to blame.[7]

Despite the reduced prices, the Norton Sound roe-herring fishery had come of age and had quickly become of great importance to the region's residents. In 1980, more local residents fished for herring than for salmon. Alaska's Legislature and the Alaska Board of Fisheries did what they could to facilitate opportunities for local fishermen.

Prior to the 1980 season, the Alaska Board of Fisheries adopted a public proposal that began the process of eliminating purse seines in the Norton Sound roe-herring fishery. The board's goal was to encourage local participation. The board's reasoning for eliminating purse seines was twofold: First, local fishermen could afford gillnet gear. Second, gillnetters' competition with the highly efficient purse seines would be eliminated. The new regulation authorized the use of gillnets and beach seines, but a purse-seining season would be opened only if the gillnet fleet was unable to take its allowable harvest.

Because gillnet fishermen in 1980 had demonstrated that they were capable of taking the allowable harvest, in 1981 the Board of Fisheries passed a regulation that completely prohibited purse-seine gear in the Norton Sound herring fishery. Beach seines, however, could still be utilized.[8]

To make beach seines more efficient, some fishermen fastened tag lines at various points along the seine's leadline. Once a set was made, tightening the tag lines gathered the bottom of the seine together. The system didn't function nearly as well as a purse line, but it helped concentrate the fish. A drawback of beach-seine gear was that it could not select for the most desirable herring, the older-age-class fish, which have larger egg skeins. Gillnetters, on the other hand, could utilize mesh sizes that selected for these fish. This ability was important during years of low demand, when buyers wanted only the best fish.[9]

There was another change for the 1981 season. During its 1980 session, the Alaska Legislature passed what some called the "in-state herring processing bill."[10] The legislation made it unlawful for a person to remove herring from the state before it had been frozen or otherwise processed for shipment. The primary intent was to end the practice of salting herring, then loading it aboard foreign freighters, as the *Maren 1* had been doing at Togiak.

However, at the request of Norton Sound herring fishermen, Jack Fuller, a Nome resident and a member of Alaska's Legislature, included a provision that would relax until July 1, 1982, the regulations for processing Bering Sea roe herring. The provision authorized roe stripping and the dumping of carcasses, which had been prohibited during the 1979 and 1980 seasons pursuant to legislation passed in 1977 (see chapter 11). The legislature justified its action by stating that "in certain circumstances, the processing technique commonly referred to as 'stripping' provides benefits of such importance to the state economy that the benefits may outweigh the waste involved in the process."[11]

Under this law, it became legal for processors of Bering Sea herring to remove the roe from herring, then dump the carcasses. In addition to Norton Sound, the law applied to Togiak and to places where the roe-herring fisheries were relatively small but nonetheless important to the local economy, such as at Cape Romanzof. The bill's sponsors believed that by the 1983 season local processors would be capable of fully utilizing the herring. This apparently wasn't the case, because in early 1983 the legislature passed a bill that extended the legalization of roe stripping in the Bering Sea until July 1, 1984. Kyokko Suisan Alaska, a company owned by Shigeyoshi Kitano, who had been instrumental in the development of the Togiak roe-herring fishery (see chapter 13), provided the Norton Sound fishermen with information regarding stripping techniques and was a potential market for their product.[12]

Meanwhile, the Norton Sound roe-herring fishery continued to grow, with 332 fishermen making at least one delivery during the 1981 season. About half of the fishermen were residents of the Norton Sound area, the others mostly being gillnetters who accompanied the thirty tenders and processors that came north after the season at Togiak closed. Norton Sound residents referred to these fishermen as the "gypsy fleet." The total harvest for the season was 4,370 tons, almost doubling 1980's harvest. Local fishermen were responsible for approximately 30 percent of the harvest.[13]

The value of roe herring is based on the percentage of their weight that is mature roe. In a given season, fish with a roe content of 12 percent can be worth perhaps 40 percent more than fish with a roe content of 8 percent. In Norton Sound, as in Alaska's other roe-herring fisheries, the large, older fish—the ones with high roe content—tend to be the first to spawn.

In 1984, to maximize the value of the Norton Sound roe-herring fishery and at the same time reduce pressure on younger stocks, ADF&G began managing the fishery for an above-average roe recovery. The department's new policy was to focus on harvesting herring older than six or seven years of age, which meant less fishing later in the season, when herring schools were dominated by younger, smaller fish.

The change was controversial at first, but the benefits were soon recognized, and the policy was accepted as the norm. Fishermen's incomes increased, buyers appreciated the improved quality, and the shortened herring season allowed fishermen more time to get where they needed to be for the upcoming salmon season. The effect on the herring was mixed. While the reduced pressure on younger stocks helped guarantee the future of the fishery, the harvest of older stocks—the most fecund component of the herring population—impacted egg production.[14]

During the years 1982–1984, the roe-herring catch in Norton Sound didn't fluctuate dramatically and averaged about 3,800 tons. The type of fishing gear employed, however, changed significantly. Prior to the 1984 season, the catch by fishermen using beach seines was negligible, but during the 1984 season, ten beach seiners landed 327 tons of roe herring, 10 percent of the total harvest. The beach seiners were competition for the local gillnet fleet, and in the fall of 1984, the Board of Fisheries limited the length of beach seines in the Norton Sound roe-herring fishery to 100 fathoms. The board also limited the harvest by beach seiners to "not exceed 10 percent of the total herring sac roe harvest projection as published by the department."[15] In 1987, the board further restricted the length of beach seines, limiting them to seventy-five fathoms.

Also in 1984, the Board of Fisheries, pressured primarily by Native fishermen's organizations, attempted to slow the growth of the number of fishermen engaged in the Norton Sound roe-herring fishery and at the same time bolster local involvement by designating Norton Sound as a super-exclusive registration area. As such, a fisherman harvesting and selling herring in Norton Sound between February 1 and June 30 was prohibited from harvesting or selling herring or crewing on a herring vessel anywhere else in Alaska during the same year. The same restrictions applied to herring-fishing vessels. The board likewise designated the Cape Romanzof roe-herring fishery as super-exclusive. There was also a management/conservation element to the board's actions: they slowed the pace of fishing on herring stocks that were of unknown magnitude, distribution, and resiliency.

The measure seemed to have had some success because the number of fishermen declined significantly, from 272 in 1983 to 191 in 1984. The decline, however, was attributed not to the imposition of the super-exclusive registration rule but to late-season sea ice that prevented many fishermen from participating in the fishery. In 1985, the number of fishermen rebounded to 277, and the 1987 season saw the highest level of fishing effort on record: 564 fishermen made at least one delivery.

The average income of the fishermen, however, declined accordingly.[16] ADF&G biologist Charles Lean summed up the fishery's development: "The Norton Sound herring fishery has developed from a new fishery in 1979 to a fully capitalized fishery in 1987. Fleet efficiency increases annually with increased utilization of larger fishing vessels, hydraulic reels and shakers and increased experience of the fishermen."[17]

Following that crowded season, the roe-herring fishermen in western Alaska formally asked the Alaska Commercial Fisheries Entry Commission to consider limiting entry into the Norton Sound roe-herring fishery. The commission agreed to do so and immediately established a moratorium on entry. Only captains who had fished legally (some were thought to have violated the super-exclusive registration area provision introduced in 1984) and had made landings prior to January 1, 1987, were eligible to apply for interim-use permits to fish in 1988. An exception was educational gillnet permits issued in some years to the Bering Strait School District commercial fisheries vocational class.

The commission subsequently limited the number of gillnet permits to 301 and the number of beach seine permits to four, and it developed a point system based on priorities and criteria for awarding the permits. By 1993, the commission had begun awarding limited-entry permits to qualified individuals. Fishermen who were qualified to fish under the moratorium yet whose limited-entry applications were still under review were issued interim-use permits and allowed to fish while their eligibility was being evaluated.

Though participation in the fishery had over the years declined steeply—the last year there were more than 301 gillnetters in Norton Sound was 1992—the process of awarding limited-entry permits was still ongoing in 2004 but seemed to have ended in 2005, the last year of a major roe-herring fishery in the sound. And for beach seiners, limited entry had been a moot point since 1998. There had been little market interest that year in herring caught in beach seines, and the last beach seine harvest was in 2000.[18]

The following paragraphs chronicle, beginning in 1988, the Norton Sound roe-herring fishery.

1988

During the 1988 season, 348 fishermen landed 4,672 tons of herring. The ex-vessel price for herring was high—an average of $835 per ton—and the catch was worth a record $3.9 million.

1989

Eight processors and thirty-six tenders were on the grounds in 1989. The fishing effort and catch were similar to those of 1988, but the average ex-vessel price had declined to $482 per ton, and the value of the fishery declined to $2.3 million. The decline in value was attributed to the death of Japanese Emperor Hirohito in January, which put a damper on the celebration of the Japanese New Year. ADF&G estimated that abandoned gillnets and a lost beach seine in 1989 resulted in the waste of about thirty tons of herring.[19]

1990

Despite some 300 square miles of sea ice that broke loose late in the season and drifted into some of Norton Sound's most productive gill-netting areas, the 365 fishermen who participated in the 1990 fishery landed a record 6,439 tons of herring. The average ex-vessel price had increased about 16 percent, to $559 per ton, and the fishery was worth $3.6 million. Among the waste that season was a dozen abandoned gillnets, each estimated to have been carrying five tons of herring.

1991

Similar to the previous year, the 1991 season was pretty much routine save for a substantial decline in the ex-vessel price of herring. The price that year averaged $423 per ton, yielding 279 fishermen $2.4 million for the 5,671 tons of herring they harvested.[20]

1992

Based on aerial surveys, ADF&G projected 1992 to be a banner year. Prior to the season, at least five seafood companies (Icicle Seafoods, Lafayette Fisheries, New West Fisheries, Sno-Pac Products, and Trident Seafoods) expressed interest in participating in the fishery. It was not to be.

The roe-herring harvests in Prince William Sound and at Togiak, completed in May, had been large, saturating the market. Moreover, on June 10, pack ice covered more than 80 percent of Norton Sound. Mobilizing there would have been problematic, to say the least. By mid-June, all the packers had decided to pass on Norton Sound, and there was no commercial roe-herring fishery there in 1992.

Based on income estimates from the previous five years, the Norton Sound's 350 roe-herring fishermen—about 60 percent of whom lived on the sound—and their crewmembers were deprived of about $3 million in revenue.[21] "They're really depressed around here today," said Dora Smith, of Golovin, in mid-June 1992. "We were depending on that fishery to pay our bills this winter."[22]

Citing the failure of the herring fishery, in July 1992 the US Small Business Administration declared Norton Sound an economic disaster area. Herring fishermen in twenty-one villages became eligible for low-interest thirty-year loans.[23]

1993

In large part because of projections of low prices, the 1993 fishing effort was significantly smaller than it had been in 1991. A total of 264 fishermen—257 gillnetters and seven beach seiners—participated. Their total harvest was 5,029 tons, not including about forty-five tons that was wasted in abandoned beach seine sets. The ex-vessel base price for the fish was about $300 per ton, and the estimated ex-vessel value of the fishery was $1.4 million.[24]

1994

Even fewer fishermen—212 gillnetters and three beach seiners—participated in the 1994 fishery. But the processing effort was robust, comprising eleven processors that were accompanied by forty-six tenders.

The season was a disappointment, however, because an offshore mass of cold water slowed the migration of spawning herring into Norton Sound. In fishermen's parlance, it was a "scratch" fishery, characterized by catches just large enough to keep fishermen on the water. In all, only 960 tons of herring were landed. The average ex-vessel base price of the catch, about $295 per ton, was similar to what it was in 1993. The value of the catch was $271,000, less than 20 percent of what it had been in 1993. In its year-end regional management report, ADF&G noted that "many fishermen and some buyers [are] reconsidering their future in this fishery."[25]

1995

Despite ADF&G's observation, the number of fishermen who chose to participate in the 1995 Norton Sound roe-herring fishery—209 gillnetters and six beach seiners—was about the same as in 1994, as was the number of processors.

They made a good choice. Norton Sound sea ice cleared out early, and there was an abundance of herring. Moreover, the market was improved, and the base ex-vessel price of herring had increased to $600 per ton. Fishermen landed a record 6,763 tons of herring that had a total ex-vessel value of $4.2 million—a more than fifteen-fold increase over the value of the 1994 catch and a new record.[26]

1996

Given the success of the 1995 season and an increase in the strength of the roe-herring market, it was no surprise that the roe-herring fishing effort in Norton Sound increased in 1996. Six beach seiners and 281 gillnetters participated. Likewise, processing capacity increased; nine companies were on the grounds, and their fleet of processors and tenders totaled seventy-five vessels. Fishermen landed 6,170 tons of herring, for which the base ex-vessel price was $800 per ton. The fishery's total ex-vessel value, $4.6 million, broke the previous year's record.[27]

1997

ADF&G had projected that there would be substantially fewer herring in Norton Sound in 1997. Also, as evidenced by earlier fisheries in 1997, the price of roe had declined substantially. The prospect of fewer fish and a lower price diminished interest among fishermen and processors alike. While the number of beach seiners remained the same (six), the number of gillnetters declined to 214. There was the same number of processors (nine), but their complement of tenders declined to forty-six.

Moreover, storms delayed the arrival of many boats in Norton Sound. The 1997 season was the last in which beach seines were a significant component in the roe-herring harvest in the sound.

The total roe-herring harvest in 1997 was 3,971 tons. For their catch, fishermen were paid a base price of $154 per ton, bringing the total ex-vessel value of the fishery to $612,000—about 13 percent of the value of the previous year's catch.[28]

1998

ADF&G projected that there would be an abundance of herring in Norton Sound in 1998, but the market for herring roe remained depressed. Only two processors and thirty-five fishermen—all gillnetters—elected to participate. The department opened the fishery on May 22 and—in a novel move—left it to the processors to manage it. It was in the processors' best interest to purchase herring in quantities and on a schedule that most efficiently utilized their processing capacity. ADF&G closely monitored the fishery, and overall the self-regulation system worked well under these unique circumstances.

The season ended on June 9, and the total harvest was 2,624 tons—not bad for the number of fishermen involved—but the estimated ex-vessel base price for roe herring was a paltry $72 per ton, and the total ex-vessel value of the fishery was just $203,000.[29]

1999

The 1999 roe-herring market was expected to be poor, as it had been the previous two years, but weak showings of herring and low harvests in Alaska's southern districts brought the price up as the fisheries progressed to the Bering Sea. Four companies sent five processors and thirteen tenders to Norton Sound. The vessels were delayed by heavy sea ice and did not begin buying fish until June 14. The fishing fleet, which consisted completely of gillnetters, numbered 119. Though portions of Norton Sound remained ice choked to the last days of the fishery, fishermen managed to land 2,693 tons of herring before the season closed on June 22. The ex-vessel base price for the fish ranged from $200 to $250, and the ex-vessel value of the catch was $615,000—fully three times what it had been in 1998.[30]

2000

> Recent drops in sac roe market interest results in an increasing demand for high-quality product. Fishers are required to be much more selective and managers have responded by increasing fishing time to allow fishers more time to locate acceptable herring. Managers are required to work in close cooperation with the Norton Sound herring fishing industry.—Alaska Department of Fish and Game, 2000[31]

Four companies sent five processors and eighteen tenders to Norton Sound for the 2000 roe-herring season. Ninety-four gillnetters landed 4,463 tons of herring. Three beach seiners landed eighty-one tons but found it difficult to sell their catch because the fish were smaller and less valuable than those caught in gillnets. This was the last year beach seines were utilized in the Norton Sound roe-herring fishery.

The base ex-vessel price for roe herring was $200 per ton, and the ex-vessel value of the fishery was $894,000.[32]

2001

Interest in Norton Sound's roe herring was diminished in 2001. Three companies (Icicle Seafoods, NorQuest Seafoods, and Trident Seafoods) sent processors and thirteen tenders to the sound. The ex-vessel base price had fallen to $133 per ton, and seventy-three gillnet fishermen landed 2,245 tons, worth $348,000.[33] The continuing poor prices that the fishermen were receiving were a reflection on Japanese consumers—as they had been for years—losing interest in kazunoko in favor of Western-style foods.

2002

Interest in Norton Sound's roe herring diminished further in 2002. Two buyers sent two processors and seven tenders to the sound. The fishing effort comprised forty-six gillnetters who together landed only 1,059 tons of herring. With an ex-vessel base price of $155, the fishery was worth $160,000, an average of about $3,500 for each fisherman who made a landing.[34]

2003

Only one processor, NorQuest Seafoods' *Aleutian Falcon*, four tenders, and thirty-one gillnetters were on the grounds in 2003. The ex-vessel base price for the 1,587 tons of roe herring landed was $150 per ton, and the total ex-vessel value of the fishery was $117,000.[35]

2004

Processors opted to forgo participation in the Norton Sound roe-herring fishery in 2004, leaving fishermen without a market.[36]

2005

Interest in Norton Sound roe herring picked up in 2005. Two companies, Icicle Seafoods (floating processor *Discovery Star*) and Norton Sound Seafoods (plants at Unalakleet and Nome) sent ten tenders to the sound. The catch by the fifty-six gillnetters totaled 1,951 tons and had an ex-vessel value of $321,580.[37]

2006–2018

During the years 2006 through 2018, the herring harvest in Norton Sound averaged 234 tons, ranging from one ton in 2014 to 807 tons in 2011. The fish were harvested for roe as well as for bait and food. Throughout this period, Norton Sound Seafood Products was the only herring buyer on the sound. However, in years in which herring were harvested for roe (2006, 2010, 2011, and 2013), the Norton Sound Economic Development Corporation, the parent company of Norton Sound Seafood Products, contracted with Icicle Seafoods to bring a floating processor and tenders to the sound.[38]

15

FOOD HERRING IN THE MODERN ERA

In Petersburg in 1973, the herring story wasn't only about roe; there was also a food-herring component. Petersburg Fisheries, a mostly fishermen-owned company that, since its inception in 1965, endeavored to expand, diversify, and provide opportunities for fishermen and plant workers, installed two Swedish-made herring filleting machines. The machines produced butterfly-style fillets, splitting the fish along the ventral (bottom) side and removing the entrails, backbone, and other bones, which resulted in the fillets from each side being attached at the dorsal (top) side. Each machine could fillet about 2,000 fish per hour. The fillets were frozen into blocks and shipped to secondary processors, including in Europe, for later use for smoking or pickling.[1] The herring were prime for filleting in late fall and early winter, creating a demand for "winter" herring.

Each fall, Petersburg Fisheries salted a few hundred pounds of the herring fillets in five-gallon plastic buckets for local consumption, mostly to be used to make pickled herring—the kind done in jars with vinegar and spices.

Petersburg Fisheries dismantled its herring filleting operation in 1976. I was in Petersburg in the fall of 1977 and, to supply the local demand for salted herring fillets, took up—in a miniscule manner—the herring filleting operation. Herring were abundant in Wrangell Narrows, and without any permit from ADF&G, I set a gillnet about sixty feet long along a float in Petersburg's boat harbor. Each morning, I would empty it of herring, typically enough to fill a couple of five-gallon buckets. Then I hand-filleted the fish in the kitchen of the house I was housesitting. I salted the fillets in five-gallon plastic buckets on the back porch, let them sit for a few days, then resalted them in fresh salt. By Thanksgiving, I had accumulated about thirty-five buckets of salted herring, which I—again, without any sort of permit or license—sold to very appreciative locals.

In 2004, Alaska governor Frank Murkowski created the Alaska Global Food Aid Program in an effort to expand the market and develop new customers for Alaska seafood products within both domestic and international federal food aid programs. Since 2007, the program has been administered by the Alaska Seafood Marketing Institute. During the program's early years, it focused on supplying canned salmon.

However, at the request of the Department of Agriculture, which administers federal food aid programs, and its "customers" for an increased variety of shelf-stable fish products, canned herring was identified as a product that might be suitable. In late 2008, the initial steps of learning to can herring commenced. During the summer of 2009, the program exported a shipping container of canned salmon to Uganda, and included in it was a small batch of canned Alaska herring.

In 2010, the Alaska Department of Commerce, Community, and Economic Development, within which the Alaska Seafood Marketing Institute is situated, asked Alaska's Legislature for $300,000 to begin and partially fund the Western Alaska Canned Chum & Herring Demonstration Project. Part of the money would be used to purchase and install a nobbing machine that would remove herrings' heads, tails, and guts in preparation for canning them in the same cans in which salmon are canned. The purpose of the project was to increase the utilization of herring and salmon from Alaska's Bering Sea coast that were (and still are), for a variety of reasons, underutilized. Moreover, economic development in the region was badly needed. Canning both salmon and herring would add to the viability of a canning facility.

Alaska's Legislature granted $200,000 to the project, and the nobbing machine was purchased. However, because experimentation/development must happen in the off season, and there are no canneries or year-round processing facilities along the coast north of Bristol Bay, the machine was installed at the Ocean Beauty Seafoods (OBI) plant in Kodiak, where the appropriate canning equipment was available. Unfortunately, the "filler" machine that put the herring into cans was designed for putting chunks of salmon into cans. It was not compatible with the herring that were processed in the nobbing machine, and the weights of the cans' contents varied substantially.[2] Herring fillets, however, would prove to be compatible with the filling machines.

In 2013, the Department of Commerce, Community, and Economic Development transferred the project, now officially the Canned Salmon, Herring and Protein Powder Project, to the Alaska Seafood Marketing Institute. To fund it, the department requested a $300,000 appropriation from Alaska's Legislature.[3] The money was granted, and a German-made (Baader) herring filleting machine was purchased to replace the nobbing machine in Kodiak. To be marketable, herring fillets must be on the large side, and Togiak herring are the largest in Alaska. Herring at Togiak are harvested almost entirely for their roe, so the male herring have little (or negative) value. These male herring were targeted for filleting and canning as means of adding value to the fishery.

The program's canned herring fillets were tested with consumers around the world, from Africa to Central and South America, and were well received. Notable were the several shipping container loads sent to Liberia, where in a clinical study the fish were shown to dramatically improve the nutritional profile and overall health of people in a community that was suffering from the AIDS epidemic.

Despite this success, international politics changed the policies of federal food aid programs and diminished the demand for canned herring. Because of this reduced demand, coupled with the high cost of producing canned herring in Alaska, the effort to further develop this product ceased.

As an alternative, the Alaska Seafood Marketing Institute turned its focus to frozen herring fillets and soon after started the Alaska Herring Development Project. Its goal was to help the Alaska seafood industry "create a sustainable market for Alaska Herring as a human food."[4] North Pacific Seafoods, which has a processing plant at Naknek, in Bristol Bay, requested to use the fillet machine to produce herring fillets for potential customers in Asia. At the same time, it could supply the Alaska Herring Development Project. The filleting machine, along with a new descaling machine, was transferred to the Naknek plant, which began producing individually quick-frozen herring fillets in the spring of 2015. (It must be noted that herring fillets do not freeze well due to high oil content.[5] The oil begins to turn rancid after a relatively short while at standard freezer temperatures.)

In June 2015, to introduce herring to the American public, the Alaska Seafood Marketing Institute began sponsoring Alaska Herring Week in Seattle. Eight restaurants showcased herring during the event, and the number grew to thirty-three for the 2016 Alaska Herring Week. North Pacific Seafoods that year donated 5,000 pounds of herring fillets. In 2017, fully fifty-four restaurants and four grocery stores were involved. Eight of the participating chefs were winners of the prestigious James Beard Award.[6] "It was in everything from pâté to tacos and piled high on open-faced sandwiches," wrote Suzanna Caldwell, of the *Anchorage Daily News*, of the 2017 event.[7]

That same year, in a report on specialty seafood products commissioned by the Alaska Seafood Marketing Institute, the Juneau-based consulting firm McDowell Group wrote that "The biggest challenges [sic] for fillet production in Alaska is competing with low prices of Canadian suppliers. In addition, the softer flesh quality of Alaska herring, high transportation costs, and the lack of infrastructure in Alaska to handle herring fillet production are major obstacles."[8]

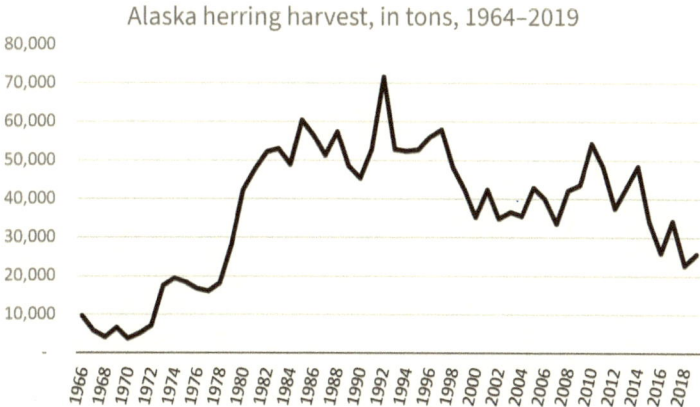

Figure 15.1. Data sourced from ADF&G management reports.

Alaska Herring Week did not happen during 2018 because the North Pacific Seafoods plant was being overhauled during the spring herring fishery at Togiak, and an alternative source of fish could not be located.[9] But there was a bigger problem: despite the exposure over the course of three years, once Herring Week was over, restaurants did not retain herring on their menus. For this reason, Herring Week did not happen in 2019, and the COVID-19 pandemic precluded such an event in 2020. The Alaska Seafood Marketing Institute, however, intends to sponsor Alaska Herring Weeks in several major cities once the food-service industry recovers from the disruptions caused by the COVID-19 pandemic.

Meanwhile, since 2011, one small Seattle-based company, Deckhand Seafoods, has been smoking headed-and-gutted herring caught at Togiak and canning them.[10]

Part III
Herring Spawn on Kelp

Figure 16.1. Jumbo-grade kazunoko kombu. (John Swanson)

16

GENESIS OF ALASKA'S HERRING SPAWN-ON-KELP FISHERY

> The whims of both mother nature and the marketplace have determined whether any season was a success or failure for the hundreds of fishermen who participate in the wild kelp fishery. In order to be marketable, the final product must contain clean kelp of a particular species, have herring spawn of adequate density, and occur within a relatively short span of time before the eggs begin to hatch.—Alaska Department of Fish and Game, 1982[1]

Herring spawn on kelp, *kazunoko kombu* (also *komochi kombu*) in Japanese, is a traditional delicacy that is typically served as an appetizer, especially during the New Year's celebration in Japan. In 1988, high-grade kazunoko kombu demanded the highest price of any fisheries product in Alaska—up to $25 per pound to the fishermen. In more recent years, however, the value of this product has declined.[2] In 2019, the average price for the spawn on kelp produced in Alaska was $8.15 per pound.[3]

Macrocystis pyrifera (giant kelp), a large, bladed kelp that grows in Southeast Alaska, especially the region's southern reaches, is the preferred kelp. High-grade kazunoko kombu must have an even herring egg coverage at least three layers thick (more than 350 eggs per square inch) on both sides of a kelp blade, and the blade must measure about four inches by fourteen inches, or larger.

In Prince William Sound, ribbon kelp (*Laminaria* sp.) was the most desirable native species, but sieve kelp (*Agarum* sp.), hair kelp (*Desmarestia* sp.), and rockweed (*Fucus* sp.) were also harvested. (Rockweed and hair kelp are not true kelps, but ADF&G lumps them in with true kelps in its management of the spawn-on-kelp fishery.) At Togiak, in Bristol Bay, herring spawn on rockweed and sugar wrack (*Saccharina* sp.) was harvested, and in Norton Sound, the harvest was spawn on rockweed. In addition, many pound operators brought *Macrocystis* kelp from Southeast Alaska to Prince William Sound.

For shipment and storage, kazunoko kombu is packed in brine in plastic containers and kept refrigerated.

In the early years of the fishery, herring spawn on kelp was harvested by handpickers who worked the kelp beds from small boats, using knives and grappling hooks; by scuba divers using knives; and for rockweed, an intertidal species, by pickers using rakes. The product

Figure 16.2. Grading and packing kazunoko kombu at Icicle Seafoods plant at Petersburg, April 2018. (John Swanson)

of this "wild" (or "natural") spawn-on-kelp fishery varied greatly in quality, and there was a lot of waste.

To produce a higher-quality product and reduce waste, fishermen began harvesting desirable kelp prior to the anticipated herring spawning period and suspending it from lines in areas where herring were expected to spawn. Once spawning was complete, the egg-laden kelp was harvested.

To gain even more control, fishermen hung the kelp in a netting enclosure, then they captured about-to-spawn herring with purse seines and transferred them into the *closed pound*. Once the herring spawned, they were released and the egg-laden kelp was harvested.

The evolution of the herring spawn-on-kelp fishery in Alaska has been the replacement of the wild fishery with pound fisheries that, though they produce a superior product, employ fewer fishermen than the wild fisheries and have concentrated the wealth derived from the fishery.

The spawn-on-kelp fishery developed parallel to the roe-herring fishery. In 1958, the Japanese, whose own herring stocks had collapsed, were looking for a replacement supply of herring roe and of herring spawn on kelp. There must have been some discussion of developing a spawn-on-kelp fishery in Southeast Alaska because the territorial

ADF&G recommended in 1958 that the commercial taking of herring spawn for export purposes be prohibited until herring populations in Southeast Alaska were "at a high peak of abundance."[4]

Alaska became a state on January 3, 1959. The Alaska Statehood Act, however, specified that Alaska would not obtain jurisdiction over its fish and wildlife resources until it could convince the federal government that adequate provision for the administration, management, and conservation of those resources had been made.[5]

Nevertheless, Alaska's Legislature had since its inception as a territorial body maintained an interest in Southeast Alaska's herring resource. As discussed in part I, most of the legislature's actions were focused on the elimination of herring reduction plants, and the Statehood Act's jurisdictional provision didn't deter the new state legislature from passing a bill to protect Southeast Alaska's herring. The legislation, passed on March 11, 1959, and, echoing ADF&G's 1958 recommendation, made it illegal to commercially take herring spawn in state waters. The legislation did not prohibit Alaska residents from taking herring spawn for personal consumption or for "barter or exchange for the necessities of life."[6] Until the state gained jurisdiction over its fisheries, however, the legislation was meaningless. There were at the time no federal regulations limiting where herring spawn on kelp could be taken, by what means it could be taken, or the quantity that could be taken.[7]

In passing this legislation, Alaska's Legislature may have been attempting to prevent a planned herring spawn-on-kelp fishery in the vicinity of the communities of Craig and Klawock, on the west shore of Prince of Wales Island, in southern Southeast Alaska, where herring spawning normally occurs in middle to late March. If that was the intent, it was unsuccessful. In its annual report for 1959, ADF&G reported the commercial harvest of 107,900 pounds of herring spawn on kelp, which was valued at $5,395.[8] To gather the kelp, the "kelpers" there used grappling hooks, which substantially tore up the kelp beds.[9]

At least some of this product—salted in barrels—may have been sold to Mutual Fish Company, a Japanese American wholesale/retail seafood business in Seattle. Harry Yoshimura, whose family owned the business, recalled that the company purchased herring spawn on kelp from an Alaska fisherman at about that time.[10]

Alaska gained jurisdiction over its fish and wildlife resources in January 1960 and quickly began the process of managing them according to the sustained-yield principle, as mandated in the state's constitution.[11]

Its new authority in hand, Alaska's Legislature, recognizing the damage that had been done to the kelp beds and to the herring resource, yet desirous of fostering a new industry, in early 1960 amended its

1959 law. The amended legislation authorized a commercial herring spawn-on-kelp fishery, but with restrictions: individuals or businesses wanting to harvest herring spawn on kelp were required to obtain a permit from ADF&G and could take the product only at "places and times and in such amounts and manner as may be permitted by the Board of Fish and Game." The legislation became effective on March 17, 1960, in time to allow a herring spawn-on-kelp fishery in the Craig/Klawock area.[12]

There was, however, no wild (natural) spawn-on-kelp fishery in the area in 1960 or 1961. Instead, there was an open-pound fishery at Craig in 1960, probably the first such operation in Alaska. Two individuals there gathered *Macrocystis* kelp and suspended it on the grounds open-pound style. According to a Fish and Wildlife Service report, approximately two-thirds of the weight of the forty-six tons of the spawn on kelp harvested that spring consisted of roe. The remaining approximately one-third was the weight of the kelp. The report also stated that biologists "claimed" the amount of roe removed in this operation had no measurable effect on herring production.[13]

According to the federal Bureau of Commercial Fisheries, the first export to Japan of herring spawn on kelp occurred in 1962. The bureau's report did not elaborate, but ADF&G's catch summary for 1962 listed 46,150 pounds of "herring kelp, salted" worth $22,500 as being produced in Southeast Alaska.[14]

The wild herring spawn-on-kelp fishery at Craig/Klawock grew substantially in 1963. During the two-week season that opened on April 3, an estimated 150 kelpers harvested 200,176 pounds of product. The kelp was processed in Petersburg. Alaskan Glacier Seafoods, which primarily operated a shrimp-canning plant, was likely involved. Dave Ohmer, whose father operated the plant, remembers barrels of spawn on kelp stored in the alley next to the plant. The other kelp processor at Petersburg was Kito Enterprises, which likely processed its product at the Petersburg Cold Storage plant.[15]

Three important aspects of the spawn-on-kelp fishery were that the participants were largely Alaskans, the fishery occurred at the time of year when other fishing income was lowest, and gear and vessel requirements were minimal. (Most kelpers worked out of open skiffs.) The Bureau of Commercial Fisheries estimated that a larger percentage of the income from this fishery was spent in the villages and small communities of Southeast Alaska than for any other fishery.[16]

The herring spawn-on-kelp season at Craig/Klawock in 1964 lasted twelve hours. During that time, 200 kelpers harvested 228,404 pounds of product, a little more than the 100-ton quota that ADF&G had established by emergency order. The number of processors on the grounds that year had swelled to seven.[17]

At Sitka in 1964, a spawn-on-kelp fishery had a modest beginning. There, fifteen kelpers harvested fifteen tons of product over three days. In anticipation of an increase in effort the following year, the Alaska Board of Fish & Game during its fall 1964 meeting established a seventy-five-ton quota for the Sitka area's 1965 season.[18]

In 1965, the Sitka area herring spawn-on-kelp fishery, in the words of ADF&G, "exploded into a full-fledged Commercial fishery."[19] The season comprised two openings. The first was on the west side of Sitka Sound and lasted two hours. About 100 fishermen participated. The second was on the sound's east side. It lasted three hours and involved about 200 fishermen. The total harvest for the two areas was 63.1 tons, a bit short of the quota. The department described the competition among kelpers in both areas as "keen but orderly."[20]

ADF&G's 1965 Sitka area report noted that there had been 20.75 linear miles of herring spawn in the area. The report also noted that herring had not spawned in Silver Bay for the past several years, which the department believed was probably attributable to effluents from the Japanese-owned Alaska Pulp Company pulp mill that had begun operating there in late 1959.[21]

In 1966, herring-spawn-on-kelp fisheries occurred at three locations in Southeast Alaska: Craig/Klawock, Hydaburg, and Sitka.

At Craig/Klawock, where ADF&G had set the quota at 100 tons, 638 kelpers worked from 330 skiffs and 71 larger vessels. (Some of the larger vessels were used only to transport harvested kelp to processors.) Most of the kelpers were Alaskans, but one vessel, the *Fred D. Parr*, with a crew of twenty-eight, came from Seattle. The kelpers had to work fast because the single opening, on April 4, lasted only ninety minutes. When it was over, 146 tons of spawn on kelp had been harvested, greatly exceeding ADF&G's quota. Eighteen processing companies paid kelpers about $0.45 per pound for their harvests. The average kelper's payday was about $207.

ADF&G's quota for the spawn-on-kelp fishery at Hydaburg was fifty tons. The fishery there, at McFarland Island, opened on April 25 and, with a one-hour duration, was even shorter than the Craig/Klawock fishery. During that time, 435 fishermen working from 320 skiffs and 61 larger vessels harvested 132 tons of product, more than doubling the quota. The thirteen processors on the grounds paid $0.65 per pound for the product, and the average kelper's payday was $394.[22]

The fishery at Sitka, for which ADF&G had the previous year established a quota of seventy-five tons, had an interesting twist. Bob

Wyman, owner of the Katlian Packing Company, saw an opportunity for increased production by transplanting kelp from an area where herring didn't spawn to an area where they usually spawned. Prior to the kelping season's opening, Wyman paid divers to harvest about twenty-five tons of *Macrocystis* kelp near Kruzof Island, in Sitka Sound, where herring did not spawn. The divers passed the kelp up to workers on a barge, who hung it around barge's perimeter. Wyman then towed the barge to an area near Middle Island that was a favored spawning area for herring but was full of eelgrass. There, his divers "transplanted" the *Macrocystis*, putting the kelp's holdfasts, the gnarled structures that attach kelp to rocks, into bags that were weighed down with rocks.

Wyman's scheme didn't sit well with some Sitkans. They resented the fact that the transplanted kelp was for Wyman's exclusive use, and they complained to ADF&G. The department sought legal advice and subsequently decided that the kelp could be harvested by anyone who chose to do so.

The Sitka opening lasted ninety minutes. There, 423 kelpers working from 249 skiffs and thirty-nine large vessels harvested fifty-eight tons of spawn on kelp, including Wyman's. Wyman later estimated that 90 percent of the kelp he had transplanted was harvested by others. Shortly after the season closed, ADF&G announced an emergency regulation that prohibited the taking of herring spawn on transplanted kelp.

The total spawn-on-kelp production in Southeast Alaska in 1966 was 274 tons (trimmed weight, apparently). ADF&G estimated its ex-vessel value to be about $600,000. Another 2.5 tons was produced in southcentral Alaska.[23]

Despite his frustrating experience in 1966, the ever-enterprising Bob Wyman in 1968 cut a quantity of *Macrocystis* kelp in Sitka Sound and chartered an old herring seiner, the *Victory*, to take it to Kodiak. There, the kelp was not transplanted but was hung open-pound-style from ropes strung between anchored buoys. The intensity of the herring spawn at Wyman's chosen location, however, was insufficient to obtain a satisfactory product.

Undeterred, Wyman subsequently sent a load of *Macrocystis* kelp aboard the floating processor *Alaska Star* to Togiak, in Bristol Bay. The kelp spoiled en route and was discarded overboard. As will be discussed below, transporting *Macrocystis* kelp from Southeast Alaska to Prince William Sound for spawn-on-kelp operations would become an enterprise of considerable magnitude.[24]

While the spawn-on-kelp fishery at Hydaburg was closed after the 1966 season due to reduced spawn deposition, the 1967 season at Craig/

Klawock was even more intense than the previous season, lasting only twenty minutes. On the grounds were 617 skiffs and 87 purse seiners and packers, and during the brief opening 1,230 kelpers—among them the Petersburg women's bowling team—harvested 101.5 tons of product for which the grounds price was $0.70 per pound. An ADF&G spokesman later said the fishery was "as close to being unmanageable as it could get." Sadly, three individuals, a man from Metlakatla and two brothers from Hydaburg, drowned in connection with the fishery.[25]

At Sitka, the season lasted forty-five minutes. During that time, 850 kelpers harvested 81.3 tons. Buyers on the grounds paid kelpers $1 to $1.25 per pound, which meant the average kelper at Sitka earned about $200. The first-wholesale value of the product was about $2 per pound.

Because of the high market value of herring spawn on kelp—pound for pound, spawn on kelp then had a higher ex-vessel value than halibut or sockeye salmon—its ease of collection, and the simple processing requirements, the temptation to harvest it illegally was great. State troopers in 1967 apprehended at least three buyers and a large number of kelpers for trafficking in illegally taken product.[26]

In its 1967 annual report on the Sitka area, ADF&G noted that the herring spawn-on-kelp fishery had "mushroomed into a full-fledged, nearly uncontrollable, commercial fishery."[27] To reduce the "gold rush" nature of the spawn-on-kelp fishery in Southeast Alaska, in 1968 the Board of Fish & Game adopted a regulation that required kelpers in the region to register for a single area. The quota for the area would then be divided among the registered kelpers.[28]

The regulation never got tested. Spawning herring at Craig/Klawock in 1968 were far less numerous than they had been in previous years. Where there had been an average of twelve lineal miles of spawn during the early years of the fishery, by 1968 it had declined to only half that amount. Moreover, the fish that did show up in 1968 behaved atypically. Spawn deposition in previous years occurred over about a ten-day period, but in 1968, spawning at Craig/Klawock was sporadic over a thirty-day period, and the amount of spawn–on–*Macrocystis* kelp available was not sufficient to provide for a fishery. At Sitka, there was a normal spawn, but the herring avoided the available *Macrocystis* kelp beds, and the fishery there was canceled.[29] There was, however, an interest among buyers for limited amounts of herring spawn on hair kelp, and ADF&G used its emergency powers to authorize its harvest in the Sitka area. Production totaled thirty-six tons.[30]

The Craig/Klawock and Sitka wild spawn-on-kelp fisheries were never reopened, but in 1992 closed-pound spawn-on-kelp operations began at Craig/Klawock. This fishery is discussed at length below.

For several seasons prior to 1967, ADF&G had issued herring spawn-on-kelp permits in its lower Cook Inlet area, which extended east almost to Prince William Sound. The fishermen there were apparently looking for *Macrocystis* kelp, which does not grow in southcentral Alaska, so their ventures failed. In 1967, however, one fisherman found spawn-laden kelp of an undetermined species in Aialik Bay (just west of Resurrection Bay) and used grappling hooks to harvest it. His production was 19,800 pounds, which was packed in wooden barrels that each held about 200 pounds. About half the production was reportedly sold for $1 per pound, the remainder for somewhat less.[31] This harvest seems to have been a one-time event.

In 1968, while the herring spawn-on-kelp fishery temporarily waned in Southeast Alaska, a similar fishery had its inception in Bristol Bay. The fishery took place at Eagle Bay, a small bay just east of Togiak, and produced 54,600 pounds of herring egg–laden rockweed (*Fucus*) kelp. It is discussed below.

The following year, 1969, marked the beginning of the herring spawn-on-kelp fishery in Prince William Sound. It started in Landlocked Bay and Johnson Cove, in the northeastern reaches of the sound, where fourteen kelpers working from skiffs mostly used grappling hooks to harvest ninety-five tons of herring egg–laden kelp growing along the shore. Ribbon kelp was the most desirable, but small amounts of sieve kelp and hair kelp were also harvested. Subsequent reports indicated that the product was of a poor grade and was contaminated with silt and sand.[32]

Quality control improved, and the herring spawn-on-kelp fishery in Prince William Sound quickly expanded and became something of a first-rite-of-spring free-for-all for the local population. By 1971, more than 150 kelpers, some of whom were divers, were involved in the annual harvest of herring spawn on kelp.

I was among the kelpers in Prince William Sound in the spring of 1972. The fishery opened at 6:00 a.m. on May 1, and within four hours at Landlocked Bay, my skipper and I had grappled about a thousand pounds of kelp. It was of marginal quality, and we spent the rest of that day and most of the next day trying to get a tender to buy it. Ultimately, we were able to sell 190 pounds of our harvest—the portion of our catch that met minimum quality standards—for $0.50 per pound. We discarded the remainder and ended our brief career as kelpers.

Prince William Sound's herring spawn-on-kelp fishery peaked in 1975, when 437 kelpers harvested 459 tons. In the process, Bidarka Point, a traditional subsistence harvest area for Natives living in the nearby village of Tatitlek, on the eastern shore of Prince William Sound, was completely denuded of kelp. Concerned about depletion of kelp beds, in 1977 the Alaska Board of Fisheries outlawed grappling hooks and required that kelp be cut by a "hand-held unpowered blade-cutting device"—a knife—and that kelp blades be cut at least four inches above the stipe (stalk), leaving the holdfast and the regenerative portion of the plant intact. Due to the board's action, future harvesting would be done primarily by knife-wielding divers. In 1978, the board established a guideline harvest level of 200 tons. Production in 1978, however, was only seventy-one tons.[33]

HERRING POUNDS

Open Pounds

To reduce waste and gain more control over the situation and to produce a more uniform and higher-quality product, fishermen began employing what were referred to as open pounds. They suspended kelp blades from weighted lines stretched between anchored buoys in an area where herring were expected to soon spawn. Once spawning was complete, the egg-covered blades were removed from the water and layered with salt in wooden boxes or plastic buckets.

ADF&G defines an open pound as "suspending kelp from a floating frame structure in an area where herring are spawning. The herring are not impounded by a net but instead are allowed to naturally spawn on the suspended kelp."[34]

Open pounds were employed primarily in Prince William Sound. Nevertheless, probably the first open pound in Alaska was in Resurrection Bay in 1972. ADF&G's annual report for that year described the operation:

> Seward Fisheries imported from Japan an unidentified species of kelp. Presumably, the kelp was of the variety that the Japanese savor most. It came in the form of dried strips three inches wide and about two feet long. There were 1,200 of these strips in all. The idea was to tie the strips on a line and suspend the line between two floats in an area where the herring were spawning in hopes that the herring would spawn on the kelp. A permit was obtained from the Commissioner to place 100 of these strips in the boat harbor on an experimental basis only, with the remaining strips to be placed in Resurrection Bay outside the small-boat harbor. Seward Fisheries was to choose the outside locations. The strips were prepared, first by soaking in a brine solution and

then in straight sea water. The prepared kelp was then tied to a halibut line and hung on the dock in hopes that spawning herring would make use of it.

Unfortunately, no spawning herring were found. The kelp fermented, and the project was abandoned.[35]

Closed Pounds

> The impetus behind the development of the pound-type fishery has been the desire to eliminate some of the uncertainties that have plagued the wild kelp fishery. The pound technique involves the confinement of mature herring in a small enclosure (pound) along with carefully selected kelp, and hopefully forces the fish to deposit their eggs.—Alaska Department of Fish and Game, 1982[36]

> In concept, pounding, also called ponding, is simply a matter of constructing a frame, fitting it with webbing, waiting, hanging kelp, seining herring for the pound, waiting, then, hopefully, harvesting a consistently superior product.—D. B. Pleschner, *Pacific Fishing*, 1986[37]

To gain even better control over the process of obtaining spawn on kelp and to produce a uniformly higher-quality product, Alaskans borrowed technology that had been employed for several years by British Columbia spawn-on-kelp harvesters (who had borrowed it from local First Nations). They suspended carefully selected blades of kelp from lines stretched near the surface across a floating rectangular log-frame structure from which an enclosure of webbing was suspended—a closed pound. They then caught almost-ready-to-spawn herring with purse seines and transferred them live into the pound. The fish were confined there until they spawned, after which they were released. Several days later, after the eggs had cemented themselves to the kelp, the spawn-on-kelp was harvested. If all worked according to plan, the resulting product, because of its uniformity and the care with which it was handled, was far superior to the wild variety or the open-pound variety.

ADF&G defines a closed pound as "a single, floating, rectangular frame structure with suspended webbing that is used to enclose herring for a period of time in order to produce spawn-on-kelp."[38]

The first successful employment of closed pounds in Alaska was in Prince William Sound in 1980, but the *Exxon Valdez* oil spill in 1989 marked the beginning of the end of impound operations in Prince William Sound. Closed pounds were operated at several locations in

Southeast Alaska, and they continue to be operated near the Prince of Wales Island communities of Craig and Klawock.

For management purposes, ADF&G allocates a portion of an area's herring guideline harvest level to the spawn-on-kelp sector. The department also calculates the annual removal of reproductive capacity from the herring population resulting from spawn-on-kelp harvests. It assumes that the average roe recovery from about-to-spawn herring is 10 percent and that 80 percent of the spawn-on-kelp harvest weight consists of eggs. The department estimated that the 1975 Prince William Sound spawn-on-kelp harvest of 459 tons (the record) removed 3,668 tons of reproductive capacity from the spawning population.[39]

Stress mortality on impounded herring is also a consideration. ADF&G assumes 100 percent mortality for herring captured in the spawn-on-kelp fishery, but the actual percentage is almost certainly lower. "Not all survive—we don't have a good handle on mortality—but a good portion survive to come back the following year," said Dave Gordon, ADF&G's fishery biologist at Sitka, in 2006.[40] Operators influence mortality. Some stuff their pounds with herring and end up with profuse roe coverage but kill a lot of fish in the process. An advantage of ADF&G's assumption of 100 percent mortality is that any error favors the resource.

The process of transferring herring from the seine into the pound is accomplished with a pushable/towable "tow pound" or "net pen" that, like the seine, is fitted with net doors (figure 16.3). Once fish are captured in the seine, the net pen is brought alongside the seine, and the door on the seine is laced together with the door on the net pen. Then the seine is "dried up," forcing the fish to swim through the opening into the net pen. The web door of the pen is then raised and closed, completing the transfer. The net pen is then towed to the pound site, where the herring are transferred from the net pen into the pound in a similar fashion.

The first attempt to employ closed pounds in Alaska was by two individuals in Prince William Sound in 1979. Due to a lack of time to construct the pounds, neither operation was brought to fruition, but the men had imported a quantity of *Macrocystis* kelp from Southeast Alaska, and one of them attached the kelp to lines suspended between anchored buoys, open-pound style, in Valdez Arm, and some production did occur.[41]

The closed-pound system was perceived to have considerable potential, and the following year ADF&G issued fourteen permits for closed pounds in Prince William Sound. Four pounds were built, but only two,

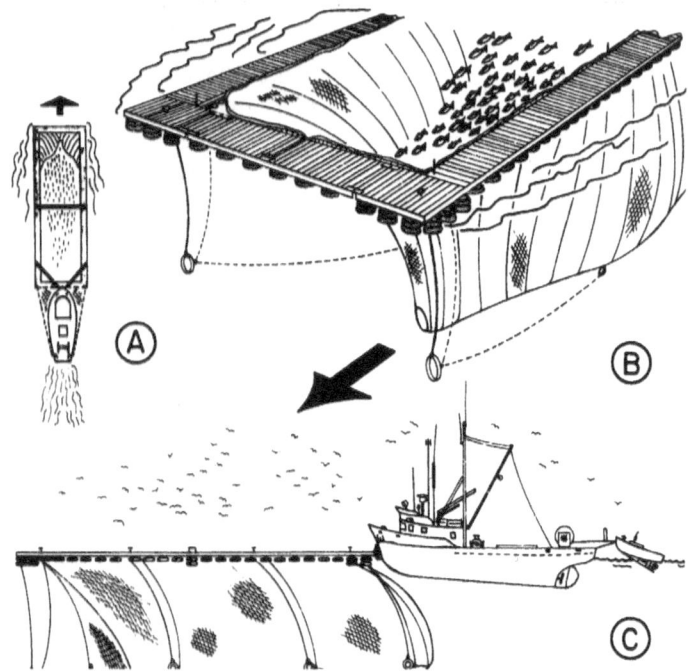

Figure 16.3. Tow pound. (*Pacific Fishing*, February 1985)

both in Landlocked Bay, were utilized. One pound was hung with locally sourced ribbon kelp and the other with *Macrocystis* kelp imported from Southeast Alaska. Production from the two pounds totaled 8.6 tons of egg-laden ribbon kelp and 0.4 tons of egg-laden *Macrocystis* kelp. The operator of the pound stocked with ribbon kelp considered the quality of his egg-laden product to be excellent, with much denser egg coverage than any of the kelp obtained from the wild harvest.

Interest in wild spawn-on-kelp, however, remained strong. The harvest in 1980 involved ten buyers and a whopping 469 divers. The harvest totaled 306 tons, which was about eighty-five tons more than the guideline harvest level. ADF&G attributed the overage in part to illegal harvesting and stockpiling of kelp during the closure between the periods open to harvesting and to the sale of "subsistence harvest" kelp.[42]

17

PRINCE WILLIAM SOUND HERRING SPAWN-ON-KELP FISHERIES, 1981–1999

The following is a chronology of the Prince William Sound spawn-on-kelp fishery for the years 1981 through 1999.

1981

Participation and production in the herring-pound fishery grew quickly. In 1981, eighteen pounds were deployed in Landlocked Bay, which ADF&G designated as the pounding site. The department allocated sixteen tons of spawn-laden kelp (1,800 pounds of finished product per permit holder) and provided a 200-ton herring quota that was divided equally among the pound operators. Eleven of the operators harvested 8.8 metric tons of what appeared to be a fine product.

Additionally, several "pounders" (pound operators) imported *Macrocystis* kelp from the Sitka area and suspended it open-pound style in areas that spawning herring were known to utilize. The experiment was a failure because the spawning herring behaved abnormally that season, and no production resulted.

In the wild fishery, 214 divers harvested fifty-five metric tons of kelp during a twelve-hour opening on April 24. It was the smallest harvest since the fishery began. ADF&G that year noted a "cyclic pattern in the amount and quality of the harvest . . . [in] the kelp fishery due to fluctuations in herring abundance and changes in timing, location and density of spawn in traditional harvest areas."[1]

1982

In 1982, to protect wild spawn-on-kelp harvest areas, ADF&G restricted closed-pound operations to portions of Landlocked Bay and Boulder Bay. Eleven pounds were deployed, and four individuals imported an estimated six tons of *Macrocystis* kelp from Southeast Alaska. A small portion of this was suspended in two of the pounds. The remainder had either deteriorated beyond usability or was suspended in open pounds in traditional spawning areas. Production, all of which came from seven pounds, totaled 22.8 tons of ribbon kelp and 0.4 tons of *Macrocystis* kelp.

The wild spawn-on-kelp fishery involved 151 divers and eight buyers. Production during seventy-three hours of kelping totaled 155 tons.[2]

1983

Thirty closed pounds were operated in Prince William Sound in 1983. Complicating matters was the lack of herring in the traditional areas, and the pound operators were forced to relocate north to Galena Bay—for most operators a tow of over fifteen miles. Their production, which consisted mostly of ribbon kelp, was twenty-eight tons, and ADF&G estimated its value was probably in the neighborhood of $250,000. (The value of the fishery was difficult to measure because most pounders marketed their own product.)

The lack of herring in the traditional areas also forced the harvesters of wild spawn-on-kelp, who mostly worked from small open skiffs, to relocate to the Naked Island area, thirty miles southwest of Galena Bay. Fortunately, the weather was good. The wild-harvest season consisted of a single twelve-hour opening on April 27 during which 186 divers working from more than 100 boats harvested 150 tons of primarily ribbon and sieve kelp. Nine buyers paid approximately $650,000 for the total harvest.

An encouraging development in the Prince William Sound spawn-on-kelp fishery in 1983 was the success of several fishermen in producing spawn on open-pounded kelp. The fishermen had imported *Macrocystis* kelp from Southeast Alaska and strung it in the midst of natural spawning prior to a wild-kelp opening, then harvested it when the fishery opened. Their production was almost twenty tons, about 13 percent of the sound's wild-kelp harvest that year.[3]

1984

Herring were generally abundant in Prince William Sound in 1984, but because of the distribution and long duration of spawning, egg coverage on kelp was never of sufficient quantity to justify a commercial harvest, and there was no wild spawn-on-kelp fishery in the sound.

There were, however, both closed-pound and open-pound fisheries. *Macrocystis* kelp, 61.5 tons of which were imported into the sound, figured big in the fisheries. Closed-pound operators imported approximately 16.5 tons of *Macrocystis*, and, in addition to a quantity of ribbon kelp, used it in the forty-five closed pounds that were deployed. Thirty-seven of the pounds produced, and their total harvest was 28.5 metric tons, 64 percent of which was *Macrocystis*, the remainder being ribbon kelp. The total estimated value was $270,000.

The open-pound fishery was based entirely on forty-five tons of *Macrocystis* kelp imported from Southeast Alaska. (This large quantity of kelp was likely transported in a boat, rather than as air cargo, as would become the standard practice.) Unfortunately, at least twenty tons were lost because of poor handling and poor timing. The remaining approximately twenty-five tons was suspended in open pounds spread over an area from Valdez Arm west to Esther Island. As was the case with the wild fishery, herring spawn was inadequate, and none of the open-pounded *Macrocystis* was harvested.[4]

1985

The nonexistent 1984 wild spawn-on-kelp season in Prince William Sound reduced interest in the fishery, and in 1985 only seventy-nine divers participated. Due to poor quality of kelp, which was mostly attributed to storms that embedded a layer of silt between the herring spawn and the kelp blades, buyers turned down a significant portion of the harvest. The salable harvest totaled thirty-eight tons, about half ribbon kelp and half sieve kelp. It had an ex-vessel value of approximately $36,300.

Fifty-nine closed pounds were deployed, of which fifty produced a total of forty tons of spawn on kelp. Twenty-eight tons were *Macrocystis* kelp, and the remainder was ribbon kelp. This was the last year significant quantities of ribbon kelp were used in the fishery. The *Macrocystis* kelp–based product was worth $8 per pound; the ribbon kelp–based product was worth $5 per pound. The total value of the fishery was approximately $570,000.[5]

1986

The wild spawn-on-kelp fishery in 1986 attracted only twenty-nine divers. Their production, almost all ribbon kelp, totaled forty-seven tons, about a third of the sector's GHL. Overall, the quality of the product was above average, and three processors paid a total of approximately $160,000 for the harvest.

Primarily to compensate for the vagaries of herring spawning, ADF&G relaxed its regulation and permitted both the wild fisheries and closed-pound fisheries to occur in all districts in the sound. The number of pounds continued to increase; eighty-two were deployed in 1986. ADF&G regulations specified minimum dimensions for a pound: fifty by twenty-five by eight feet, or the equivalent, with a surface area of at least 900 square feet and twenty-five feet on a side. Herring could be held in a pound for no longer than six (later, eight) days.

In the early days, pound frames were made of notched logs. By 1986, most pounders used a modular "walk-around" system constructed of foam blocks sandwiched between lengths of plywood that were roughly two feet wide.

A unique aspect of the closed-pound fishery was its cooperative nature. To gain an economy of scale to facilitate pound construction and operations, pounders divided themselves into groups—some called them clubs—comprising five to ten individuals. (In 1990, there were 128 pounders divided into twenty groups.) The fee to enter a group could be substantial. As it was forming, one group charged $10,000, using the funds to purchase the pounds, nets, etc., that would be used in the group's first year of operations. As the roe-herring season approached, two or three members would travel to Southeast Alaska to get a supply of kelp while the others assembled, located, and prepped the pounds for the kelp.

Securing the kelp was no simple feat. ADF&G regulations required kelp harvesters to obtain a permit that specified the area to be harvested and the amount allowed. The permit also listed cutting restrictions that were necessary to maintain healthy kelp beds.

Getting the kelp from Southeast Alaska to Prince William Sound could take a week. In one group, the Glacier Kelp Group, several members would travel to Craig, in southern Southeast Alaska, and charter a fishing boat to take them to nearby Sea Otter Sound. There, they would cut kelp and pack it into waxed cardboard boxes of the sort used to ship fresh fish. Once back in Craig, they would load the boxes into a refrigerated truck that would deliver it to the Ketchikan airport. From there it would be shipped to Anchorage on Alaska Airlines. In Anchorage, the crew would load the boxes into a rented van and then drive 300 miles to Valdez. There, the kelp would be loaded aboard a member's boat and delivered to the group's pounds. In total, pound operators in Prince William Sound imported sixteen tons of *Macrocystis* in 1986, and the quantity would increase to more than twice that amount within a couple of years.

Groups would also contract seiners—some seiners were also pound operators—to capture herring for the members' pounds.

ADF&G established a guideline harvest level of sixty tons for the 1986 pound fishery, and this amount was divided equally as a production limit on each pound. Each pounder's allocation was 1,463 pounds of spawn on kelp, which represented 9.1 tons of herring and 296 pounds of kelp. *Macrocystis* kelp was utilized exclusively, and the quality of the final product was, in the words of ADF&G, "far superior to the previous year." It was valued at $8 per pound, putting the estimated total value of the seventy-two-ton harvest—which substantially exceeded the guideline harvest level—at $1.2 million.[6]

1987

Participation in the closed-pound spawn-on-kelp fishery had been increasing and seemed likely to continue to do so. A potential limit, however, was a decline in herring abundance. To stabilize the fishery at an optimal number of participants, in December 1986 the Alaska Commercial Fishery Entry Commission limited entry into the Prince William Sound closed-pound spawn-on-kelp fishery. By January 1, 1987, the commission had issued 129 permanent and interim-use permits. The commission's effort, though, was moot because in less than a decade, a combination of the *Exxon Valdez* oil spill and widespread disease in herring would cause the fishery to collapse.

During the 1987 season, 108 pounds produced 61.3 tons of product. Almost all of the kelp used was *Macrocystis*, a minor portion being locally sourced ribbon kelp. The overall quality of the product, however, was marred by the fact that the bulk of the herring impounded were three-year-olds, which had lower fecundity than the older fish normally utilized. Egg coverage was below normal, and most of the product was of low grade. Nevertheless, the market was strong, and the season's production averaged $15 per pound, making the fishery worth $1.8 million.

In the wild spawn-on-kelp fishery, ADF&G reduced the GHL to eighty-five tons due to declining herring abundance. Sixty divers operating from thirty vessels harvested eighty-eight tons of product that was valued at $284,250.[7]

1988

In 1988, ADF&G restricted the closed-pound fishery to the traditional areas of Valdez Arm and Port Fidalgo. Nevertheless, the pound fishermen's 1988 season harvest of 124 tons set a record, more than doubling the previous year's harvest. Moreover, the quality of the product surpassed that of any previous year. Fishermen received an estimated average of $18 per pound for their product, and its value totaled $4.46 million—about $37,500 for each of the 119 pounds that produced (122 pounds were deployed). By contrast, the 125 divers in the wild fishery were paid only $1.20 per pound for their product, and their ninety-seven-ton harvest yielded just $232,000.[8]

1989

ADF&G closed the spawn-on-kelp fisheries, as well as the roe-herring fishery, for the 1989 season due to the potential for contamination of catches from the *Exxon Valdez* oil spill.

1990

The 1990 spawn-on-kelp fisheries were conducted much as if the *Exxon Valdez* oil spill had never happened. The GHL for pounders was 118 tons; for the wild harvest it was 104 tons. Production in the 122 pounds that produced (six had no production) was 101 tons, and with an estimated average price of $11.40 per pound, it was worth about $2.3 million.

Seven buyers and 128 divers were involved in the wild harvest. Production was 119 tons of mostly hair and ribbon kelp that had an estimated ex-vessel value of $213,840.[9]

1991

Under the management plan for pound operators in effect before 1991, each permit holder was allowed to harvest a specified poundage of product. Measuring that poundage was an issue because the kelp was trimmed after it was harvested, and there was also some shrinkage of the product. The finished weight was less than the raw (on-grounds) weight. To address this disparity as well as overharvest by some permit holders, in 1991 the department adopted a management plan that distributed individual harvest quota to each permit holder by specifying the number of kelp blades that could be hung in each pound. The 1991 limit was 1,200.

There were 126 pounds in Prince William Sound that year. The GHL for the fishery was 220 tons, and the raw weight of their production was 202.4 tons, comfortably below the GHL. After trimming and shrinkage, the weight was approximately 160 tons. The quality of the product was lower than in past years, which was attributed to the *Macrocystis* kelp having thin, narrow blades. This drove the average price received for the processed weight down to $9 per pound, putting the value of the fishery at about $2.9 million.

In the wild fishery, marketable kelp of the traditional species (ribbon, hair, and sieve) was scarce. However, a market for herring egg–laden rockweed, an intertidal species, developed. As will be discussed below, rockweed had been harvested at Togiak since 1968. Most divers in Prince William Sound that year abandoned their diving gear in favor of handpicking rockweed on the beach at low tide. The forty-eight pickers harvested 108 tons of product, significantly less than the wild sector's 195-ton GHL. Ninety-nine percent of the harvest was rockweed, for which the average price was $0.80 per pound. The pickers' payday totaled $172,160.[10]

1992

After the 1991 season, several fishermen petitioned the Commercial Fisheries Entry Commission to limit entry into the wild fishery. The possibility that the fishery might be limited motivated individuals who had not previously participated to do so in order to establish a history that might result in their eventually receiving a limited-entry permit. ADF&G issued 385 permits, more than 100 more than were issued in 1991. Of these, 217 participated in the fishery. The total harvest by divers and handpickers was 252 tons, a bit above the 243-ton GHL. Seventy-six percent of the harvest was rockweed, and 21 percent was ribbon kelp. Prices were the lowest ever: $0.40 for rockweed and $0.70 for ribbon kelp. Processors turned away boatloads of kelp that did not meet their standards. This kelp may have been sold to other processors or dumped. ADF&G regulations did not prohibit discarding product.

The pound fishery in 1992 was similar to the fishery in 1991, except the quality of the product was even lower. Production from the

127 pounds that were installed totaled 210 tons, well short of the 276-ton GHL. The average price paid was $8 per pound, a dollar per pound less than in 1991.[11]

1993

The prospects for the 1993 season were rosy. ADF&G's model estimated a biomass for 1993 of 134,133 tons, the highest projection on record. Likewise, the GHL for the pound fishery was the highest on record: 305 tons. Reflecting the anticipated abundance of herring, ADF&G increased the number of blades allowed per pound to 1,950. Permit holders deployed 124 pounds.

On April 10, after test sets indicated mature fish were present, ADF&G opened, in a limited area, the seine fishery that would supply the pounds. Test fishing continued, but an adequate supply of herring failed to show, and the department closed the fishery on April 20. Despite the projected high abundance, herring were scarce throughout the sound. ADF&G did not open the general roe-herring seine fishery, although it did open the roe-herring gillnet fishery.

Nevertheless, while fishing to supply the pounds was open, nineteen enterprising pounders in Galena Bay introduced token numbers of herring into their pounds. They then towed the pounds to areas where herring were spawning. There, they reintroduced kelp and dropped the webbing surrounding their pounds, ostensibly to release the herring. At the same time, this allowed spawning herring to swim in and deposit their eggs. Other pounders followed suit. ADF&G didn't condone this practice—a variation on the open pounds that had been used in earlier years—but there was no regulation that prohibited pounders from dropping their web once herring had been introduced.

It was worth the effort, because when it was over, fifty-two pounds produced 106 tons of untrimmed product, worth $2 million.

In the wild spawn-on-kelp fishery, eighty-three fishermen harvested 163 tons of product, all of which was rockweed. Its ex-vessel value was $178,860.[12]

1994

At its February 1994 meeting, the Board of Fisheries increased the minimum herring biomass threshold from 8,400 tons to 22,000 tons.

Peak biomass of spawning herring that spring was 19,647 tons, below the threshold level for opening the sound's herring fisheries. But there was more: the age structure of the fish indicated that the stock was becoming senescent. Moreover, many of the fish caught in test sets were infected with a virus (see chapter 11).[13]

1995

ADF&G aerial surveys in April and early May in 1995 determined the biomass of spawning herring in the sound to be 7,100 tons, the lowest ever recorded and below the threshold level for opening any of the commercial herring fisheries there.[14]

1996

Peak biomass of spawning herring had increased to 10,600 tons, but this was still below the threshold level, and all spring herring fisheries were canceled. More than half the biomass was age eight or older, though there was an encouraging showing of age-3 and age-4 recruits.[15]

1997

The biomass of spawning herring in Prince William Sound in 1997 exceeded the threshold level, and for kelpers the season was a semblance of normalcy. During two four-hour periods in late April, forty-four kelpers harvested 26.4 tons of herring roe on rockweed, worth an estimated $32,000.

In the pound fishery, ADF&G gave permit holders the option of choosing to operate an open pound with a kelp quota of 640 blades or a closed pound with a kelp quota of 410 blades. Eighty-four fishermen elected to operate open pounds, and seven elected to operate closed pounds. To accommodate the open-pound operations, the season was open from April 10 to May 6. Operators of open pounds reported marginal success, and the total production was 26.7 tons, worth an estimated $427,000.[16]

1998

One buyer and thirty-five kelpers participated in the wild fishery. Production totaled 29,337 pounds of rockweed and 5,458 pounds of ribbon kelp, the latter of which was harvested by divers. This was the final wild herring spawn-on-kelp fishery in Prince William Sound.

In the pound fishery, ADF&G once more gave permit holders options. They could operate an open pound with a kelp quota of 660 blades or a closed pound with a kelp quota of 425 blades. Moreover, open pounds would be allowed to be placed in the Montague Island area, where most of the sound's herring had spawned in recent years.

The season opened on April 11 for the deployment of open pounds and April 12 for seining herring for introduction into closed pounds. Twenty-two permit holders deployed closed pounds and eleven deployed open pounds. Unfortunately, much of herring initially captured for impoundment in the closed pounds were spawn-outs, and nine of the operators of closed pounds switched to open pounds. (The thirteen operators who stuck with the closed pounds later secured suitable herring.) Production for the closed pounds totaled 4,780 pounds; for the open pounds it was 8,200 pounds.[17]

1999

For the 1999 season, ADF&G gave fishermen the option of operating an open pound with a kelp quota of 680 blades or a closed pound with a kelp quota of 435 blades. Additionally, open pounds would again be allowed in the Montague Island area.

Because of an overall low abundance of herring, however, there was little interest in the fishery. Two permit holders together operated the single open pound that was deployed, and seven permit holders combined to operate two closed pounds. (Four permit holders operated one pound; three permit holders operated the other.)

Seining for the closed pounds opened on April 16 and closed two days later. Then, on April 20, ADF&G announced that due to a widespread lack of herring, the 1999 roe-herring fisheries in Prince William Sound—including the wild spawn-on-kelp fishery—were canceled. All herring impounded in the closed pounds had to be released and all kelp harvested by April 25. Likewise, kelp in the open pounds had to be harvested by that same date.

The total spawn-on-kelp harvest for the two closed pounds was 10,000 pounds, while the harvest for the single open pound was 12,307 pounds.

And so ended the herring spawn-on-kelp fishery in Prince William Sound. None of the sound's spawn-on-kelp, roe herring, or food/bait fisheries have reopened.[18]

18

SOUTHEAST ALASKA HERRING SPAWN-ON-KELP POUND FISHERIES

HOONAH SOUND

Until 1990, there had not been a commercial spawn-on-kelp fishery in Southeast Alaska since the wild fishery at Craig/Klawock was closed after the 1968 season. At its January 1989 meeting, the Alaska Board of Fisheries approved a closed-pound fishery that would begin in the spring of 1990 in Hoonah Sound, on Chichagof Island, approximately thirty-five miles northeast of Sitka. Important in the board's favorable consideration was that the herring there were not being otherwise utilized.

Though the spawning population of herring in Hoonah Sound is relatively small, a record spawn of seventeen nautical miles was observed there in 1989, and ADF&G estimated the spawning population that year was approximately eight million pounds. Herring spawning in Hoonah Sound normally occurs during the last two weeks of April. The Board of Fisheries established an annual herring-harvest limit of 150 tons for the new fishery, with the requirement that the fish be divided equally among pound operators. (The board dropped the equal-distribution requirement prior to the 1994 season.)[1]

The Hoonah Sound fishery sparked considerable interest. Approximately 400 fishermen applied for permits, though only 104—most of whom were inexperienced—deployed pounds by the April 5 cutoff date.

ADF&G's primary method of managing the fishery and achieving harvest objectives was to specify the maximum number of kelp blades that could be placed in a pound. The number was based on a ratio of one ton of spawn on kelp produced for each 12.5 tons of herring allocated to the fishery. For example, a herring allocation of 125 tons would be expected to produce approximately ten tons of spawn on kelp. The department assumed that approximately 25 percent of the weight of the final spawn-on-kelp product was kelp. Thus, a harvest target of ten tons of spawn on kelp would require approximately 2.5 tons of kelp. One ton of kelp was considered to equal 7,000 blades, and the department divided the number of blades in the kelp allocation equally among pound operators. For the 1990 season, each pound operator was allocated 240 blades.[2]

By regulation, the *Macrocystis* kelp utilized was obtained under permits issued by ADF&G at locations where herring historically did

Figure 18.1. Stringing *Macrocystis* kelp for closed-pound, Craig/Klawock. (Tom Swanson)

not spawn. Sitka Sound was then the approximate northern boundary of naturally growing *Macrocystis* in commercial quantities, but it was off-limits to harvest, so the kelp mostly came from the Prince of Wales Island vicinity. Today, *Macrocystis* kelp is abundant at least as far north as Cape Cross, at the south entrance to Cross Sound.[3]

Regarding the 150 tons of herring allocated to the fishery, ADF&G initially allowed operators to introduce 2,300 pounds of herring into each pound. Seiners, however, weren't particularly supportive, and it became clear that many pounds would not be stocked before spawning ended. To the benefit of those who had the wherewithal to stock their pounds, the department raised the amount to 6,000 pounds of herring and the number of kelp blades to 300.

The spawn-on-kelp harvest totaled 9.7 tons. When it was all over, Bob DeJong, ADF&G biologist in charge of the fishery, characterized the season as "a learning experience" for both fishermen and the department.

And lucrative the fishery was not. The ex-vessel price for the production averaged $8.46 per pound, and the total value of the harvest was $201,000. The pound operators' ex-vessel income averaged a little over $2,000.[4]

Revenues were similar in 1991, but they increased in 1994, when the ex-vessel spawn-on-kelp price hit a record $25.74 per pound and the harvest was 32.7 tons. The fishery was worth $1.7 million, and ex-vessel income for the 109 pound operators averaged $15,444.[5]

During the 1990 and 1991 seasons, seined herring were towed to the pound site in the seine. Substantial numbers of fish were injured or killed in the process. In 1992, ADF&G outlawed transporting herring in a seine and required that transport to the pound site be done in a towable net pen or in the pound itself. The department advised pound operators to tow slowly to avoid crushing herring. Later, it mandated that "transport to the pound site may be done only with a pushable net pen or a separate tow pound and not with the seine." The department also stipulated that "transfer of herring is only allowed by temporarily joining nets together and allowing the herring to swim freely into the transfer net or pound."[6]

Not all fishermen complied with this requirement, and the fish paid a price. Quoting ADF&G's summary of the 1994 spawn-on-kelp fishery at Hoonah Sound:

> Most rough handling occurred during the first day of fishing when large sets were made and herring were drug in seine nets prior to transfer into the herring pounds. Although this practice violates permit stipulations, neither managers nor the enforcement officer were on the grounds when this practice was occurring. In many pounds, herring were heavily descaled, in addition to overcrowded. In a few cases, herring were so badly descaled that herring eggs could be seen adhering to the backs of impounded fish.[7]

The excessive quantity of herring utilized in the fishery was another concern. Based on the amount of product sold, ADF&G determined that the quantity of herring captured at Hoonah Sound during the 1994 and 1995 seasons was four times the 150-ton limit that had been in effect since the fishery was established. The amount captured in 1995 represented fully two-thirds of the area's spawning population. To rein in this excess, in January 1997 the Board of Fisheries adopted regulations that eliminated the 150-ton limit and authorized ADF&G to manage the fishery based on biomass estimates, with a guideline harvest level (GHL) of 10 to 20 percent of the estimated biomass.[8]

In January 1995, the Alaska Commercial Fisheries Entry Commission adopted a regulation placing the herring spawn-on-kelp pound fishery in Hoonah Sound under limited entry, beginning with the 1996 season. The commission subsequently limited the fishery to 108 permits.[9] It would take years, however, for the commission to determine final eligibility, and in 1995 there were 152 applicants for the Hoonah Sound fishery. One hundred and thirty-two pounds were built, of which 121 produced a marketable product. Both the harvest and the ex-vessel price of spawn on kelp had declined, and the fishery was worth $1.2 million.

The limited-entry system was irrelevant for the 1996 Hoonah Sound season: the forecasted biomass of mature herring was below ADF&G's 1,000-ton threshold for Hoonah Sound, so no fishery took place. Fortunately, spawning herring were far more abundant in 1997, and pounders harvested sixty-five tons of spawn on kelp. The harvest in 1998 was eighty-six tons, and in 1999 it was seventy-two tons.

ADF&G's GHL for the Hoonah Sound spawn-on-kelp fishery in 1999 was 778 tons. In 2000, the department forecasted a reduced amount of mature herring, and the GHL decreased by more than half, to 359 tons—not enough to fill the number of pounds that would likely be deployed. Ninety-six pounds were on the grounds in 1999, and 106 permit holders had applied to participate in the fishery in 2000. For management purposes, ADF&G assumed each pound required twenty tons of herring.

To provide for a fishery under the reduced GHL, ADF&G in 2000 authorized "double-permit" closed pounds. Two permit holders could operate a single pound, and the department provided an incentive for them to do so. Up to 300 blades of *Macrocystis* kelp could be suspended in a double-permit pound, as opposed to 110 blades in a single-permit pound.

The quantity of herring introduced into a double-permit pound, however, was generally about the same as for a single-permit pound. This meant that roughly the same number of eggs would be spread over more blades of kelp, lowering the grade of the final product. At least for some operators, however, the greater harvest weight enabled by the additional blades of kelp more than offset the lower grade of spawn on kelp produced, and they calculated that the gross revenue from one double-permit pound was greater than for two single-permit pounds. Moreover, double-permit pounds offered operational efficiencies over single-permit pounds.

Ninety-two permit holders chose to consolidate their operations. For the season, there were forty-six double-permit pounds and two single-permit pounds. Spawn-on-kelp production totaled thirty-six tons.[10]

In 2002, production exceeded 100 tons for the first time, and for the years 2002–2012 it averaged 195 tons, ranging from 137 tons in 2002 to 290 tons in 2010. The most lucrative year was 2008, when the harvest was worth $5.1 million. One hundred pounders harvested a marketable product, and their average gross income from the 223-ton harvest was $51,000.

At the same time, in 2008 Hoonah Sound experienced the highest recorded spawning biomass ever. There was an estimated 14.5 nautical miles of spawn and, based on ADF&G spawn deposition surveys, an escapement of 19,975 tons. Unfortunately, and for reasons unknown, herring biomass in this area had begun decreasing, though the fishery's record production—290 tons—was two years later, in 2010. The harvest in 2011 declined to 193.7 tons.

The 2012 season, with a harvest of 186.5 tons, was worth a bit more than $4 million, and was the third most lucrative season in Hoonah Sound. But the herring resource at Hoonah Sound was collapsing. ADF&G surveys of herring prior to the 2013 season indicated the biomass of herring had declined by almost half, and the season was canceled. The herring biomass decline continued, and the Hoonah Sound fishery was never reopened after the 2012 season. The fishery had lasted twenty-three years. No spawn has been documented in the sound since 2015.[11]

CRAIG/KLAWOCK

In 1988, to provide employment opportunities for the area's residents, Natives in Craig and Klawock, on the west coast of Prince of Wales Island, promoted the development of a spawn-on-kelp closed-pound fishery such as the one ongoing then in Prince William Sound. Given that the *Macrocystis* kelp utilized in the Prince William Sound fishery was sourced in the vicinity of Craig and Klawock and that an abundance of herring typically spawned in the area, it seemed like a natural for development. The group held several well-attended workshops in which experienced operators from Canada and Prince William Sound explained the closed-pound process.[12]

That fall, the Klawock Cooperative Association, a federally recognized Native organization, petitioned the Board of Fisheries to establish a commercial spawn-on-kelp fishery in the Craig/Klawock area. ADF&G surveys in 1988 indicated a healthy herring resource. That spring, approximately twenty-seven nautical miles of spawn had been observed, and the department estimated that approximately 32.7 million pounds of herring had spawned there.

The Board of Fisheries discussed the request at its January 1989 meeting but denied it.[13] In January 1991, however, the board addressed a similar proposal by the Klawock Heenya Corporation, a for-profit Native corporation established under the Alaska Native Claims Settlement Act of 1971. The corporation noted that the only utilization of Prince of Wales Island herring stocks was for bait, adding that the bait fishery provided the lowest-value commercial use of the fish. Bait fishermen, however, opposed the proposal, fearing that a spawn-on-kelp fishery would impact their catches. The board, noting that herring stocks had been increasing, authorized a small pound fishery on the west coast of Prince of Wales Island beginning in 1992. Herring typically spawn there in mid- to late March.

The decision limited the pound fishery to 15 percent of the total herring quota available for harvest in the area if that amount provided at least 100 tons of herring for the pound fishery. That meant the area quota would have to be at least 670 tons if the pound fishery was to get its 100-ton minimum.[14] The bait fishery received 85 percent of the area's total herring quota.

Herring were reasonably abundant on the west coast of Prince of Wales Island in 1992, and the herring quota for the spawn-on-kelp fishery was 403 tons. And there was a lot of interest: 531 individuals applied for permits, and 243 constructed pounds, many with logs salvaged off local beaches. A few of those who constructed pounds had been involved in the Hoonah Sound pound fishery and knew what they were doing, but for most, it was a brand-new experience.

Fishing began on March 18. Brian Kandoll was one of the newcomers and was working with several others like himself. All were excited about the fishery's prospects. They had experience gillnetting for herring and seining for salmon, and for them, seining herring was the fun part. What they didn't realize, however, was that the key to producing a good product was catching the herring at the right time.

The group easily caught enough herring for their needs and introduced them into their pounds. They expected the fish to soon start spawning, but four or five days later, the herring were still swimming in circles. ADF&G's regulations limited the time herring could be impounded to seven days so, spawn or no spawn, the fish would have to be released in two or three days.

In an act of desperation, the group attempted to stimulate spawning. They took their seine boat to Fish Egg Island, where a natural spawn was occurring. There, they pumped milty water into the boat's hold, then ran back to their pounds and pumped the milty water into them. This, they hoped, would nudge the impounded female herring into spawning. Whether the milty water caused it is debatable, but a few of the fish did eventually spawn. Unfortunately, there weren't enough eggs on the kelp to render it salable.

Probably the most unconventional operation that first year was an effort to seine herring using a Bayliner pleasure boat about twenty-six feet long and a jet ski. The Bayliner was the seiner and the jet ski was the power skiff. There was no deck winch or power block (see chapter 2) on the Bayliner, though there were a couple of blocks (pulleys) hanging amidships. The operation's seine was shallow, only about fifteen feet deep. Once the seine was set and closed, the ends of the purse lines were run through the blocks and fastened to the stern of the jet ski. The little vessel then rushed off at full throttle, and the net pursed in a matter of seconds. The crew of the Bayliner then hauled the net by hand. Kandoll, who continues to participate in the Craig/Klawock pound fishery, recalled that, though fun to watch, this operation didn't catch any herring, and the following year the Bayliner/jet ski seiners either used a conventional seiner or left the fishery.

The 1992 Craig/Klawock pound fishery product was harvested on April 4, and production totaled 25.7 tons. The quality, however, was low, which ADF&G attributed to the harvest of immature herring, poor kelp quality, and inexperienced handling of both kelp and herring. The average price was $3.50 per pound—half the price received by the Hoonah Sound pounders. The total ex-vessel value of the Craig/Klawock fishery was $180,000, and the average ex-vessel value for fishermen was a paltry $785. On a positive note, 22.6 nautical miles of spawn was observed along the area's shoreline.[15]

Herring were expected to be less abundant in 1993, and ADF&G reduced the herring quota for pounders to 240 tons. Despite this and the small fishermen's payday in 1992, interest in the fishery remained high, and 209 pounds were deployed. It was all downhill from there. A storm destroyed some of the pounds, and the fish showed up late and not in the anticipated quantities. Only twenty-three pounds (11 percent of those deployed) produced spawn on kelp, the total of which was 5.7 tons. And herring spawn along the area's shoreline had declined from the previous year by more than half, to 8.9 nautical miles.[16]

Reflecting this decline, ADF&G reduced the 1994 season quota to 135 tons. The number of pounds deployed fell to 147, but at least some operators were getting more skilled at managing them. And, fortunately for the eighty-three operators who made landings, the price of spawn on kelp had more than doubled since 1993, to an average of $11 per pound. Even with the reduced herring quota, production increased to 16.5 tons. The harvest was worth $364,000.

Though the herring quota fell again in 1995, production again increased, to 25.4 tons. Moreover, the ex-vessel price of spawn on kelp also increased, to $19 per pound, bringing the ex-vessel value of the fishery to a million dollars. And, importantly, the operators' skill had continued to increase. Of the 159 pounds deployed, fully 146 (92 percent) produced a marketable product.[17] Also in 1995, the Alaska Commercial Fisheries Entry Commission limited entry into the Craig/Klawock herring spawn-on-kelp pound fishery, effective for the 1996 season. The commission subsequently limited the fishery to 229 permits.[18]

The 1996 season again experienced a reduction in the herring quota, to 100 tons. Nevertheless, 162 pounds were deployed, and spawn-on-kelp production increased by almost half, to 37.3 tons. The ex-vessel price had increased slightly, to $20 per pound, and the 1996 fishery was worth $1.5 million.

Pursuant to the Board of Fisheries' 1991 action that established the Craig/Klawock spawn-on-kelp fishery, the fishery had been allocated 15 percent of the total herring harvest quota for the area. Bait fishermen were allocated 85 percent of the total harvest quota, but they were not utilizing all of it. For the 1994–1995 winter bait fishery, their quota was 617 tons, but bait fishermen harvested only 124 tons. The bait quota for the 1995–1996 season was 558 tons, but only thirty-four tons were harvested. Bait herring that season had an ex-vessel value of about $13,600, less than 1 percent of the value of the 1996 spawn-on-kelp fishery.

Recognizing the growing value of the spawn-on-kelp fishery, in January 1997 the Board of Fisheries increased the allocation of herring to the fishery to 40 percent of the area's total herring quota. Additionally, unharvested quota from the winter bait-herring fishery, which by regulation ended on the last day of February, would be transferred to the spawn-on-kelp fishery.[19]

For the 1997 spawn-on-kelp fishery, the original quota was 100 tons—the minimum required for the fishery to proceed—but an additional 100 tons of unharvested bait quota was later added. Unfortunately for pound operators, the market for spawn on kelp was weak, and its price tumbled to $6 per pound. Recognizing the weak market, some operators chose to opt out of the fishery, and the number of pounds declined to 119. The twenty-three tons of spawn on kelp harvested that year had a value of only $270,000, less than 20 percent of its value the previous year. The average price of spawn on kelp fell to $3.39 per pound in 1998 and would remain below $5 per pound until 2006.[20]

For the 2000 season, the guideline harvest level for the Craig/Klawock pound fishery was 280 tons. Fifty pounds were deployed. Herring arrived in abundance, but, unfortunately, not in the areas where seiners were allowed to fish. Fishermen were unable to fill their pounds, and the season's production was zero.[21]

The 2001 season was back to business, though only thirty-one pounds were operated, the fewest since the fishery's inception in 1992. The reason for the lack of interest was twofold. First was uncertainty over obtaining a supply of herring. The quota for the fishery was 913 tons, the highest ever, but there was no guarantee the fish would be in areas open to seining. The second reason was likely the poor market for spawn on kelp. In the end, all thirty-one pounds produced a salable product, but its average worth was only $2.70 per pound, the lowest since the fishery's inception. The total harvest of 27.2 tons was worth $146,859. By ADF&G's calculation, the fishermen's average income from the fishery was $2,880.[22]

Herring abundance began increasing in 2004, but that didn't translate into increased production until 2008, when the value of the spawn on kelp produced was also high—an average of $10.33 per pound. Production that year in sixty-six pounds was 148.5 tons, which was worth a record $3.07 million.

Herring remained abundant, and in 2012 the quota was 6,847 tons, the highest since the fishery's inception. The price, too, was high—an average of $10.69 per pound—and the 98.1 tons of spawn on kelp produced in the fishery's thirty-five pounds was worth $2.1 million, with the average fisherman taking home nearly $33,000.

New for 2013 was a regulation that allowed two closed pounds to be combined into a single structure, as had been done in the Hoonah Sound fishery in 2001. And, as in Hoonah Sound, ADF&G provided an incentive for fishermen to do so. The kelp allocation for a single-permit pound in 2013 was 600 blades, but for a double-permit pound it was 750 blades per permit holder, or 1,500 blades.

As explained in the discussion of the Hoonah Sound fishery, the additional blades would, at least in theory, enable a greater amount of product to be produced, but with a lower rate of egg deposition, the grade of the product would likely be lower.

Prospects for the 2013 season also looked good, with a herring quota of 4,808 tons. Probably reflecting the anticipated easy availability of herring and the comparatively lucrative 2012 season, eighty pounds were deployed, a number of them of the double-permit variety. Production that year totaled 137.7 tons, a substantial increase from 2012. Moreover, the average price of the spawn on kelp produced had increased to $12 per pound. The fishery's worth set a new record, $3.1 million, with the average operator taking home almost $24,000.[23]

The 2014–2016 seasons were business as usual, but the pertinent data is unavailable to the public because of ADF&G's confidentiality policy. At least three processors must participate in a fishery for the

Figure 18.2. John and Logan Swanson at John Swanson's herring pound, Hoonah Sound, 2008. (John Swanson)

data to be made public, but only two participated in the Craig/Klawock fishery during those years.

The 2017 season was anything but business as usual. Until 2017, ADF&G had controlled production in the pound fishery by allocating kelp blades. At the same time, the department worked under the assumption that each pound required twenty tons of herring. Herring were projected to be scarce in 2017, and ADF&G set the quota at 349 tons. By the department's calculation, this was enough for only about seventeen pounds. In 2016, forty-six pounds had been deployed; in 2015, there had been seventy-six.[24]

To equitably distribute herring among pound operators, ADF&G ruled that each pound would be shared by six permit holders. Mark Jensen, Petersburg's mayor and a participant in the fishery, wondered how it would all work. "It's hard to get two fishermen to agree on something and now you're talking about 120 permit holders and 20 pounds," Jensen told a reporter.[25]

To compensate somewhat for the reduced amount of herring, ADF&G increased the number of blades allowed to 500 blades per permit, so a pound shared by six permit holders could suspend 3,000 blades. For the 2016 season, the department had authorized 300 blades per single-permit pound.

In the end, nineteen pounds were installed on the grounds. To everyone's surprise, however, herring showed up in about twice the volume expected. Unfortunately, the fish were in nontraditional areas that were closed to commercial harvest. Recognizing that its forecast was not accurate, ADF&G quickly opened additional waters to fishing. The areas opened were a considerable distance from the pounds, and though it was a challenge, all but a couple of operators were able to relocate their pounds and successfully add herring. The season's production totaled 69.9 tons, which was worth $933,000. ADF&G calculated the average income per permit holder to have been a little over $8,000.[26]

The herring quota for 2018, 1,602 tons, was a large increase from the previous year. Given the availability of herring, there was no requirement to operate multiple-permit pounds, though some fishermen chose to do so. Sixty-six pounds were placed on the grounds in 2018, and the harvest was 205 tons, a record. The product was worth $3.26 million—also a record but one that would be broken in 2019.

The herring quota in 2019, 2,911 tons, was a major increase over that of the previous year. Though seventy-three pounds were installed in 2019, an increase of seven over the previous year, production declined slightly, to 202 tons. The average per-pound price for the spawn on kelp produced, however, increased 20 cents over the 2018 price, to $8.15, and the fishery was worth $3.3 million. ADF&G calculated the average income per permit holder to have been almost $24,000. And, boding well for the future, in mid-April, the department documented fifty-six nautical miles of spawn along the coastline at Craig/Klawock, the most ever recorded and almost twice the previous record.[27]

TENAKEE INLET

In January 2003, the Alaska Board of Fisheries adopted regulations to provide for a spawn-on-kelp fishery in Tenakee Inlet, on the Chatham Strait shore of Chichagof Island. That fishery occurred for the first time in April 2003, with fifty-five participants who mostly utilized double pounds. Production was 47.6 tons, which had an ex-vessel value of $580,500. Participation in 2004 increased to eighty-five individuals, and the harvest was ninety-five tons, worth $981,464. No one knew it at the time, but this represented the peak income for the fishery. Ninety-eight individuals participated in the 2005 fishery, and their production was 101 tons. Unfortunately, the price of spawn on kelp had declined, and the value of the harvest was $512,900. Even worse, the spawning stock of herring at Tenakee Inlet, which had historically exhibited cycles of abundance, was on the decline. There was no spawn-on-kelp fishery in 2006–2008, and after that there were harvests in only two years, 2009 and 2014. No herring spawn was documented in Tenakee Inlet in 2016, and in the spring of 2019, ADF&G documented 0.5 nautical miles of what it characterized as "very light" spawn.[28]

ERNEST SOUND

Ernest Sound is about twenty-five miles south of Wrangell. Like the Tenakee Inlet fishery, the Ernest Sound spawn-on-kelp fishery was authorized by the Alaska Board of Fisheries in January 2003. The first fishery occurred in 2004, when sixty-four participants harvested 56.1 tons of product that had an ex-vessel value of $514,912. The following three years, 2005–2007, herring abundance was below ADF&G's threshold level, so no fishery occurred. Subsequently—due either to a scarcity of herring or a lack of interest in the fishery—there were only four more spawn-on-kelp harvests in Ernest Sound (2008, 2009, 2013, and 2014). The peak year was 2013, when the ex-vessel price of spawn on kelp was high (an average of $12.25 per pound), and eighty-one participants harvested 64.3 tons of product. The harvest's total value that year was $1,574,729—an average of $19,441 per participant.[29]

19

TOGIAK AND NORTON SOUND HERRING SPAWN-ON-KELP FISHERIES

TOGIAK

In late May 1968, at Eagle Bay, a small bay east of Togiak Bay, six to eight local villagers using their hands and garden rakes worked during low tides to harvest 56,400 pounds of herring egg–laden rockweed (*Fucus* sp.), an intertidal species. Typically, they could work for about two hours on either side of the low tide. Their product was then packed in brine in wooden barrels and was valued at $0.45 per pound, delivered to Dillingham. The species, however, was not especially desirable on the Japanese market, and the egg deposition was not optimal. A year later, much of the product remained unsold.[1]

Despite this inauspicious beginning, kelpers and processors in the following years were able to produce a marketable product, and the fishery grew, though modestly, and expanded to Metervik Bay and other nearby bays. For the next five seasons, only one processor was involved and only limited harvests that averaged less than 40,000 pounds resulted.[2]

Production increased in 1974, when four processors were involved and twenty-six kelpers, most of whom were Natives from Togiak and Twin Hills, harvested 126,000 pounds. They were reported to have been paid $0.16 per pound for the product, which put the total value of the harvest at about $20,000.[3] An important characteristic of the spawn-on-kelp fishery at Togiak was that the kelpers were primarily local Natives.

The fishery—humorously referred to by some as "stalking the wily kelps"—continued to grow, and in 1978, 160 kelpers harvested 329,858 pounds of product, which were sold to eleven processors. The harvest was worth approximately $120,000.[4]

The number of kelpers declined in 1979, but the number of processors increased, to sixteen. Production totaled 414,000 pounds, worth nearly $250,000. But due to the intense harvesting, the kelp beds were paying a price. In Metervik Bay, more than 33 percent of the available kelp biomass was harvested in just three low tides. And on the western shoreline of Kulukak Bay, 44 percent of the available kelp biomass was harvested.

Figure 19.1. Gathering rockweed, Bristol Bay, 1978. (Alaska Department of Fish and Game Historical Photograph Collection, 1950–1991, Alaska Resources Library and Information Services, on behalf of the Alaska Department of Fish and Game)

To prevent further overharvest, the Alaska Board of Fisheries approved a regulation that would close the kelp fishery in the Togiak District once 10 percent of the kelp biomass was taken. To protect individual kelp beds, the district was divided into eleven areas, and harvest quotas were established for each.[5]

The regulation was implemented for the 1980 season, and the maximum allowable harvest for the district was set at 934,000 pounds. However, due primarily to a scarcity of spawn, the harvest was only 190,000 pounds. The product was graded fair to poor, and seventy-eight kelpers who made landings were paid an average of $0.50 per pound.[6]

The following is a chronology of the Togiak spawn-on-kelp fishery for the years 1981 through 2002.

1981

The harvest at Togiak in 1981 was 380,000 pounds, for which the 108 kelpers were paid an average of $0.66 per pound, which put the total value of the fishery at about $250,000. ADF&G figured that less than a half metric ton of kelp was wasted.[7]

1982

Interest in the fishery was growing and in 1982, 214 kelpers, nearly double the number of the previous year, participated. The season's first opening began on May 21, and during its thirty-nine-hour duration kelpers harvested 236,000 pounds of spawn-laden kelp, about 76 percent of the quota. A second opening was expected to follow, but a storm moved into the area, stirring up sediment and contaminating kelp. Moreover, some of the eggs on the kelp had begun to "eye up" (the eyes of the embryonic fish within the egg became visible), making it unsalable. ADF&G then closed the season. It wasn't much consolation, but the price kelpers were paid for the harvest had increased to $0.75 per pound, and the fishery was worth $176,000—an average of a little over $800 per kelper.[8]

1983

Only 125 kelpers participated in the 1983 fishery, which was something of a duplicate of the 1982 season. Kelpers harvested 271,000 pounds but were unable to harvest the remainder of the quota due to a storm that stirred up sediment. A consolation was that the price paid to kelpers had increased to $1.05 per pound, and the fishery was worth approximately $284,000—about $2,300 per kelper. The harvesting power of the kelpers was demonstrated when they harvested 125,000 pounds of product under very poor conditions during one twenty-four-hour opening.[9]

1984

In 1984, the Board of Fisheries adopted a management plan that established a harvest quota of 350,000 pounds for the Togiak spawn-on-kelp fishery. To protect kelp beds, a two- or three-year rotation of harvest areas was incorporated in the plan.

The high price kelpers had received for their harvest in 1983 enticed additional participation in the fishery. Fully 330 kelpers were on the grounds in 1984. Together during three openings in late May, they harvested 406,586 pounds of spawn on kelp, substantially exceeding their 350,000-pound quota. The product, however, had light egg coverage, and the price paid for it was an average of only $0.50 per pound. This more than offset the high production, and the total value of the fishery declined to $203,000.[10]

1985

Because of its low quality, much of the spawn on kelp purchased in 1984 had not yet been sold as the 1985 season approached. At a preseason meeting, buyers informed the kelpers that they would be unwilling to purchase anything but number one–grade (the highest grade) spawn on kelp. Samples subsequently taken were all graded number two, and it was apparent that allowing a harvest at that time would result in a large amount of unsalable product that would likely be dumped. Everyone agreed to delay the fishery in the hope that subsequent spawning would improve the quality of the product. The weather, however, didn't cooperate. High seas during a gale that moved through the area contaminated the kelp beds with silt. It was clear that if the season was opened, the majority of the product harvested would be dumped, and so ADF&G elected to not open the season.[11]

1986

In 1986, 204 kelpers harvested 374,000 pounds of spawn on kelp. Although this was somewhat over the quota, it wasn't enough to raise any alarm at ADF&G. The harvest was worth $187,000.[12]

1987

The 1987 season was marred when alcohol-induced gunfire erupted at a fish camp on the beach at Middle Bay, just east of Togiak Bay, where a spawn-on-kelp opening was about to occur. Three people, all from Togiak, were killed, and a fourth was wounded. Due to the homicide, many kelpers left the area, and only a small amount of product was harvested at Middle Bay. The total harvest for the season by 187 kelpers was 303,307 pounds, worth $166,000.[13]

1988

Ten processors were on the grounds for the 1988 Togiak spawn-on-kelp fishery, twice the number as the previous year. The low tide on May 21 was at about 1:00 a.m., and during a six-hour opening centered around that tide, 259 kelpers harvested 489,320 pounds of spawn on kelp, greatly exceeding their quota. The harvest, which came entirely from two small areas, was valued at $346,000—an average of about $1,335 per kelper.[14]

1989

The year 1989 represented the peak in production in the Togiak spawn-on-kelp fishery. Kelpers, of which there were an estimated 487—a record—harvested 559,780 pounds of product during a four-hour opening (the shortest ever) on May 14, in a small area just east of Togiak Bay. A dozen processors were involved in the fishery, and they paid kelpers a total of $448,000.

This was the second consecutive year that the harvest greatly exceeded the quota, and ADF&G biologists expressed concern over the "very short, intense openings and a reduction in the staff's ability to control harvests within specified levels." Participation in the fishery had more than doubled over the past three years.[15]

1990

To address ADF&G's concern regarding being able to effectively manage the Togiak spawn-on-kelp fishery, prior to the 1990 season, the Alaska Commercial Fisheries Entry Commission held hearings on the issue and subsequently limited entry into the fishery. Only those who had participated in the fishery for one of the three years 1986–1988 were eligible. For the 1990 season, the commission issued 479 interim-use permits. Eligibility would be determined over the coming years. By 1996, the commission had issued 283 permanent permits.

The 1990 Togiak spawn-on-kelp season, which lasted three hours, was the shortest ever. The harvest was 413,844 pounds, which for the third consecutive year substantially exceeded the fishery's 350,000-pound quota. The harvest was valued at $360,000.

During the opening, ADF&G counted 481 kelpers on the beaches, two more than the number of permits the Commercial Fisheries Entry Commission had issued. The disparity was because some permit-holding kelpers were also herring fishermen, and their deckhands assisted them. Whether the deckhands could legally do so was a gray area.[16]

1991

The level of effort in the 1991 fishery was the greatest on record. ADF&G's aerial count was 532 participants, though, as was the situation in 1990, a number were deckhands who didn't possess interim-use permits. In October 1991, the Board of Fisheries limited the role of non–permit holders to assisting with the transport of kelp only after the close of the fishing period.

The 1991 Togiak area spawn-on-kelp harvest was taken in one 2.5-hour opening and totaled 348,357 pounds. It was the first time in four seasons that kelpers stayed within their quota. And at $1.10 per pound, the price paid to fishermen was the highest ever. The total value of the fishery was $383,000.[17]

1992

The Board of Fisheries' prohibition of nonpermitted crewmembers from harvesting kelp reduced the effort by 38 percent, to 386. Kelpers harvested 363,600 pounds of spawn on kelp in 3.25 hours during two openings. The five buyers on the grounds paid an average of $0.70 for the product, putting the value of the fishery at $254,000—an average of about $660 per kelper.[18]

1993

Interest in Togiak's spawn on kelp declined in 1993, and only two buyers were on the grounds. The number of kelpers also fell, to 173, the lowest number since 1981. During two early-May openings that totaled seven hours, kelpers harvested 383,000 pounds of product for which they were paid $0.70 per pound.[19]

1994

The 1994 fishery was similar to the 1993 fishery. Two buyers and 184 permit holders participated. The harvest, which totaled 308,440 pounds, fell short of the quota because heavy onshore winds just prior to the second opening contaminated the kelp with silt. The price paid to kelpers was unchanged from the past two seasons—$0.70 per pound—and the total value of the fishery was $212,000.[20]

1995

The demand for spawn on kelp was strong in 1995, and five buyers were on the Togiak grounds. The fishery consisted of three periods that totaled 14.5 hours, during which 188 kelpers harvested 281,600 pounds of product. Production fell short of the quota because ADF&G's management plan specified a two- or three-year rotation for kelp-harvest areas. ADF&G terminated the 1995 harvest because continuing it would leave only a few kelp areas available for harvest the following year.

The average price paid for the spawn on kelp was a record $1.29 per pound, which made the total value of the fishery $362,000.[21]

1996

For the 1996 season, three buyers and 200 kelpers were on the grounds. During two six-hour openings, kelpers harvested 455,800 pounds of product, 30 percent over the harvest guideline prescribed in the management plan. Contributing to the overharvest were numerous violations, including non-permit holders harvesting kelp, and, during the second opening, which occurred at night (as did the first opening), harvesting kelp before the opening and after the closure.

Buyers paid up to $1.25 per pound, similar to the price paid in 1995. ADF&G valued the fishery at $510,000, the highest ever and almost twice the 1978–1995 average.[22]

1997

Samples of kelp taken in early May showed only light coatings of eggs, and the two companies registered to process spawn on kelp determined that the product was not of marketable quality. Additional samples taken several days later confirmed this assessment. Concerned about potential waste due to the discarding of harvested kelp that might be rejected by buyers, ADF&G elected to cancel the season.[23]

1998

Though the market was unstable, two processors indicated an interest in purchasing spawn on kelp at Togiak in 1998. Nature, however, didn't cooperate. High seas during successive storms deposited silt and other debris in areas where herring were spawning, rendering the kelp that might be gathered unmarketable.[24]

1999–2002

The fishery was back on track in 1999, though the presence of only one buyer, the floating processor *Yardarm Knot,* signaled only modest interest in the product. During two late-May openings that totaled eight hours, 130 kelpers under good weather and tide conditions harvested 419,563 pounds of product, 20 percent over their quota. The entire harvest was valued at $315,000.[25]

The single processor registered to purchase spawn on kelp at Togiak in 2000 found no kelp of marketable quality, due in part to a storm that stirred up sediment that adhered to the kelp. The processor left the district, and soon thereafter ADF&G announced that there would be no fishery. For the 2001 season, no companies registered to process spawn on kelp at Togiak.[26]

In 2002, one buyer was interested in purchasing thirty metric tons of spawn on kelp at Togiak. ADF&G opened the fishery, and during the single two-hour opening that comprised the season, approximately sixty-five kelpers delivered 67,793 pounds of product.[27]

Since 2002, buyers have expressed no interest in purchasing spawn on kelp at Togiak.

NORTON SOUND

A small-scale spawn-on-rockweed (*Fucus* sp.) fishery existed in Norton Sound from 1977 to 1984. Harvests ranged from less than one ton (1977) to approximately forty-six tons (1981). Also, during the 1984 season, one enterprising operator of a Prince William Sound herring pound transported a ton of *Macrocystis* kelp to Norton Sound. He suspended it open-pound style in an area near Elim where herring spawned, and he harvested approximately three tons of product.

Some individuals wanted to expand the spawn-on-kelp fishery, but a local controversy developed regarding harvesting spawn on kelp while simultaneously harvesting spawning fish. The possibility of expanding the harvest of spawn on kelp tipped the balance, and the local fisheries advisory committees petitioned the Alaska Board of Fisheries to close all spawn-on-kelp fisheries in Norton Sound following the 1984 season. The board did so.[28] The harvesting of spawn on kelp in Norton Sound, however, was not over.

In early 1998, it was widely recognized that the market for roe herring was poor, and few processors expressed interest in Norton Sound (see chapter 14). This meant that some fishermen would have no market for their fish and would have to sit the season out. To provide additional opportunities for fishermen, Frank Rue, ADF&G's commissioner, approved emergency regulations that allowed a herring spawn-on-rockweed fishery in Norton Sound. And, at almost the same time, the Board of Fisheries approved an experimental herring spawn-on-*Macrocystis*-kelp fishery in the sound.

The effort in the wild spawn-on-kelp fishery was minuscule: three fishermen gathered a total of one ton of kelp. However, twenty-two fishermen expressed interest in open-pound *Macrocystis* operations and at least eighteen deployed kelp. The kelp was shipped via air cargo from Southeast Alaska to Nome. From there, it was flown in small planes to Elim, on Norton Sound's north shore. The first shipment arrived in good condition on May 24, just before the ice broke up. Fishermen deployed the kelp open-pound style, some from buoyed lines, some from jury-rigged floating frames. The harvest totaled 16,083 pounds, which would stand as the record for this fishery.[29]

At its January 1999 meeting, the Board of Fisheries approved open-pound spawn-on-*Macrocystis*-kelp operations in Norton Sound. At the same time, the market for herring roe had improved. Among fishermen in Norton Sound, however, there was no interest in wild kelp and only small interest in *Macrocystis*. Six fishermen had *Macrocystis* operations, but a late spring and associated ice floes complicated operations and only two harvested product. Their total production was 7,482 pounds, less than half of what it had been the previous year. And in future years, harvests would continue to decline. Production in 2000 was 4,500 pounds. The quality of the spawn on kelp that year was good, though, and it had an estimated ex-vessel value of $36,354.[30]

Production declined to 4,400 pounds in 2001, and, due to poor market conditions, there was no production in 2002. Production resumed in 2003, but the harvest by the single operator was only 1,750 pounds. This marked the end of herring spawn-on-*Macrocystis*-kelp harvests in Norton Sound. There were, however, small amounts—less than one ton—of wild herring spawn-on-rockweed harvests in Norton Sound in 2006–2008.[31]

EPILOGUE

The future ain't what is used to be.—Yogi Berra[1]

Herring, a foundation species of the North Pacific Ocean ecosystem, are underappreciated. Beginning in the 1880s and extending until the 1960s, fishermen caught untold billions of these beautiful, wild marine creatures that were then "reduced" to make basic industrial commodities: fertilizer, fish meal, and oil. Alaska's reduction fishery collapsed in the 1960s because of depressed herring stocks and competition from the Peruvian anchoveta fishery.

The reduction fishery, however, was almost immediately replaced by the roe-herring fishery. In it, fishermen caught billions more herring so the roe, a traditional luxury food item in Japan, could be extracted from the females. Disposing of the female carcasses and the male fish—about 90 percent of the weight of the catch—was (and is) simply a cost of doing business.

But as Japanese consumers increasingly eschew traditional foods in favor of a more Western-style diet, the demand for herring roe has declined. And the impact on Alaska fishermen and processors is substantial: "It's a big year [herring were expected to be abundant] for Alaska roe herring fisheries, but lackluster interest by both harvesters and processors is an ongoing story," wrote Laine Welch, a fisheries reporter, just as Alaska's 2021 roe-herring fishery was about to get under way.[2] In contrast to the crowded, hypercompetitive fisheries of the past, at Sitka, only eighteen seiners made landings in 2021, while at Togiak, traditionally Alaska's largest herring fishery, just ten seiners and three gillnetters did so.[3]

It must be noted, too, that large quantities of herring are caught not to satisfy a market but as bycatch in other fisheries, especially the pollock trawl fishery in the Bering Sea. As such, the herring have a negative value. Pollock trawlers sometimes fish in areas that are less productive of their target species to avoid catching herring because exceeding a herring bycatch limit can trigger pollock fishing closures. In 2020, by late May, pollock trawlers had caught almost 3,000 metric tons of herring, substantially exceeding their bycatch limit. In response, the National Marine Fisheries Service temporarily closed some areas of the Bering Sea to directed trawling for pollock.[4] Despite such penalties, however, herring bycatch is likely to be a persistent drain on the resource.

The good news is that Alaska herring stocks as they now exist are generally doing well. For Sitka Sound in 2021, ADF&G forecasted a mature herring return of 210,453 tons, the second largest ever. During the season, the department mapped 102.3 nautical miles of herring spawn in the sound, one of the highest on record and far above the forty-year (1981–2020) average of 59.1 nautical miles.[5] At Togiak, ADF&G forecasted a mature herring return of 236,742 tons, the highest since 1993, when the department implemented its age-structured assessment model.[6] Observations during the season indicated that this forecast was sound.[7] The Kodiak area, too, though it supports a fishery far smaller than those at Sitka and Togiak, saw an abundance of fish, and the ADF&G guideline harvest level there, 7,895 tons, was the highest ever.[8] And in long-suffering Prince William Sound, herring were nowhere near their historical abundance, but the population showed an increase over previous years.[9]

Welcome as all this news is, it pales when placed in the context of the historical extent and abundance of herring in Alaska and along the entire Pacific Northwest coast.

The ongoing decline of the roe-herring fisheries, with no other major commercial exploitation on the horizon, presents an opportunity for herring to rebuild their populations to near pre-exploitation levels along the entire Pacific Coast, an opportunity that has not been possible for more than a century. The question remains, however, of how climate change and ocean acidification will impact herring and the plankton and small crustaceans that constitute the bulk of herring's diet.

And if a major fishery for Alaska herring does develop, it would be prudent to manage it not for the maximum sustained yield but rather in a manner that recognizes the great importance of this forage fish to the marine ecosystem.

Ensuring that herring thrive should be everyone's goal. For if they do, the oceans, and the world we live in, will benefit. A good step to foster the health of the species would be to seek sustainability certification of Alaska's herring fishery by an internationally respected organization such as the Marine Stewardship Council. The council has certified several northeast Atlantic Ocean herring fisheries that have an annual combined harvest of more than a million metric tons.[10]

The United States Congress, too, may act to improve the conservation of herring. In April 2019, Representative Debbie Dingell (D-Michigan) introduced the Forage Fish Conservation Act. The legislation would have amended the Magnuson-Stevens Fishery Conservation and Management Act, the primary law that governs the management of US marine fisheries, to provide an official definition for forage fish and require the secretary of commerce to improve monitoring and management of these fishes. Representative Dingell's bill failed to pass out of committee, but in December 2020, Senator Richard Blumenthal (D-Connecticut) introduced similar legislation in the US Senate. That legislation, too, failed to gain traction, but in April 2021, with Senator Roy Blunt (R-Missouri) as a co-sponsor, Senator Blumenthal reintroduced the Forage Fish Conservation Act.[11]

And in the House of Representatives, in July 2021, representatives Jared Huffman (D-California) and Ed Case (D-Hawaii) introduced the Sustaining America's Fisheries for the Future Act, legislation to update and reauthorize the Magnuson-Stevens Act. Included in this proposed legislation is a requirement that NOAA Fisheries develop a definition of forage fish that recognizes, among other things, the importance of forage fish in the marine ecosystem.[12]

At the very least, these Congressional proposals have increased the recognition of the importance of forage fish. If either is signed into law, it would likely improve the management of these valuable fishes.

The Herring Song

Vee like 'Laska herring,
    Fat Alaska Herring;
He iss de finest fiss dot swims.
    Vee like crab an' rockfish,
    Lutefish from stockfish—
Vee tink our Scotch-cured matjes
    Fine.
Take your humpback salmon,
Fresh from a trap an' frisky,
But gif me SILD—
An' Ay don't gif a damn if Ay don't
    have whiskey.
    Vee like 'Laska herring,
    Fat Alaska herring—
He iss de finest fiss dot swims.[13]
(Sild is the Norwegian word for herring.)

Figure 20.1. Pacific herring.

NOTES

DEDICATION

1. University of Alaska, UA Journey: "From the finding aid for Clarence Louis Anderson papers 1894–1966 circa 1920–1960" (University of Washington Libraries Special Collections), accessed October 25, 2020, https://www.alaska.edu/uajourney/notable-people/juneau/clarence-j.-anderson/; Ira Rothwell, "Alaska's First Director of Fisheries," Alaska Department of Fisheries, *Annual Report No. 1 for the Year 1949*, 21; C. L. Anderson, "The Herring Industry at Aalesund, Norway," *Pacific Fisherman* (June 1922): 10–12.

2. "Oil and Meal Industry Enters Most Active Season in Years," *Pacific Fisherman* (May 1935): 60.

3. "Anderson Is Director of Alaska Fisheries Dept.," *Pacific Fisherman* (June 1949): 25; Rothwell, "Alaska's First Director of Fisheries," 12.

4. Bob King, *Sustaining Alaska's Fisheries: Fifty Years of Statehood* (Juneau: Alaska Department of Fish and Game, 2009), 57.

5. Alaska Department of Fish and Game Newsletter, *Alaska Sportsman*, July 1961.

6. "C. L. Anderson Honored," *Pacific Fisherman* (November 1962): 12.

7. "C. L. Anderson Retires with Work Well Done," *Pacific Fisherman* (December 1961): 37.

INTRODUCTION

1. Donald J. Orth, *Dictionary of Alaska Place Names*, US Geological Survey Professional Paper No. 567 (Washington, DC: GPO, 1967, reprinted with minor revisions in 1971), 417–418, 852; Alaska Department of Fish and Game, *Bristol Bay Area Annual Management Report*, 1975, 26–27.

2. Tarleton H. Bean, "Food Fishes of Alaska," 50th Cong., 2d Sess., *Fur-Seal Fisheries of Alaska*, Ho. Rept. No. 3833, January 29, 1889, XLI.

3. John N. Cobb, "Opportunities in the Fish Canning Industry," *Pacific Fisherman* (Yearbook 1920): 114, 116, 121.

4. Barton Warren Evermann, *Alaska Fisheries and Fur Industries in 1913*, Bureau of Fisheries Doc. No. 797 (Washington, DC: GPO, 1914), 125–129.

5. Robert H. Armstrong, *Alaska's Fish: A Guide to Selected Species* (Portland, OR: Alaska Northwest Books, 1996), 7.

6. Thomas F. Thornton et al., *Herring Synthesis: Documenting and Modeling Herring Spawning Areas within Socio-Ecological Systems Over Time in the Southeastern Gulf of Alaska*, North Pacific Research Board Project No. 728 (June 2010), 3.

7. William F. Royce, *Introduction to the Practice of Fishery Science* (San Diego: Academic Press, 1996), 114.

8. Alaska Department of Fish and Game, *Progress Report for the Years 1963 and 1964*, ADF&G Rept. No. 13, 70–71.

Notes

9. June Allen, "Herring Fishery Fate Unknown," *Ketchikan Daily News*, February 20–21, 1993.

10. Michael A. Perez, *Calorimetry Measurements of Energy Value of Some Alaskan Fishes and Squids*, NOAA Tech. Memorandum NMFS-AFSC-32 (February 1994), 9–13.

11. E. Pikitch et al., *Little Fish, Big Impact: Managing a Crucial Link in Ocean Food Webs* (Washington, DC: Lenfest Ocean Program, 2012), 4.

12. Robert J. Browning, *Fisheries of the North Pacific: History, Species, Gear & Processes*, rev. ed. (Anchorage: Alaska Northwest Publishing Company, 1980), 33–36.

CHAPTER 1: ALASKA HERRING

1. Tom Ohaus, "Harried Herring," *Pacific Fishing* (December 1990): 148.

2. E. J. Huizer, "History of Alaska Herring Fishery," *Alaska Department of Fisheries Annual Report for 1952*, 63–76; Robert J. Browning, *Fisheries of the North Pacific: History, Species, Gear & Processes*, rev. ed. (Anchorage: Alaska Northwest Publishing Company, 1980), 33–36; NOAA Fisheries, *Pacific Herring*, accessed August 25, 2019, https://www.fisheries.noaa.gov/alaska/endangered-species-conservation/pacific-herring; William F. Royce, *Introduction to the Practice of Fishery Science* (San Diego: Academic Press, 1996), 114; Doug Woodby et al. (Alaska Department of Fish and Game), *Commercial Fisheries of Alaska*, ADF&G Special Publication No. 05-09 (June 2005), 9; Alaska Department of Fish and Game, *Species Profile, Pacific Herring (Clupea pallasii)*, accessed August 25, 2019, http://www.adfg.alaska.gov/index.cfm?ADFG=herring.main.

3. Szymon Surma, Evgeny A. Pakhomov, and Tony J. Pitcher, "Energy-Based Ecosystem Modelling Illuminates the Ecological Role of Northeast Pacific Herring," *Marine Ecology Progress Series* (2018) 588: 147–161.

4. D. E. Hay et al., "Geographic Variation in North Pacific Herring Populations: Pan-Pacific Comparisons and Implications for Climate Change Impacts," *Progress in Oceanography* (May–June 2008): 233–240.

5. Alaska Department of Fish and Game, *Species Profile, Pacific Herring (Clupea pallasii)*, accessed August 25, 2019, http://www.adfg.alaska.gov/index.cfm?ADFG=herring.main; George A. Rounsefell, "Some Observations on the Alaska Herring," Part 2, *Pacific Fisherman* (September 1928): 20–21; Jeffrey Skrade (former ADF&G fisheries biologist), personal communication, November 14, 2020; Alaska Department of Fish and Game, "Bristol Bay Herring Fishery," in *Bristol Bay Area 2002 Annual Management Report*, Regional Information Rept. No. 2A03-18 (April 2003), 119; Alaska Department of Fish and Game, *Annual Management Report, 1995, Norton Sound–Port Clarence–Kotzebue*, Regional Information Rept. No. 3A96-30 (October 1996), 91.

6. "Whales Versus Herring," *Pacific Fisherman* (April 1919): 32.

7. Doug Woodby et al., *Commercial Fisheries of Alaska*, Alaska Dept. of Fish & Game Special Publication No. 05-09 (June 2005), 9.

8. Karen Benveniste, "Bering Sea Herring for Food," *Pacific Fishing* (March 1983): 54–60; Fritz Funk, ed., *Preliminary Forecasts of Catch and Stock Abundance for 1993 Alaska Herring Fisheries*, Alaska Department of Fish and Game Regional Information Rept. No. 5J93-06 (May 1993), 18.

9. Alaska Department of Fish and Game, *Species Profile, Pacific Herring*.

10. Huizer, "History of Alaska Herring Fishery," 63–76.

Notes

11. Browning, *Fisheries of the North Pacific*, 33–36; Alaska Department of Fish and Game, *Pacific Herring*; NOAA Fisheries, *Pacific Herring*, accessed September 13, 2019, https://www.fisheries.noaa.gov/alaska/endangered-species-conservation/pacific-herring; "Larvae," Pacific Herring: Past, Present and Future, accessed September 13, 2019, http://www.pacificherring.org/explorer/life-cycle/larvae; Gerald M. Reid, *Fishery Facts-2, Alaska's Fishery Resources—The Pacific Herring*, National Marine Fisheries Service Extension Publication (June 1972), 9; William F. Royce, *Introduction to the Practice of Fishery Science* (San Diego: Academic Press, 1996), 114; Doug McNair, "Alaska Roe Herring," *Pacific Fishing* (November 1983): 103–107; Steve Will, "Herring Harvest Ends; Quality Good," *Daily Sitka Sentinel*, April 1, 1994.

12. "By-Products Bounce Back," *Pacific Fisherman* (January 25, 1959): 225–235.

13. Alaska Department of Fish and Game, *Prince William Sound Area 1991 Annual Finfish Management Report*, ADF&G Regional Information Rept. No. 2A92-09 (May 1992), 21.

14. "Sitka Sound Herring Watch: Many 3-Year-Olds Show Up," *Daily Sitka Sentinel*, March 25, 1991.

15. Edward W. Nelson, *Report on Natural History Collections Made in Alaska between the Years 1877 and 1881* (Washington, DC: GPO, 1887), 320–321.

16. Alaska Department of Fish and Game, *Pacific Herring (Clupea pallasii), Species Profile*, accessed September 13, 2019, http://www.adfg.alaska.gov/index.cfm?ADFG=herring.main; NOAA Fisheries, *Pacific Herring*, accessed September 13, 2019, https://www.fisheries.noaa.gov/alaska/endangered-species-conservation/pacific-herring; "Larvae," *Pacific Herring: Past, Present and Future*, October 7, 2013 http://www.pacificherring.org/explorer/life-cycle/larvae; Gerald M. Reid, *Fishery Facts-2, Alaska's Fishery Resources—The Pacific Herring*, National Marine Fisheries Service Extension Publication (June 1972), 9.

17. Alaska Department of Fish and Game, *Species Profile, Pacific Herring*; Alaska Department of Fish and Game, *Pacific Herring Fishery*, accessed August 9, 2020, http://www.adfg.alaska.gov/index.cfm?adfg=commercialbyfisheryherring.main#history.

18. Iain McKechnie et al., "Archaeological Data Provide Alternative Hypotheses on Pacific Herring (*Clupea pallasii*) Distribution," *Proceedings of the National Academy of Sciences* 111, no. 9 (March 4, 2014): E807–E816, https:/doi.org/10.1073/pnas.1316072111.

19. "The Fisheries of Alaska," in *Report on Population and Resources of Alaska at the Eleventh Census: 1890* (Washington, DC: GPO, 1893), 226.

20. McKechnie et al., "Archaeological Data Provide Alternative Hypotheses on Pacific Herring (*Clupea pallasii*) Distribution."

21. George Thornton Emmons and Fredrica de Laguna, eds., *The Tlingit Indians* (Seattle: University of Washington Press, 1991), 119.

22. Ivan Petroff, "Report on the Population, Industries, and Resources of Alaska [1882]," in *Seal and Salmon Fisheries and General Resources of Alaska*, vol. 4 (Washington, DC: GPO, 1898), 271; Jefferson F. Moser, *Salmon and Salmon Fisheries of Alaska: Report of the Operations of the United States Fish Commission Steamer Albatross for the Year Ending June 30, 1898*, in *Bulletin of the United States Fish Commission*, Vol. XVIII, 1898, 55th Cong. 3d Sess. Ho. Doc. No. 308 (Washington, DC: GPO, 1899), 123.

23. Moser, *Salmon and Salmon Fisheries of Alaska*, 123.

24. E. Ruhamah Scidmore, *Alaska: Its Southern Coast and the Sitkan Archipelago* (Boston: D. Lothrop Co., 1885), 248.

25. Edward W. Nelson, *Report on Natural History Collections Made in Alaska between the Years 1877 and 1881* (Washington, DC: GPO, 1887), 320–321; John N. Cobb, *The Commercial Fisheries of Alaska in 1905*, Bureau of Fisheries Doc. No. 603 (Washington, DC: GPO, 1906), 20; Alaska Department of Fish and Game, *Norton Sound–Port Clarence–Kotzebue Annual Management Report, 1979*, 123.

26. Millard C. Marsh and John N. Cobb, *Fisheries of Alaska in 1910*, Bureau of Fisheries Doc. No. 746 (Washington, DC: GPO, 1911), 45–51.

27. Edward William Nelson, *The Eskimo About Bering Strait*, Bureau of American Ethnology, 18th Annual Rept., Part 1 (Washington, DC: GPO, 1900), 267.

28. Moser, *Salmon and Salmon Fisheries of Alaska*, 123.

29. Petroff, "Report on the Population, Industries, and Resources of Alaska [1882]," 271.

30. "Fishery Bureau Bulletins," *Pacific Fisherman* (February 1906): 18; E. H. Dahlgren (US Bureau of Fisheries), "Alaska Herring Research," Part 2, *Pacific Fisherman* (April 1940): 44, 46.

31. Madonna L. Moss, "The Nutritional Value of Pacific Herring: An Ancient Cultural Keystone Species on the Northwest Coast of North America," *Journal of Archaeological Science: Reports* 5 (2016), 649–655; Anne-Marie Victor-Howe, *Subsistence Harvests and Trade of Pacific Herring Spawn on* Macrocystis *Kelp in Hydaburg, Alaska*, ADF&G Tech. Paper No. 225, February 2008.

32. Carl Spuhn, President, Alaska Oil and Guano Company, December 1910, in Alaska Fisheries, Hearing, United States Senate, Subcommittee of the Committee on Fisheries, April 11, 1912.

33. Marsh and Cobb, *Fisheries of Alaska in 1910*, 45–51.

34. E. Lester Jones, *Report of Alaska Investigations, 1914* (Washington, DC: GPO, 1915), 62.

35. Ward T. Bower, "Alaska Fishery and Fur-Seal Industries in 1930," in *Report of the United States Commissioner of Fisheries for the Fiscal Year 1931* (Washington, DC: GPO, 1932), 58–61.

36. Ward T. Bower, "Alaska Fishery and Fur-Seal Industries in 1924," Bureau of Fisheries Doc. No. 902, in *Report of the United States Commissioner of Fisheries for the Fiscal Year 1925* (Washington, DC: GPO, 1926), 134.

37. Moser, *Salmon and Salmon Fisheries of Alaska*, 121–124; "The Fish Oil and Meal Industry of the Pacific," *Pacific Fisherman* (August 1935): 57–85; "New Alaska Regulations Less Severe Than in 1940," *Pacific Fisherman* (March 1941): 11–12; "U.S. Service Slaps a Ceiling On Fishing Effort in Alaska," *Pacific Fisherman* (December 1955): 16, 18; Bernard E. Skud, Henry M. Sakuda, and Gerald M. Reid, *Statistics of the Alaska Herring Fishery, 1878–1956*, US Fish and Wildlife Service Statistical Digest No 48 (Washington, DC: GPO, 1960), 1.

38. "By-Products," *Pacific Fisherman* (January 25, 1959): 233.

39. Alaska Department of Fish and Game, *Statewide Herring Wholesale Value and Pounds Processed*, accessed September 25, 2020, https://www.adfg.alaska.gov/index.cfm?adfg=fishlicense.coar_herringproduction.

40. Bower, "Alaska Fishery and Fur-Seal Industries in 1939," 150–153; Sarah Crawford Isto, *The Fur Farms of Alaska* (Fairbanks: University of Alaska Press, 2012): 65, 95.

41. Jones, *Report of Alaska Investigations, 1914*, 58–62.
42. "New Herring Harvest Policy Is Adopted," *Daily Sitka Sentinel*, December 18, 1973.
43. Bill Atkinson, "Japan Update," *Pacific Fisherman* (September 1997): 13; Terry Johnson, "Roe Eruption," *Pacific Fishing* (April 1993): 45–51.
44. Chapter 9, *Session Laws of Alaska, 1977*; Chapter 27, *Session Laws of Alaska, 1980*; Chapter 14, *Session Laws of Alaska, 1983*.

CHAPTER 2: EARLY DEVELOPMENT OF ALASKA'S HERRING INDUSTRY

1. John N. Cobb, *The Commercial Fisheries of Alaska in 1905*, Bureau of Fisheries Doc. No. 603 (Washington, DC: GPO, 1906), 19–22.
2. Jefferson F. Moser, *Salmon and Salmon Fisheries of Alaska: Report of the Operations of the United States Fish Commission Steamer Albatross for the Year Ending June 30, 1898*, in *Bulletin of the United States Fish Commission*, Vol. XVIII, 1898, 55th Cong. 3d Sess. Ho. Doc. No. 308 (Washington, DC: GPO, 1899), 121–122; Ted C. Hinkley, "Punitive Action in Angoon," Part 2, *Alaska Sportsman* (February 1963): 14–15, 40, 42.
3. Hinkley, "Punitive Action in Angoon," 14–15, 40, 42; Dave Kiffer, "US Navy Bombed Angoon 125 Years Ago," *SitNews*, October 29, 2007, http://www.sitnews.us/Kiffer/Angoon/102907_angoon_bombed.html; Thomas F. Thornton et al., *Herring Synthesis: Documenting and Modeling Herring Spawning Areas within Socioecological Systems over Time in the Southeastern Gulf of Alaska*, North Pacific Research Board Project No. 728 (Portland, OR: Portland State University, 2010), 261–263.
4. E. Ruhamah Scidmore, *Alaska: Its Southern Coast and the Sitkan Archipelago* (Boston: D. Lothrop Co., 1885), 248.
5. Eliza Ruhama Scidmore, *Appleton's Guide-Book to Alaska* (New York: D. Appleton and Co., 1893), 88.
6. John N. Cobb, "Report on the Fisheries of Alaska [in 1906]," Bureau of Fisheries Doc. No. 618, in *Report of the Commissioner of Fisheries for the Year 1906 and Special Papers* (Washington, DC: GPO, 1907), 55.
7. Eliza Ruhamah Scidmore, "Alaska Fish and Game," *Harper's Weekly* (September 2, 1893): 838–841; H. Bruce Franklin, *The Most Important Fish in the Sea: Menhaden and America* (Seattle: Island Press, 2007), 62; "Remarkable Development of Alaska Herring Fishery," *Pacific Fisherman* (September 1925): 23–24, 26, 28, 30; "Herring Plant for Washington Bay," *Pacific Fisherman* (March 1925): 44; "Improvements at Killisnoo Plant," *Pacific Fisherman* (February 1919): 52.
8. George A. Rounsefell, "Contribution to the Biology of the Pacific Herring, *Clupea pallasii*, and the Condition of the Fishery in Alaska," in *Bulletin of the Bureau of Fisheries, 1929* (Washington, DC: GPO, 1930), 227–320.
9. Moser, *Salmon and Salmon Fisheries of Alaska*, 121–124; Carl Spuhn, in James W. Witten (General Land Office), *Report on the Agricultural Prospects, Natives, Salmon Fisheries, Coal Prospects and Development, and Timber and Lumber Interests of Alaska* (Washington, DC: GPO, 1904), 74–75.
10. Lee H. Wakefield, President, Herring Packers Association, "The Development of the Herring Industry of Alaska," reproduced in *Protection of Alaskan Fisheries*, 71st

Cong., 2d sess., *Hearings Before the Committee on Merchant Marine and Fisheries, House of Representatives, on H.R. 253 and H.R. 7238, April 10, 1930* (Washington, DC: GPO, 1930), 17–18; Millard C. Marsh and John N. Cobb, *Fisheries of Alaska in 1907*, Bureau of Fisheries Doc. No. 632 (Washington, DC: GPO, 1908), 53–55; "Extent and Productiveness of the Pacific Coast Herring Industry," *Pacific Fisherman* (July 1905): 15.

11. "Remarkable Development of Alaska Herring Fishery," *Pacific Fisherman* (September 1925): 23–24, 26, 28, 30–31.

12. "The Fish Oil and Meal Industry of the Pacific," *Pacific Fisherman* (August 1935): 57–85; "Conditions in Southeast Alaska Herring Fishery," *Pacific Fisherman* (November 1927): 12–13.

13. Gerald M. Reid, *Fishery Facts-2, Alaska's Fishery Resources—The Pacific Herring*, National Marine Fisheries Service Extension Publication (June 1972), 14–15.

14. "Alaska Herring Floor Forced Up One-Third by Fishermen," *Pacific Fisherman* (July 1951): 48.

15. "Herring Negotiations Lag; Fish Meal Denied Protection," *Pacific Fisherman* (May 1938): 67.

16. "Fats Market Bust Leaves Herring Trade Up in the Air," *Pacific Fisherman* (May 1949): 83.

17. Gerald M. Reid, *Age Composition, Weight, Length, and Sex of Herring*, Clupea pallasii, *Used for Reduction in Alaska, 1929–66*, NOAA Tech. Rept. NMFS SSRF 634 (July 1971), 1.

18. Eliza Ruhamah Scidmore, "The First District of Alaska from Prince Frederick Sound to Yakutat Bay," in *Report on Population and Resources of Alaska at the Eleventh Census: 1890* (Washington, DC: GPO, 1893), 51; Eliza Ruhama Scidmore, *Appleton's Guide-Book to Alaska* (New York: D. Appleton and Co., 1893), 88.

19. James Sheakley, *Report of the Governor of the District of Alaska to the Secretary of the Interior, 1894* (Washington, DC: GPO, 1894), 11.

20. Moser, *Salmon and Salmon Fisheries of Alaska*, 121–124.

21. Eliza Ruhamah Scidmore, "Alaska Fish and Game," *Harper's Weekly* (September 2, 1893): 838–841; "Improvements at Killisnoo Plant," *Pacific Fisherman* (February 1919): 52.

22. E. J. Huizer, "History of Alaska Herring Fishery," *Alaska Department of Fisheries Annual Report for 1952*, 63–76.

23. Rounsefell, "Contribution to the Biology of the Pacific Herring, *Clupea pallasii*," 227–320.

24. Moser, *Salmon and Salmon Fisheries of Alaska*, 121–124; "August Buschmann," *Pacific Fisherman* (August 1904): 15.

25. Moser, *Salmon and Salmon Fisheries of Alaska*, 122.

26. Moser, *Salmon and Salmon Fisheries of Alaska*, 123.

27. Cobb, *The Commercial Fisheries of Alaska in 1905*, 19–22; "Alaska," *Pacific Fisherman* (August 1904): 10–11; "The Weise Packing Co.," *Pacific Fisherman* (January 1907): 14

28. Cobb, "Report on the Fisheries of Alaska [in 1906]," 52–56.

29. Barton Warren Evermann and Edmund Lee Goldsborough, *Fishes of Alaska*, US Bureau of Fisheries Doc. No. 624 (Washington, DC: GPO, 1907), 247.

30. Millard C. Marsh and John N. Cobb, *The Fisheries of Alaska in 1909*, Bureau of Fisheries Doc. No. 730 (Washington, DC: GPO, 1910), 48–50.

31. Evermann and Goldsborough, *The Fishes of Alaska*, 233.
32. Huizer, "History of Alaska Herring Fishery," 63–76.
33. "The Food of Fishes," *Pacific Fisherman* (October 1906): 10.
34. Act for the Protection and Regulation of the Fisheries of Alaska, June 26, 1906 (34 Stat. 478), § 8.
35. "Fertilizers—Fish Oils—Whaling," *Pacific Fisherman* (January 1908): 56.
36. Marsh and Cobb, *The Fisheries of Alaska in 1909*, 48–50.
37. Act of May 7, 1906 ("Delegate Act"), (34 Stat. 169).
38. "Herring Supply Threatened," *Pacific Fisherman* (March 1911): 13.
39. *Alaska Fisheries, Hearings on S. 5856, A Bill to Amend an Act for the Protection and Regulation of the Fisheries of Alaska*, Sen. Subcommittee of the Committee on Fisheries, 62nd Cong, 2d sess., April 11, 1912 (Washington, DC: GPO, 1912), 4.
40. Charles Nagel, Secretary, Department of Commerce and Labor, March 27, 1912, in *Alaska Fisheries, Hearings on S. 5856, A Bill to Amend an Act for the Protection and Regulation of the Fisheries of Alaska*, Sen. Subcommittee of the Committee on Fisheries, 62nd Cong, 2d sess., April 11, 1912 (Washington, DC: GPO, 1912), 3–4.
41. Cobb, "Report on the Fisheries of Alaska [in 1906]," 52–56.
42. Carl Spuhn, President, Alaska Oil and Guano Company, December 1910, in Alaska Fisheries, Hearing, United States Senate, Subcommittee of the Committee on Fisheries, April 11, 1912.
43. Walter Clark, Governor, Territory of Alaska, in Alaska Fisheries, Hearing, United States Senate, Subcommittee of the Committee on Fisheries, April 11, 1912.
44. E. Lester Jones, *Report of Alaska Investigations, 1914* (Washington, DC: GPO, 1915), 58–62.
45. Act for the Protection and Regulation of the Fisheries of Alaska, June 26, 1906 (34 Stat. 478), § 1.
46. Act to Create a Legislative Assembly in the Territory of Alaska, and Confer Legislative Power Thereon, August 24, 1912 (37 Stat. 512), § 3.
47. An act to establish a system of taxation, create revenue, and provide for collection thereof for the Territory of Alaska, and for other purposes, Chapter 52, *Session Laws of Alaska, 1913* (May 1, 1913).
48. Chapter 76, *Session Laws of Alaska, 1915* (April 29, 1915).
49. Chapter 74, *Session Laws of Alaska, 1917* (May 3, 1917).
50. Ward T. Bower, "Alaska Fisheries and Fur Industries in 1918," Appendix VII to the *Report of the United States Commissioner of Fisheries for the Fiscal Year 1918* (Washington, DC: GPO, 1920), 61–63.
51. *Alaska Fish Salting & By-Products Co. v. Smith*, 6 Alaska 173, 179 (D.Alaska 1919), aff'd, 255 U.S. 44 (1921).
52. "Fish Oil and Fertilizer Industry in 1921," *Pacific Fisherman* (Yearbook 1922): 94; "Alaska Fishery Taxes Changed," *Pacific Fisherman* (June 1921): 18; Chapter 31, *Session Laws of Alaska, 1921*.
53. "Alaska By-Products Season Late," *Pacific Fisherman* (August 1918): 47.
54. Ward T. Bower, *Alaska Fishery and Fur-Seal Industries in 1920*, Bureau of Fisheries Doc. No. 909 (Washington, DC: GPO, 1921), 62–63.
55. "Hearings on Herring Waste," *Pacific Fisherman* (January 1923): 31–32.
56. "The Fish Oil and Meal Industry of the Pacific," *Pacific Fisherman* (August 1935): 57–85.

57. "Possible Legislative Action Holds Herring Men's Attention," *Pacific Fisherman* (January 1935): 45.

58. "Alaska Salt Herring," *Pacific Fisherman* (Yearbook 1938): 199, 201.

59. Committee Study Resolution—Alaskan Fisheries Hearings, *Hearings before the Special Subcommittee on Alaskan Fisheries of the Committee on Merchant Marine and Fisheries, US House of Representatives, 76th Cong., 1st Sess., and the Joint Committee on Fisheries of the Legislature of Alaska* (Washington, DC: GPO, 1939), 1–3.

60. "Congress Committee Reports on Alaska Fishery Problems," *Pacific Fisherman* (July 1940): 19–20.

61. Rounsefell, "Contribution to the Biology of the Pacific Herring, *Clupea pallasii*," 227–320; Hervy M. Petrich, "Development of the Purse Seiner," *Pacific Fisherman* (January 1936): 18–19.

62. "Fishing Boat Evolution Traces to Rise of Oil and Meal Manufacturing," *Pacific Fisherman* (August 1936): 47; "Storfold and Grondahl Building New Seiner," *Pacific Fisherman* (December 1929): 47; "New Seiner for Buchan & Heinen," *Pacific Fisherman* (December 1929): 47; "Largest Alaska Herring Seiner Goes South for Tuna Fishing," *Pacific Fisherman* (November 1927): 8–9; "Seven Herring Plants to Operate," *Pacific Fisherman* (May 1931): 93; "High Boats of the Herring Fleet," *Pacific Fisherman* (December 1930): 55; "'Coolidge' Is High Boat," *Pacific Fisherman* (December 1932): 57.

63. "1,260 Tons of Herring in One Set Establishes New Fishing Record," *Pacific Fisherman* (February 1951): 13–14.

64. "Building Steel Seiner for Chatham Strait," *Pacific Fisherman* (April 1934): 54; "Steel Seiner Leaves," *Pacific Fisherman* (June 1934): 51; "First Steel Seiner," *Pacific Fisherman* (July 1934): 19–20.

65. Rounsefell, "Contribution to the Biology of the Pacific Herring, *Clupea pallasii*," 227–320.

66. Rounsefell, "Contribution to the Biology of the Pacific Herring, *Clupea pallasii*," 227–320.

67. Stephen H. Rogers, "Herring Fishery in Southeastern Alaska," *Commercial Fisheries Review* (August 1965): 1–6.

68. "The Fish Oil and Meal Industry of the Pacific," *Pacific Fisherman* (August 1935): 57–85.

69. "Oil and Meal," *Pacific Fisherman* (Yearbook 1946): 409, 411; Lawrence N. Kolloen and Keith A. Smith, "Southeastern Alaska Herring Exploratory Herring Fishing Operations, Winter 1952/53," *Commercial Fisheries Review* (November 1953): 1–24.

70. Ward T. Bower, *Alaska Fisheries and Fur-Seal Industries: 1946*, Statistical Digest No. 17 (Washington, DC: GPO, 1948), 40–42.

71. "Boom-Point Power Offers New Angle for Seine Fishing," *Pacific Fisherman* (May 1955): 26–27.

72. "Whale Slaughter Harms Herring Industry," *Pacific Fisherman* (November 1906): 13.

73. Barton Warren Evermann, *Alaska Fisheries and Fur Industries in 1913*, Bureau of Fisheries Doc. No. 797 (Washington, DC: GPO, 1914), 129; Ward T. Bower, "Alaska Fishery and Fur-Seal Industries in 1923," Bureau of Fisheries Doc. No. 973, in *Report of the United States Commissioner of Fisheries for the Fiscal Year 1924* (Washington, DC: GPO, 1925), 108.

74. Evermann, *Alaska Fisheries and Fur Industries in 1913*, 133–134.

75. "Whales Versus Herring," *Pacific Fisherman* (April 1919): 32; Senate Joint Memorial No. 4, Alaska Senate Journal, 1919.

76. George A. Rounsefell, "Some Observations on the Alaska Herring," Part 2, *Pacific Fisherman* (September 1928): 20–21.

77. Roy Chapman Andrews, *Whale Hunting with Gun and Camera* (New York: D. Appleton and Company, 1916), 4–5, 185.

78. Andrews, *Whale Hunting with Gun and Camera*, 4–5.

79. Dale Rice (deceased whale biologist), information courtesy Janice M. Straley (University of Alaska), May 25, 2021.

80. Dale Vinnedge, *Alaska's Whaling Coast* (Charleston, SC: Arcadia Publishing, 2013), 42.

81. Briana H. Witteveen, Robert J. Foy, Kate M. Wynne, "The Effect of Predation (Current and Historical) by Humpback Whales (*Megaptera novaeangliae*) on Fish Abundance near Kodiak Island, Alaska," *Fishery Bulletin* 104 (2006): 10–20; M. J. Kirchoff, *Baranof Island: An Illustrated History* (Juneau: Alaska Cedar Press, 1990), 95; Thomas F. Thornton et al., *Herring Synthesis: Documenting and Modeling Herring Spawning Areas within Socio-Ecological Systems over Time in the Southeastern Gulf of Alaska*, North Pacific Research Board Project No. 728 (June 2010), 295.

82. 81 *Fed. Reg.* 62260, September 8, 2016.

83. Alaska Department of Fish and Game, *Species Profile, Pacific Herring (Clupea pallasii)*, accessed August 25, 2019, http://www.adfg.alaska.gov/index.cfm?ADFG= herring.main.

84. John R. Moran and Janice M. Straley, *Humpback Whales As Indicators of Herring Movements in Prince William Sound*, Exxon Valdez Oil Spill Long-term Monitoring Program, Gulf Watch Alaska, Final Report, Restoration Project 16120114-N (July 2017).

CHAPTER 3: SALTED HERRING

1. Donald K. Tressler et al., *Marine Products of Commerce; Their Acquisition, Handling, Biological Aspects, and the Science and Technology of Their Preparation and Preservation* (New York: Chemical Catalog Company, 1923), 327.

2. Jefferson F. Moser, "Salmon and Salmon Fisheries of Alaska: Report of the Operations of the United States Fish Commission Steamer Albatross for the Year Ending June 30, 1898," in *Bulletin of the United States Fish Commission*, Vol. XVIII, 1898, 55th Cong. 3d Sess. Ho. Doc. No. 308 (Washington, DC: GPO, 1899), 121–124.

3. Jefferson F. Moser, "Salmon Investigations of the Steamer *Albatross* in the Summer of 1900," in *Bulletin of the United States Fish Commission*, Vol. XXI, 1902 (Washington, DC: GPO, 1903), 258.

4. Jefferson F. Moser, "Salmon Investigations of the Steamer Albatross in the Summer of 1901," in *Bulletin of the United States Fish Commission for 1901* (Washington, DC: GPO, 1902), 261.

5. "An Important Factor in Alaska Herring Industry," *Pacific Fisherman* (February 1918): 32; E. J. Huizer, "History of Alaska Herring Fishery," *Alaska Department of Fisheries Annual Report for 1952*, 63–76.

6. John G. Brady, *Report of the Governor of the District of Alaska to the Secretary of the Interior, 1903* (Washington, DC: GPO, 1903), 22.

7. Huizer, "History of Alaska Herring Fishery," 63–76.

8. John G. Brady, *Report of the Governor of the District of Alaska to the Secretary of the Interior, 1901* (Washington, DC: GPO, 1901), 17; "A New Cannery at Yakutat," *Pacific Fisherman* (May 1904): 29.

9. John N. Cobb, *The Commercial Fisheries of Alaska in 1905*, Bureau of Fisheries Doc. No. 603 (Washington, DC: GPO, 1906), 20–21.

10. Huizer, "History of Alaska Herring Fishery," 63–76; Barton Warren Evermann, *Alaska Fisheries and Fur Industries in 1911*, Bureau of Fisheries Doc. No. 766 (Washington, DC: GPO, 1912), 59–62; "Canadian Dry Salting Regulations," *Pacific Fisherman* (September 1922): 41; Tressler et al., *Marine Products of Commerce*, 310.

11. Huizer, "History of Alaska Herring Fishery," 63–76; "Red and Black Feed," *Pacific Fisherman* (June 1929): 53; John N. Cobb, "Report on the Fisheries of Alaska [in 1906]," Bureau of Fisheries Doc. No. 618, in *Report of the Commissioner of Fisheries for the Year 1906 and Special Papers* (Washington, DC: GPO, 1907), 52–56.

12. Huizer, "History of Alaska Herring Fishery," 63–76; E. Lester Jones, *Report of Alaska Investigations, 1914* (Washington, DC: GPO, 1915), 58–62; George Rounsefell, "Fluctuations in the Supply of Herring (*Clupea Pallasii*) in Southeastern Alaska," in *Bulletin of the Bureau of Fisheries*, Vol. XLVII, Bulletin No. 2 (Washington, DC: GPO, 1931), https://babel.hathitrust.org/cgi/pt?id=coo.31924078251000&view=1up&seq=35.

13. Capt. A. W. Thomas, "Protect the Alaska Herring Fisheries," *Pacific Fisherman* (April 1919): 24.

14. Ward T. Bower, "Alaska Fishery and Fur-Seal Industries in 1928," Appendix VI to *Report of the United States Commissioner of Fisheries for the Fiscal Year 1929* (Washington, DC: GPO, 1930), 285–289.

15. Thomas, "Protect the Alaska Herring Fisheries," 24.

16. George A. Rounsefell, "Contribution to the Biology of the Pacific Herring, *Clupea pallasii*, and the Condition of the Fishery in Alaska," in *Bulletin of the Bureau of Fisheries, 1929* (Washington, DC: GPO, 1930), 227–320; Howard H. Hungerford (US Bureau of Fisheries), *Report of Operations—Kodiak-Afognak District—1933*, 10.

17. "Wakefield Interest Consolidated," *Pacific Fisherman* (January 1927): 34; "Wakefield's Herring Plans," *Pacific Fisherman* (February 1928): 36.

18. Rounsefell, "Contribution to the Biology of the Pacific Herring, *Clupea pallasii*," 227–320.

19. Barton Warren Evermann, *Alaska Fisheries and Fur Industries in 1913*, Bureau of Fisheries Doc. No. 797 (Washington, DC: GPO, 1914), 125–129.

20. Ward T. Bower and Henry T. Aller, "Alaska Fisheries and Fur Industries in 1916," Appendix II to the *Report of the United States Commissioner of Fisheries for the Fiscal Year 1916* (Washington, DC: GPO, 1917), 71–72.

21. "A New Alaska Herring Industry," *Fisheries Service Bulletin* (December 1, 1916), 6; Rounsefell, "Contribution to the Biology of the Pacific Herring, *Clupea pallasii*," 227–320; E. H. Dahlgren and L. N. Kolloen, *Outlook for the Alaska Herring Fishery in 1943*, US Fish and Wildlife Service Leaflet No. 16, March 1943; Lee H. Wakefield, President, Herring Packers Association, "The Development of the Herring Industry of Alaska," reproduced in *Protection of Alaskan Fisheries, 71st Cong., 2d sess., Hearings Before the Committee on Merchant Marine and Fisheries, House of Representatives, on H.R. 253 and H.R. 7238, April 10, 1930* (Washington, DC: GPO, 1930), 17–18; Thomas, "Protect the Alaska Herring Fisheries," 24; Bower and Aller, "Alaska Fisheries and

Fur Industries in 1916," 71–72; "Wakefield-Wilson Fisheries Consolidation," *Pacific Fisherman* (Yearbook 1918): 95–96; advertisement, Wakefield & Company, *Pacific Fisherman* (February 1920): 31.

22. "An Important Factor in Alaska Herring Industry," *Pacific Fisherman* (February 1918): 32.

23. "Southern Alaska Co. Absorbs A.P. Herring Co.," *Pacific Fisherman* (March 1919): 44.

24. "Bortz Takes Pessimistic View," *Pacific Fisherman* (June 1925): 50.

25. "Will Introduce Scotch Herring Method," *Pacific Fisherman* (May 1917): 29.

26. Ward T. Bower and Henry D. Aller, *Alaska Fisheries and Fur Industries in 1917*, Bureau of Fisheries Doc. No. 847 (Washington, DC: GPO, 1918), 49–51.

27. "Increased Production of Pickled Herring Demanded," *Pacific Fisherman* (June 1917): 29.

28. "Promotion of Alaska Fisheries," *Fisheries Service Bulletin* (May 1, 1917), 3; H. F. Moore, "Educational Campaign of the US Bureau of Fisheries," *Pacific Fisherman* (January 1918): 112–114.

29. Bower and Aller, *Alaska Fisheries and Fur Industries in 1917*, 49–51; "Promotion of Alaska Fisheries," *Fisheries Service Bulletin* (May 1, 1917), 3.

30. "Good Outlook for Alaska Scotch Herring," *Pacific Fisherman* (August 1917): 26–27; *Report of the United States Commissioner of Fisheries for the Year 1917* (Washington, DC: GPO, 1919), 48–49; Tressler et al., *Marine Products of Commerce*, 330.

31. "Alaska Fishery Exploitation," *Fisheries Service Bulletin* (June 1, 1917), 2–3.

32. "Good Outlook for Alaska Scotch Herring," *Pacific Fisherman* (August 1917): 26–27; Ward T. Bower, "Alaska Fisheries and Fur Industries in 1918," Appendix VII to the *Report of the United States Commissioner of Fisheries for the Fiscal Year 1918* (Washington, DC: GPO, 1920), 61–63.

33. "Good Outlook for Alaska Scotch Herring," *Pacific Fisherman* (August 1917): 26–27.

34. US Bureau of Fisheries, *Fisheries Service Bulletin* (August 1, 1917): 3; Bower and Aller, *Alaska Fisheries and Fur Industries in 1917*, 49–51.

35. US Bureau of Fisheries, *Fisheries Service Bulletin* (July 1, 1918): 1–2.

36. George A. Rounsefell and Edwin H. Dahlgren, "Fluctuations in the Supply of Herring, *Clupea pallasii*, in Prince William Sound, Alaska," US Bureau of Fisheries Bulletin No. 9 (May 12, 1932), in *Bulletin of the United States Bureau of Fisheries*, XLVII (Washington, DC: GPO, 1935), 263–291.

37. Rounsefell and Dahlgren, "Fluctuations in the Supply of Herring, *Clupea pallasii*," 263–291.

38. Aug. H. D. Klie, "Scotch Method of Curing Herring," *Pacific Fisherman* (January 1918): 89a–89d.

39. Robert J. Browning, *Fisheries of the North Pacific: History, Species, Gear & Processes*, rev. ed. (Anchorage: Alaska Northwest Publishing Company, 1980), 33–36.

40. Aug. H. D. Klie, "Scotch Method of Curing Herring," *Pacific Fisherman* (January 1918): 89a–89d.

41. Bower, "Alaska Fisheries and Fur Industries in 1918," 61–63; "Southern Alaska Co. Absorbs A.P. Herring Co.," *Pacific Fisherman* (March 1919): 44; George Rounsefell, "Some Observations on the Alaska Herring," Part 1, *Pacific Fisherman* (August 1928): 62; Rounsefell, "Contribution to the Biology of the Pacific Herring, *Clupea*

pallasii," 227–320; "Review of Cured Herring Industry in 1923," *Pacific Fisherman* (Yearbook, January 1924): 104, 106.

42. Marsh and Cobb, *Fisheries of Alaska in 1910*, 45–51.

43. Bower, "Alaska Fisheries and Fur Industries in 1918," 61–63; Rounsefell, "Contribution to the Biology of the Pacific Herring, *Clupea pallasii*," 227–320; "Review of Cured Herring Industry in 1923," *Pacific Fisherman* (Yearbook, January 1924): 104, 106.

44. John N. Cobb, "Increasing Our Pacific Coast Fishery Resources," *Pacific Fisherman* (November 1918): 21–22.

45. "Caution to Scotch-Cure Herring Packers," *Pacific Fisherman* (September 1918): 37–39.

46. Bower, "Alaska Fisheries and Fur Industries in 1918," 61–63.

47. "Hearings on Herring Waste," *Pacific Fisherman* (January 1923): 31–32.

48. "Requirements of Alaska Herring Industry," *Pacific Fisherman* (December 1922): 12–13.

49. Advertisement, Western Cooperage Co., *Pacific Fisherman* (January 1926): 136–137.

50. Aug. H. D. Klie, "Scotch Method of Curing Herring," *Pacific Fisherman* (January 1918): 89a–89d.

51. "Importance of Cooperage for Herring Packing," *Pacific Fisherman* (July 1919): 54.

52. J.M. Wyckoff, "History of Lumbering in Southeast Alaska," *Timberman* (June 1930): 38–40; Advertisement, Henry Imhoff Cooperage, *Ketchikan Alaska Chronicle*, May 17, 1922.

53. Klie, "Scotch Method of Curing Herring," 89a–89d.

54. "A New Scotch-Style Herring Barrel," *Pacific Fisherman* (May 1919): 49.

55. Advertisement, Western Cooperage Co., *Pacific Fisherman* (July 1924): 52; advertisement, Western Cooperage Co., *Pacific Fisherman* (January 1926): 136–137.

56. Klie, "Scotch Method of Curing Herring," 89a–89d.

57. "Adopt Standard Herring Barrel," *Pacific Fisherman* (May 1921): 34.

58. "Where Your Salt Comes From," *Pacific Fisherman* (January 1916): 83.

59. Klie, "Scotch Method of Curing Herring," 89a–89d.

60. "Where Your Salt Comes From," 83.

61. "Salt Dock for Seattle," *Pacific Fisherman* (February 1925): 48.

62. Thomas, "Protect the Alaska Herring Fisheries," 4.

63. Alaska House Joint Memorial No. 6, April 25, 1919, *Session Laws of Alaska*, 1919; "Fisheries Legislation in Alaska," *Pacific Fisherman* (May 1919): 51.

64. "Scotch vs. Alaska Herring," *Pacific Fisherman* (August 1919): 52.

65. Ward T. Bower, "Alaska Fisheries and Fur Industries in 1919," Appendix IX to *Report of the United States Commissioner of Fisheries for the Fiscal Year 1919* (Washington, DC: GPO, 1921), 54.

66. "Alaska Pickled Food Herring, 1919," *Pacific Fisherman* (Yearbook 1920): 132.

67. "Alaska Matjes Herring Wanted," *Pacific Fisherman* (April 1920): 51; "Beyer Comments on Herring Industry," *Pacific Fisherman* (August 1922): 44, 46.

68. "A Scotchman's View of Alaska Herring," *Pacific Fisherman* (April 1920): 52.

69. George A. Rounsefell, "Pacific Herring," in "Progress in Biological Inquiries, 1926," Appendix II, *Report of the United States Commissioner of Fisheries for the Fiscal Year 1927* (Washington, DC: GPO, 1928), 650–652.

70. Ward T. Bower, *Alaska Fishery and Fur-Seal Industries in 1920*, Bureau of Fisheries Doc. No. 909 (Washington, DC: GPO, 1921), 62–63; Rounsefell and Dahlgren, "Fluctuations in the Supply of Herring, Clupea *pallasii*," 263–291.

71. "Freight Rate Advance Hits Herring Industry," *Pacific Fisherman* (August 1920): 49; "Salt Fish Market," *Pacific Fisherman* (November 1920): 52; "Matjes from Port Walter," *Pacific Fisherman* (July 1921): 35.

72. "Pacific Cured Fish Association Organized," *Pacific Fisherman* (March 1921): 44.

73. "Urge Protection for Herring Industry," *Pacific Fisherman* (April 1921): 23.

74. Act for the Protection and Regulation of the Fisheries of Alaska, June 26, 1906 (34 Stat. 478), § 1; Chapter 76, *Session Laws of Alaska, 1915* (April 29, 1915).

75. Chapter 31, *Session Laws of Alaska, 1921* (May 5, 1921).

76. "Review of Salt Herring Industry in 1921," *Pacific Fisherman* (Yearbook 1922): 91–92.

77. "Demand for Herring Increases," *Pacific Fisherman* (September 1921): 35; "Review of Salt Herring Industry in 1921," 91–92.

78. "Plans for Large Herring Pack," *Pacific Fisherman* (April 1922): 43.

79. "Herring Season Begins," *Pacific Fisherman* (July 1922): 43.

80. "Plans for Large Herring Pack," 43.

81. Ward T. Bower, *Alaska Fishery and Fur-Seal Industries in 1922*, Bureau of Fisheries Doc. No. 951 (Washington, DC: GPO, 1923), 75; "1922 Alaska Herring Pack Breaks Record," *Pacific Fisherman* (Yearbook 1923): 95.

82. "Heavy Shipments of Alaska Herring," *Pacific Fisherman* (August 1922): 43.

83. "1922 Alaska Herring Pack Breaks Record," *Pacific Fisherman* (Yearbook, January 1923): 95–96.

84. "Kodiak Season a Failure," *Pacific Fisherman* (December 1922): 46–48; "1922 Alaska Herring Pack Breaks Record," *Pacific Fisherman* (Yearbook, January 1923): 95–96.

85. "O'Malley Comments on Alaska Herring," *Pacific Fisherman* (May 1923): 30; "Improper Handling Means Loss," *Pacific Fisherman* (January 1923): 32–33; "Losses in Herring Trade," *Pacific Fisherman* (April 1923): 30; Ward T. Bower, "Alaska Fishery and Fur-Seal Industries in 1923," Bureau of Fisheries Doc. No. 973, in *Report of the United States Commissioner of Fisheries for the Fiscal Year 1924* (Washington, DC: GPO, 1925), 102–103.

86. "Herring Pack Curtailed," *Pacific Fisherman* (August 1923): 42, 44.

87. "New Herring Projects Planned," *Pacific Fisherman* (January 1924): 40; "Reports Poor Herring Season," *Pacific Fisherman* (January 1924): 42.

88. "Herring Shortage Serious," *Pacific Fisherman* (October 1923): 26; "Review of Cured Herring in 1923," *Pacific Fisherman* (Yearbook 1924): 104–106.

89. "Review of Cured Herring in 1923," *Pacific Fisherman* (Yearbook 1924): 104–106.

90. "Review of Cured Herring in 1923," *Pacific Fisherman* (Yearbook 1924): 104–106.

91. "Golovin Herring Season," *Pacific Fisherman* (March 1925): 46.

92. Rounsefell, "Contribution to the Biology of the Pacific Herring, *Clupea pallasii*," 236–239.

93. "Golovin Packers Leave for Alaska," *Pacific Fisherman* (June 1919): 49.

94. "Alaska Pickled Food Herring," *Pacific Fisherman* (Yearbook 1917): 100.

95. "Alaska Pickled Food Herring, 1917," *Pacific Fisherman* (Yearbook 1918): 105.

96. "Alaska Pickled Food Herring, 1918," *Pacific Fisherman* (Yearbook 1919): 122.

97. Health Analytics and Vital Records, Division of Public Health, Alaska Department of Health and Social Services, *1918 Pandemic Influenza Mortality in Alaska*, September 20, 2018, http://dhss.alaska.gov/dph/VitalStats/Documents/PDFs/AK_1918Flu_DataBrief_092018.pdf; "Golovin Herring Industry Expanding," *Pacific Fisherman* (January 1919): 39.

98. "Golovin Packers Discouraged," *Pacific Fisherman* (February 1920): 57; "Alaska Pickled Food Herring, 1919," *Pacific Fisherman* (Yearbook 1920): 132; "Golovin Herring Arrives," *Pacific Fisherman* (November 1920): 52.

99. "Alaska Pickled Food Herring, 1921," *Pacific Fisherman* (Yearbook 1922): 91.

100. "Alaska Pickled Food Herring, 1922," *Pacific Fisherman* (Yearbook 1923): 95.

101. "Alaska Pickled Food Herring, 1923," *Pacific Fisherman* (Yearbook 1924): 104.

102. "Kruse Organizes Herring Company," *Pacific Fisherman* (April 1927): 36; Ward T. Bower, "Alaska Fisheries and Fur-Seal Industries in 1936," Administrative Rept. No. 28, Appendix II to *Report of the United States Commissioner of Fisheries for the Fiscal Year 1937* (Washington, DC: GPO, 1939), 317–320; Ward T. Bower, "Alaska Fisheries and Fur-Seal Industries in 1937," Administrative Rept. No. 31, in *Report of the United States Commissioner of Fisheries for the Fiscal Year 1938* (Washington, DC: GPO, 1940), 116–118.

103. Norman B. Wigutoff and Clarence J. Carlson, "A Survey of the Commercial Fishery Possibilities of Seward Peninsula Area, Kotzebue Sound, and Certain Inland Rivers and Lakes in Alaska," US Fish and Wildlife Service Leaflet No. 375 (May 1950), 11–12.

CHAPTER 4: EARLY ALASKA HERRING FISHERY REGULATION AND RESEARCH

1. E. H. Dahlgren (US Bureau of Fisheries), "Alaska Herring Research," Part 1, *Pacific Fisherman* (March 1940): 33–34.

2. *Fisheries of Alaska, Hearings Before the Committee on Merchant Marine and Fisheries, House of Representatives, 68th Cong. 1st sess, January 31, February 1, 6, 7 and 8, 1924* (Washington, DC: GPO, 1924), 11.

3. Executive Order of November 3, 1922, *Monthly Catalogue of Public Documents*, No. 331 (Washington, DC: Superintendent of Public Documents, July 1922), 298.

4. Ward T. Bower, "Alaska Fishery and Fur-Seal Industries in 1923," Bureau of Fisheries Doc. No. 973, in *Report of the United States Commissioner of Fisheries for the Fiscal Year 1924* (Washington, DC: GPO, 1925), 53, 62–63; US Bureau of Fisheries, "Alaska Service Notes," *Fisheries Service Bulletin* (November 1, 1923): 4–5.

5. Bower, "Alaska Fishery and Fur-Seal Industries in 1923," 48–51; *Leasing of Salmon Trap Sites, Hearings before a Subcommittee of the Senate Committee on Interstate and Foreign Commerce and a Subcommittee of the House Merchant Marine and Fisheries Committee on S. 1446 and H.R. 3859, Bills to Authorize the Leasing of Salmon Trap Sites in Alaskan Coastal Waters, and for Other Purposes, 80th Cong. 2d Sess.* (Washington, DC: GPO, 1948), 81.

6. "Fish Oil and Fertilizer Industry in 1923," *Pacific Fisherman* (Yearbook 1924):113–114, 116.

7. Bower, "Alaska Fishery and Fur-Seal Industries in 1923," 49–50.

8. Act for the Protection of the Fisheries of Alaska ("White Act"), June 6, 1924 (435 Stat. 464).

9. Executive Order of June 7, 1924.

10. Act for the Protection of the Fisheries of Alaska ("White Act"), June 6, 1924 (435 Stat. 464), § 1.

11. Ward T. Bower, "Alaska Fishery and Fur-Seal Industries in 1924," Bureau of Fisheries Doc. No. 902, in *Report of the United States Commissioner of Fisheries for the Fiscal Year 1925* (Washington, DC: GPO, 1926), 75–77.

12. "Additional Alaska Salmon Fishery Regulations," *Fisheries Service Bulletin* (October 1, 1924), 6.

13. "Review of Cured Herring Industry in 1924," *Pacific Fisherman* (Yearbook, January 1925): 136, 138, 140, 143–144.

14. "Imlach Runs Three Plants," *Pacific Fisherman* (March 1925): 44.

15. *Committee Study Resolution—Alaskan Fisheries Hearings, Hearings before the Special Subcommittee on Alaskan Fisheries of the Committee on Merchant Marine and Fisheries, US House of Representatives, 76th Cong., 1st Sess., and the Joint Committee on Fisheries of the Legislature of Alaska* (Washington, DC: GPO, 1939), 391.

16. "Herring Regulations Beneficial," *Pacific Fisherman* (February 1924): 28.

17. Ed Wyman (family owned Katlian Packing Co.), personal communication, February 4, 2020.

18. Ward T. Bower, "Alaska Fishery and Fur-Seal Industries in 1926," Bureau of Fisheries Doc. No. 1023, Appendix IV to *the Report of the United States Commissioner of Fisheries for the Fiscal Year 1927* (Washington, DC: GPO, 1928) 232, 236, 237, 239, 245.

19. Gerald M. Reid, *Age Composition, Weight, Length, and Sex of Herring*, Clupea pallasii, *Used for Reduction in Alaska, 1929–66*, NOAA Tech. Rept. NMFS SSRF 634 (July 1971), 1.

20. C. L. Anderson, "The Herring Fishery at Aalesund, Norway," *Pacific Fisherman* (June 1922): 10–12; "Wakefield's Alaska Operations," *Pacific Fisherman* (September 1925): 96.

21. "Anderson's Views on Herring," *Pacific Fisherman* (February 1925): 44, 46.

22. George A. Rounsefell, "Contribution to the Biology of the Pacific Herring, *Clupea pallasii*, and the Condition of the Fishery in Alaska," in *Bulletin of the Bureau of Fisheries, 1929* (Washington, DC: GPO, 1930), 227–320.

23. George A. Rounsefell, "Report of Progress in Alaska Herring Investigation," *Pacific Fisherman* (December 1926): 20–21; Rounsefell, "Contribution to the Biology of the Pacific Herring, *Clupea pallasii*," 227–320.

24. "Will Study Herring Fisheries," *Pacific Fisherman* (March 1925): 9; "Investigation of Herring in Alaska," *Fisheries Service Bulletin* (April 1, 1925): 2; Ward T. Bower, "Alaska Fishery and Fur-Seal Industries in 1925," Bureau of Fisheries Doc. No. 1008, Appendix III to *Report of the United States Commissioner of Fisheries for the Fiscal Year 1926* (Washington, DC: GPO, 1927), 128–132.

25. Rounsefell, "Report of Progress in Alaska Herring Investigation," 20–21.

26. Reid, *Age Composition, Weight, Length, and Sex of Herring*, 1.

27. "Alaska Herring Controversy Shows Need of Investigation," *Pacific Fisherman* (November 1926): 7–8.

28. "Alaska Herring Controversy Shows Need of Investigation," *Pacific Fisherman* (November 1926): 7–8.

29. George A. Rounsefell, "Contribution to the Biology of the Pacific Herring, *Clupea pallasii*, and the Condition of the Fishery in Alaska," in *Bulletin of the Bureau of Fisheries, 1929* (Washington, DC: GPO, 1930), 317.

30. George Rounsefell, "Fluctuations in the Supply of Herring (Clupea Pallasii) in Southeastern Alaska," in *Bulletin of the Bureau of Fisheries*, Vol. XLVII, Bulletin No. 2 (Washington, DC: GPO, 1931), https://babel.hathitrust.org/cgi/pt?id=coo.31924078251000&view=1up&seq=35.

31. Rounsefell, "Fluctuations in the Supply of Herring (Clupea Pallasii) in Southeastern Alaska."

32. "Seven Herring Plants to Operate," *Pacific Fisherman* (May 1931): 93.

33. Rounsefell, "Fluctuations in the Supply of Herring (Clupea Pallasii) in Southeastern Alaska."

34. Ward T. Bower, *Alaska Fishery and Fur-Seal Industries in 1931*, in *Report of the United States Commissioner of Fisheries for the Fiscal Year 1932* (Washington, DC: GPO, 1933), 64–67.

35. Ward T. Bower, "Alaska Fishery and Fur-Seal Industries in 1932," Appendix I to *Report of the United States Commissioner of Fisheries for the Fiscal Year 1933* (Washington, DC: GPO, 1934), 12.

36. Henry O'Malley, in *Relating to the Alaskan Fisheries, Hearings on H.R. 497, A Bill to Amend Section 8 of Chapter 3547, Thirty-Fourth Statute at Large, Part 1, Entitled "An Act for the Protection and Regulation of the Fisheries of Alaska," Approved June 26, 1906*, Ho. Committee on Merchant Marine, Radio, and Fisheries, 72nd Cong, 1st sess., February 11, 1932 (Washington, DC: GPO, 1932), 1; *Protection of Alaskan Fisheries, Hearings on H.R. 253 and H.R. 7238*, Ho. Committee on the Merchant Marine and Fisheries, 71st Cong., 2d sess., April 10, 1930 (Washington, DC: GPO, 1930), 18–19.

37. "Herring Operators Oppose Anti-Reduction Bill," *Pacific Fisherman* (February 1932): 61.

38. "O'Malley States His Position on Fish Reduction to Congress," *Pacific Fisherman* (April 1932): 66.

CHAPTER 5: ALASKA'S HERRING INDUSTRY EXPANDS: 1924–1931

1. George A. Rounsefell, "Contribution to the Biology of the Pacific Herring, *Clupea pallasii*, and the Condition of the Fishery in Alaska," in *Bulletin of the Bureau of Fisheries, 1929* (Washington, DC: GPO, 1930), 227–320.

2. "Fish Oil and Meal Market," *Pacific Fisherman* (June 1921): 51.

3. "Fish Oil and Fertilizer Industry in 1923," *Pacific Fisherman* (Yearbook 1924): 113.

4. "Fish Oil and Fertilizer Industry in 1923," 113; "Fish Oil and Fertilizer Industry in 1922," *Pacific Fisherman* (Yearbook 1923): 100; June Allen, "A Biography of Alaska's Herring: A Little Fish of Huge Importance," *SitNews*, March 14, 2004, http://www.sitnews.net/JuneAllen/Herring/031404_herring.html.

5. Ward T. Bower, "Alaska Fishery and Fur-Seal Industries in 1924," Bureau of Fisheries Doc. No. 902, in *Report of the United States Commissioner of Fisheries for the*

Fiscal Year 1925 (Washington, DC: GPO, 1926), 135–136; "Remarkable Development of Alaska Herring Fishery," *Pacific Fisherman* (September 1925): 23–24, 26, 28, 30.

6. Bower, "Alaska Fishery and Fur-Seal Industries in 1924," 135–136; "Remarkable Development of Alaska Herring Fishery," *Pacific Fisherman* (September 1925): 23–24, 26, 28, 30.

7. "The Alaska Herring Pack of 1925 Second Largest on Record," *Pacific Fisherman* (January 1926): 142, 144, 146, 148, 150.

8. "Remarkable Growth of the Alaska Herring Industry," *Pacific Fisherman* (June 1925): 8–9, 24–25.

9. "Remarkable Development of Alaska Herring Fishery," *Pacific Fisherman* (September 1925): 23–24, 26, 28, 30.

10. Ward T. Bower, "Alaska Fishery and Fur-Seal Industries in 1925," Bureau of Fisheries Doc. No. 1008, Appendix III to *Report of the United States Commissioner of Fisheries for the Fiscal Year 1926* (Washington, DC: GPO, 1927), 128–132.

11. Rounsefell, "Contribution to the Biology of the Pacific Herring, *Clupea pallasii*," 227–320.

12. "Pacific Fish Oil Production Makes a New Record," *Pacific Fisherman* (Yearbook 1927): 212, 215–216, 218.

13. "Fish Oil and Meal Output Declines," *Pacific Fisherman* (Yearbook 1931): 217, 219, 221–222.

14. Elmer Higgins, "Progress in Biological Inquiries, 1927," in *Report of the United States Commissioner of Fisheries for the Fiscal Year 1928* (Washington, DC: GPO, 1929), 218–219.

15. Dan McNeil, "My Herring-Choker," *Pacific Fisherman* (October 1924): 44.

16. "The Alaska Herring Pack of 1925 Second Largest on Record," *Pacific Fisherman* (January 1926): 142, 144, 146, 148, 150.

17. "Remarkable Growth of the Alaska Herring Industry," *Pacific Fisherman* (June 1925): 8–9, 24–25.

18. "Remarkable Growth of the Alaska Herring Industry," *Pacific Fisherman* (June 1925): 8–9, 24–25; Statement of Ottar Hofstad, October 11, 1939, in *Committee Study Resolution—Alaskan Fisheries Hearings, Hearings before the Special Subcommittee on Alaskan Fisheries of the Committee on Merchant Marine and Fisheries, US House of Representatives, 76th Cong., 1st Sess., and the Joint Committee on Fisheries of the Legislature of Alaska* (Washington, DC: GPO, 1939), 263.

19. "Remarkable Growth of the Alaska Herring Industry," *Pacific Fisherman* (June 1925): 8–9, 24–25.

20. "Remarkable Growth of the Alaska Herring Industry," *Pacific Fisherman* (June 1925): 8–9, 24–25; Bower, "Alaska Fishery and Fur-Seal Industries in 1925," 128–132; Ward T. Bower, "Alaska Fishery and Fur-Seal Industries in 1931," in *Report of the United States Commissioner of Fisheries for the Fiscal Year 1932* (Washington, DC: GPO, 1933), 64–67.

21. "The Alaska Herring Pack of 1925 Second Largest on Record," *Pacific Fisherman* (January 1926): 142, 144, 146, 148, 150.

22. Ward T. Bower, "Alaska Fishery and Fur-Seal Industries in 1927," in *Report of the United States Commissioner of Fisheries for the Fiscal Year 1928*, Part 1 (Washington, DC: GPO, 1929), 132–137.

23. "Alaska Herring Output Shows Substantial Increase," *Pacific Fisherman* (Yearbook 1929): 143, 145, 147, 149; " 'Alice Cooke' Is Destroyed by Fire on Prince William Sound," *Pacific Fisherman* (December 1931): 73.

24. "Refrigeration Plant Installed in Floating Herring Saltery," *Pacific Fisherman* (May 1928): 12–15; Ward T. Bower, "Alaska Fishery and Fur-Seal Industries in 1928," Appendix VI to *Report of the United States Commissioner of Fisheries for the Fiscal Year 1929* (Washington, DC: GPO, 1930), 285–289; "Shoreplants: The Backbone of the Fish Oil and Meal Business," *Pacific Fisherman* (August 1936): 31–32; Ward T. Bower, "Alaska Fisheries and Fur-Seal Industries in 1937," Administrative Rept. No. 31, in *Report of the United States Commissioner of Fisheries for the Fiscal Year 1938* (Washington, DC: GPO, 1940), 116–118.

25. "Schooner 'John A' to Blue Fox Bay," *Pacific Fisherman* (June 1936): 67.

26. "Problem of Marketing the Herring Pack," *Pacific Fisherman* (January 1926): 42.

27. "Alaska Cured Herring Pack Smallest in Years," *Pacific Fisherman* (January 1927): 196, 198, 200, 202.

28. Bower, "Alaska Fishery and Fur-Seal Industries in 1926," 294–297.

29. "Pacific Fish Oil Production Makes a New Record," *Pacific Fisherman* (Yearbook, January 1927): 212, 215–216, 218; Bower, "Alaska Fishery and Fur-Seal Industries in 1926," 294–297.

30. "Remarkable Development of Alaska Herring Fishery," *Pacific Fisherman* (September 1925): 23–24, 26, 28, 30.

31. Bower, "Alaska Fishery and Fur-Seal Industries in 1926," 294–297; "Alaska Cured Herring Pack Smallest in Years," *Pacific Fisherman* (January 1927): 196, 198, 200, 202.

32. "Herring Packers Organizing," *Pacific Fisherman* (November 1926): 28.

33. "Alaska Herring Packers Organize," *Pacific Fisherman* (December 1926): 19–20.

34. "Conditions in Southeast Alaska Herring Fishery," *Pacific Fisherman* (November 1927): 12–13.

35. "Herring Packers Organizing," *Pacific Fisherman* (November 1926): 28.

36. "Alaska Herring Packers Organize," *Pacific Fisherman* (December 1926): 19–20.

37. "Alaska Herring Packers Organize," *Pacific Fisherman* (December 1926): 19–20.

38. "Herring Assn. Establishes Office," *Pacific Fisherman* (January 1927): 34.

39. Alaska Senate Joint Memorial No. 9, February 25, 1953, *Session Laws of Alaska, 1953*.

40. "Storfold & Grondahl Incorporated," *Pacific Fisherman* (February 1925): 40; "Storfold & Grondahl Branch Out," *Pacific Fisherman* (March 1925): 42, 44; "Storfold & Grondahl Branch Out," *Pacific Fisherman* (February 1926): 58.

41. "Alaska Herring Production Smallest in Years," *Pacific Fisherman* (Yearbook, 1928): 178, 180, 182.

42. Rounsefell, "Contribution to the Biology of the Pacific Herring, *Clupea pallasii*," 227–320.

43. Bower, "Alaska Fishery and Fur-Seal Industries in 1927," in *Report of the United States Commissioner of Fisheries for the Fiscal Year 1928*, Part 1 (Washington, DC: GPO, 1929), 132–137; "Alaska Herring Production Smallest in Years," *Pacific Fisherman* (Yearbook, 1928): 178, 180, 182; "Pacific Coast Fish Oil Production Sets Another Record," *Pacific Fisherman* (Yearbook, 1928): 198, 200, 202, 204, 206.

44. "Interest of Herring Industry Centers on Unalaska," *Pacific Fisherman* (October 1928): 24, 34.

45. Bower, "Alaska Fishery and Fur-Seal Industries in 1928," 285–289.
46. "Interest of Herring Industry Centers on Unalaska," *Pacific Fisherman* (October 1928): 24, 34; "Alaska Herring Output Shows Substantial Increase," *Pacific Fisherman* (Yearbook 1929): 143, 145, 147, 149; Bower, "Alaska Fishery and Fur-Seal Industries in 1928," 285–289.
47. "Big Herring Run at Unalaska," *Pacific Fisherman* (September 1928): 40.
48. Bower, "Alaska Fishery and Fur-Seal Industries in 1928," 285–289.
49. "Alaska Herring Output Shows Substantial Increase," *Pacific Fisherman* (Yearbook 1929): 143, 145, 147, 149; Bower, "Alaska Fishery and Fur-Seal Industries in 1928," 285–289.
50. "Some Observations on the Alaska Herring," *Pacific Fisherman* (August 1928): 62.
51. Bower, "Alaska Fishery and Fur-Seal Industries in 1928," 285–289; "New Records Established in Fish Oil and Meal Output," *Pacific Fisherman* (Yearbook 1929): 162, 164, 166–167.
52. Ward T. Bower, "Alaska Fishery and Fur-Seal Industries in 1929," in *Report of the United States Commissioner of Fisheries for the Fiscal Year 1930* (Washington, DC: GPO, 1931), 296–300.
53. "Fish Oil and Meal Production is Largest in History," *Pacific Fisherman* (Yearbook 1930): 209–210, 212, 214; Bower, "Alaska Fishery and Fur-Seal Industries in 1929," 296–300.
54. "Fish Oil and Meal Output Declines," *Pacific Fisherman* (Yearbook 1931): 217, 219, 221–222.
55. "Herring Oil Sold for 10 Cents; Sardine Operations Delayed," *Pacific Fisherman* (September 1932): 63.
56. "Seven Herring Plants to Operate," *Pacific Fisherman* (May 1931): 93.
57. "Herring Oil Sold for 10 Cents; Sardine Operations Delayed," *Pacific Fisherman* (September 1932): 63.
58. "Alaska Herring Pack Smallest in History of the Industry," *Pacific Fisherman* (Yearbook 1930): 192, 194, 197; Ward T. Bower, "Alaska Fishery and Fur-Seal Industries in 1930," in *Report of the United States Commissioner of Fisheries for the Fiscal Year 1931* (Washington, DC: GPO, 1932), 58–61.
59. "Good Herring in Demand," *Pacific Fisherman* (March 1930): 50; "Only Quality Goods in Call," *Pacific Fisherman* (May 1930): 49.
60. "Dutch Harbor Packing," *Pacific Fisherman* (August 1930): 56.
61. "Cured Herring Pack of Alaska Gains, But Still Is Light," *Pacific Fisherman* (Yearbook 1931): 210, 213; Bower, "Alaska Fishery and Fur-Seal Industries in 1929," 296–300; Bower, "Alaska Fishery and Fur-Seal Industries in 1930," 58–61.

CHAPTER 6: A CHRONICLE OF ALASKA'S HERRING INDUSTRY: 1932–1948

1. "Many Small Herring," *Pacific Fisherman* (November 1933): 45.
2. Ward T. Bower, "Alaska Fisheries and Fur-Seal Industries in 1932," Appendix I, in *Report of the United States Commissioner of Fisheries for the Fiscal Year 1933* (Washington, DC: GPO, 1934), 49–51.

3. "Cured Herring," *Pacific Fisherman* (Yearbook 1933): 177, 179.

4. Bower, "Alaska Fisheries and Fur-Seal Industries in 1932," 49–51.

5. "Alaska Herring Position Good Save for Threat Due to Cheap Currencies," *Pacific Fisherman* (January 1933): 41.

6. "Beer Law Gives Herring Trade a Thrill; Import Threat Checks Financing," *Pacific Fisherman* (April 1933): 51.

7. "Hoch Der Herring," *Pacific Fisherman* (May 1933): 53.

8. Ward T. Bower, "Alaska Fishery and Fur-Seal Industries in 1933," in *Report of the United States Commissioner of Fisheries for the Fiscal Year 1934* (Washington, DC: GPO, 1936), 254; "Herring Position Improving; Code Matters Commanding Attention," *Pacific Fisherman* (February 1934): 53.

9. "Alaska Producers Take Firmer Grip on Herring Marketing Situation," *Pacific Fisherman* (October 1933): 51.

10. "Packers Form Cooperative Concern to Market Alaska Herring," *Pacific Fisherman* (November 1933): 45; "Cured Herring," *Pacific Fisherman* (Yearbook 1934): 199–201.

11. "Buchan & Heinen Again High Plant," *Pacific Fisherman* (October 1934): 42; "Olaf Floe Once More Wins Herring Belt," *Pacific Fisherman* (November 1936): 63; "Olaf Floe Wins Third Leg on Belt," *Pacific Fisherman* (November 1937): 82.

12. Bower, "Alaska Fishery and Fur-Seal Industries in 1933," 283–286; "Cured Herring," *Pacific Fisherman* (Yearbook 1934): 199–201; "Fish Oil and Meal," *Pacific Fisherman* (Yearbook 1934): 203, 205–209.

13. "Alaska Herring Movement Proves Very Satisfactory; Outlook Good," *Pacific Fisherman* (January 1934): 45.

14. "Herring Gibbers Go North in Style," *Pacific Fisherman* (July 1934): 53.

15. "Early Yield of Herring Products Fair; Transportation Problem Is Serious," *Pacific Fisherman* (July 1934): 55; "Seamen's Truce Opens Shipping to North Ports," *Seattle Times*, June 10, 1934.

16. "Alaska Season Best in History," *Pacific Fisherman* (Yearbook 1935): 205; "Alaska Herring," *Pacific Fisherman* (Yearbook 1935): 195; Ward T. Bower, "Alaska Fishery and Fur-Seal Industries in 1934," in *Report of the United States Commissioner of Fisheries for the Fiscal Year 1935* (Washington, DC: GPO, 1936), 44–47.

17. "Buchan & Heinen Again High Plant," *Pacific Fisherman* (October 1934): 42.

18. "New Record Reached by Fish Oil and Meal," *Pacific Fisherman* (Yearbook 1936): 228–229, 231.

19. E. H. Dahlgren and L. N. Kolloen, *Outlook for the Alaska Herring Fishery in 1943*, US Fish and Wildlife Service Leaflet No. 16, March 1943.

20. "New Record Reached by Fish Oil and Meal," *Pacific Fisherman* (Yearbook 1936): 228–229, 231.

21. Ward T. Bower, "Alaska Fishery and Fur-Seal Industries in 1935, Appendix I," in *Report of the United States Commissioner of Fisheries for the Fiscal Year 1936* (Washington, DC: GPO, 1938), 39–42; "New Records Reached by Fish Oil and Meal," *Pacific Fisherman* (Yearbook 1936): 227, 229, 231.

22. "Herring Pack Moves Out Well, But at Disappointing Prices," *Pacific Fisherman* (October 1935): 49; "Alaska Cured Herring," *Pacific Fisherman* (Yearbook 1936): 217, 219, 221.

23. "Olaf Floe Wears the Herring Belt," *Pacific Fisherman* (November 1935): 51.

24. Leo Baunach, "The International Fishermen and Allied Workers of America," accessed January 20, 2020, https://depts.washington.edu/dock/IFAWA_pt2.shtml; "Herring, 1936," *Pacific Fisherman* (Yearbook 1937): 229, 231, 233.

25. *Committee Study Resolution—Alaskan Fisheries Hearings, Hearings before the Special Subcommittee on Alaskan Fisheries of the Committee on Merchant Marine and Fisheries, US House of Representatives, 76th Cong., 1st Sess., and the Joint Committee on Fisheries of the Legislature of Alaska* (Washington, DC: GPO, 1939), 845, 847.

26. "Alaska Herring Situation Opens Up; Several Oil and Meal Plants to Run," *Pacific Fisherman* (June 1936): 69.

27. "Fish Price Problem Puts Brake on Oil and Meal Preparations," *Pacific Fisherman* (April 1936): 61; "Herring Price Situation Develops Dangerous Character," *Pacific Fisherman* (May 1936): 57; "Alaska Herring Situation Opens Up," *Pacific Fisherman* (June 1936): 69; Ward T. Bower, "Alaska Fisheries and Fur-Seal Industries in 1936," Administrative Rept. No. 28, Appendix II, in *Report of the United States Commissioner of Fisheries for the Fiscal Year 1937* (Washington, DC: GPO, 1939), 317–320.

28. J. M. Kniseley, "Color in Herring Oil," *Pacific Fisherman* (Yearbook 1936): 239, 241.

29. "Shoreplants: The Backbone of the Fish Oil and Meal Business," *Pacific Fisherman* (August 1936): 31–32.

30. Bower, "Alaska Fisheries and Fur-Seal Industries in 1936," 317–320.

31. "Herring, 1936," *Pacific Fisherman* (Yearbook 1937): 229, 231, 233.

32. "Herring, 1936," *Pacific Fisherman* (Yearbook 1937): 229, 231, 233; "Herring Men Are Bitter," *Pacific Fisherman* (November 1936): 63; "Alaska Yield Is Slightly Off," *Pacific Fisherman* (Yearbook 1937): 243; Bower, "Alaska Fisheries and Fur-Seal Industries in 1936," 317–320.

33. "Olaf Floe Once More Wins Herring Belt," *Pacific Fisherman* (November 1936): 63.

34. "Herring Men Are Bitter," *Pacific Fisherman* (November 1936): 63.

35. "January Fish Reduction Proves Rather Disappointing," *Pacific Fisherman* (February 1937): 55.

36. "Alaska Salt Herring," *Pacific Fisherman* (Yearbook 1938): 199, 201.

37. "January Fish Reduction Proves Rather Disappointing," *Pacific Fisherman* (February 1937): 55; "Alaska Herring Industry Faces Course Beset by Reefs and Tide Rips," *Pacific Fisherman* (March 1937): 67.

38. Alaska House Joint Memorial No. 50, March 9, 1937, *Session Laws of Alaska, 1937*.

39. "Alaska Herring Industry Faces Course Beset by Reefs and Tide Rips," *Pacific Fisherman* (March 1937): 67; "Alaska Herring Operations Delayed But Not Destroyed," *Pacific Fisherman* (June 1937): 67.

40. "Negotiations on 1937 Fish and Plant Costs Hold Reduction Industry Spotlight," *Pacific Fisherman* (April 1937): 75; "Trade Negotiations Run Hard Aground, with Tide of Fats Prices Falling," *Pacific Fisherman* (April 1937): 73; "Alaska Herring Operations Delayed But Not Destroyed," *Pacific Fisherman* (June 1937): 67.

41. "Alaska Salt Herring," *Pacific Fisherman* (Yearbook 1938): 199, 201.

42. "Fish Oil and Meal," *Pacific Fisherman* (Yearbook 1938): 261, 263, 265; Ward T. Bower, "Alaska Fisheries and Fur-Seal Industries in 1937," Administrative Rept. No. 31, in *Report of the United States Commissioner of Fisheries for the Fiscal Year 1938* (Washington, DC: GPO, 1940), 116–118; "By-Products," *Pacific Fisherman* (January 25, 1959): 225–235.

43. "By-Products," *Pacific Fisherman* (January 25, 1959): 233.
44. "Herring Seiner 'Limit' Founders with Loss of Eight Fishermen," *Pacific Fisherman* (November 1937): 81; Captain Warren Good and Michael Burwell, *Alaska Shipwrecks, 1750–2015* (Warren Good, 2018), 287.
45. "Olaf Floe Wins Third Leg on Belt," *Pacific Fisherman* (November 1937): 82.
46. *Alaska—Its Resources and Development*, 75th Cong., 3d Sess., Ho. Doc. No. 485 (Washington, DC: GPO, 1938), 59.
47. "Herring Season Change Bring Protests from Central Districts," *Pacific Fisherman* (April 1938): 77.
48. "Fishermen's Unions Adopt Consolidation Plans," *Pacific Fisherman* (April 1937): 71; Baunach, "International Fishermen and Allied Workers of America."
49. "Herring Negotiations Lag; Fish Meal Denied Protection," *Pacific Fisherman* (May 1938): 67.
50. "Herring Industry Still in Irons as Season Opens in Alaska," *Pacific Fisherman* (June 1938): 43.
51. "Three-Fourths Alaska Herring Reduction Capacity to Be Active This Season," *Pacific Fisherman* (July 1938): 51.
52. "Three-Fourths Alaska Reduction Capacity to Be Active This Season," 51; Ward T. Bower, "Alaska Fisheries and Fur-Seal Industries in 1938," Administrative Rept. No. 36, Appendix II, in *Report of the United States Commissioner of Fisheries for the Fiscal Year 1939* (Washington, DC: GPO, 1940), 89.
53. "Reduction Fishing Favorable at Most Points Along Coast," *Pacific Fisherman* (September 1938): 67; "Fish Oil and Meal," *Pacific Fisherman* (Yearbook 1938): 289, 291, 293, 295, 297, 299.
54. "Herring," *Pacific Fisherman* (Yearbook 1939): 231, 233.
55. Bower, "Alaska Fisheries and Fur-Seal Industries in 1938," 137–140; "Fish Oil and Meal," *Pacific Fisherman* (Yearbook 1938): 289, 291, 293, 295, 297, 299.
56. "Imports Responsible for Herring Decline," *Pacific Fisherman* (Yearbook 1939): 233.
57. Homer E. Gregory and Kathleen Barnes, *North Pacific Fisheries* (New York: American Council, Institute of Pacific Relations, 1939), 242.
58. Ward T. Bower, "Alaska Fishery and Fur-Seal Industries in 1939," Administrative Rept. No. 40, in *Report of the United States Commissioner of Fisheries for the Fiscal Year 1939* (Washington, DC: GPO, 1950), 137, 139, 150–153.
59. "Alaska Herring Industry Faces Course Beset by Reefs and Tide Rips," *Pacific Fisherman* (March 1937): 67.
60. Bower, "Alaska Fisheries and Fur-Seal Industries in 1937," 116–118.
61. Bower, "Alaska Fishery and Fur-Seal Industries in 1939," 150–153.
62. Bower, "Alaska Fishery and Fur-Seal Industries in 1939," 150–153.
63. E. J. Huizer, "History of Alaska Herring Fishery," *Alaska Department of Fisheries Annual Report for 1952*, 63–76; E. H. Dahlgren (US Bureau of Fisheries), "Alaska Herring Research: Part 2," *Pacific Fisherman* (April 1940): 44, 46; E. H. Dahlgren and L. N. Kolloen, *Outlook for the Alaska Herring Fishery in 1943* (US Fish and Wildlife Service Leaflet No. 16, March 1943).
64. Bower, "Alaska Fishery and Fur-Seal Industries in 1939," 150–153.
65. "Southeast Alaska Herring Operators Denied Joint Operation," *Pacific Fisherman* (May 1940): 28.

66. Reorganization Act of 1939, April 3, 1939 (53 Stat. 561); Reorganization Plan No. 2 of 1939, May 9, 1939 (53 Stat. 1431).

67. Department of the Interior news release, "Ward T. Bower Retires from Federal Service," April 3, 1947; *Report of the Governor of Alaska to the Secretary of the Interior, 1944* (Washington, DC: GPO, 1945).

68. "Drastic Herring Quota Cuts Forecast for 1949," *Pacific Fisherman* (March 1949): 39; "Herring Hearing Reveals Improved Human Climate," *Pacific Fisherman* (January 1, 1959): 16.

69. Bower, "Alaska Fishery and Fur-Seal Industries in 1939," 150–153.

70. "Herring," *Pacific Fisherman* (Yearbook 1940): 297, 299.

71. Lawrence N. Kolloen, *The Decline and Rehabilitation of the Southeastern Alaska Herring Fishery* (US Fish and Wildlife Service Fishery Leaflet No. 252, July 1947); "Southeast Alaska Herring Operators Denied Joint Operation," *Pacific Fisherman* (May 1940): 28.

72. "North Pacific Coast Reduction Fishing Light in August," *Pacific Fisherman* (September 1940): 71–72; Ward T. Bower, *Alaska Fishery and Fur Seal Industries: 1940*, Statistical Digest No. 2 (Washington, DC: GPO, 1942): 44–47.

73. "Southeast Alaska Closed in 1940 by Bureau," *Pacific Fisherman* (Yearbook 1940): 299.

74. "North Coast Pilchard and Herring Price Agreements Reached in July," *Pacific Fisherman* (August 1940): 65.

75. "Cured Herring," *Pacific Fisherman* (Yearbook 1941): 259–260; Bower, *Alaska Fisheries and Fur Seal Industries: 1940*, 44–47.

76. "Alaska," *Pacific Fisherman* (Yearbook 1941): 285, 287, 289, 291.

77. "New Electric Detector Picks Out Tagged Herring Plant," *Pacific Fisherman* (July 1935): 60.

78. E. H. Dahlgren (US Bureau of Fisheries), "Alaska Herring Research: Part 1," *Pacific Fisherman* (March 1940): 33–34.

79. E. H. Dahlgren (US Bureau of Fisheries), "Alaska Herring Research: Part 3," *Pacific Fisherman* (May 1940): 49–50.

80. Ward T. Bower, *Alaska Fishery and Fur Seal Industries: 1941*, Statistical Digest No. 5 (Washington, DC: GPO, 1943), 39–42.

81. "Alaska Fishing Regulations Are Somewhat Relaxed," *Pacific Fisherman* (June 1941): 38.

82. Bower, *Alaska Fishery and Fur Seal Industries: 1941*, 39–42; "Cured Herring," *Pacific Fisherman* (Yearbook 1942): 259–260.

83. Harold L. Ickes, "The Nation Turns to Its Fisheries," *Pacific Fisherman* (January 1943): 13; "Concentration Program Announced for Alaska Canned Salmon Industry," *Pacific Fisherman* (April 1943): 11, 14.

84. "Canned Herring," *Pacific Fisherman* (Yearbook 1942): 261–262.

85. "Alaska Herring Industry Protests Drastic Production Curtailment," *Pacific Fisherman* (April 1942): 71.

86. Ward T. Bower, *Alaska Fishery and Fur Seal Industries: 1942*, Statistical Digest No. 8 (Washington, DC: GPO, 1944), 35–37; "Fish Oil and Meal," *Pacific Fisherman* (Yearbook 1943): 333, 335, 337, 339; "No Fish Oil Restrictions Contemplated," *Pacific Fisherman* (October 1942): 51; "Salt Herring," *Pacific Fisherman* (Yearbook 1943): 361.

87. "No Fish Oil Restrictions Contemplated," *Pacific Fisherman* (October 1942): 51.
88. Untitled article, *Pacific Fisherman* (April 1943): 19.
89. "Herring Operations Favored By New Alaska Regulations," *Pacific Fisherman* (March 1943): 41; Ward T. Bower, *Alaska Fisheries and Fur Industries: 1943*, Statistical Digest No. 10 (Washington, DC: GPO, 1944), 34–36.
90. "Alaska Herring," *Pacific Fisherman* (Yearbook 1944): 227, 331; Bower, *Alaska Fisheries and Fur Industries: 1943*, 34–36.
91. "Salt Herring, 1943," *Pacific Fisherman* (Yearbook 1944): 355.
92. "Alaska Herring," *Pacific Fisherman* (Yearbook 1944): 227, 331; Bower, *Alaska Fisheries and Fur Industries: 1943*, 34–36.
93. "Alaska Herring Industry Contributes Needed Supplies," *Pacific Fisherman* (Yearbook 1943): 331.
94. "Herring Quotas Raised 75%," *Pacific Fisherman* (April 1944): 67.
95. "Herring Plans in Outline," *Pacific Fisherman* (May 1944): 65.
96. "Alaska Herring Continue Come-Back," *Pacific Fisherman* (November 1944): 85; "Alaska Herring," *Pacific Fisherman* (Yearbook 1945): 385.
97. "Pacific Fish Meal and Oil," *Pacific Fisherman* (Yearbook 1945): 363.
98. "Alaska Herring Outlook Rosy," *Pacific Fisherman* (June 1944): 75.
99. "Oil and Meal," *Pacific Fisherman* (Yearbook 1946): 409, 411.
100. Ward T. Bower, *Alaska Fishery and Fur-Seal Industries: 1945*, Statistical Digest No. 15 (Washington, DC: GPO, 1948), 37–38.
101. "1945–Alaska Cured Herring–1945," *Pacific Fisherman* (Yearbook 1946): 283.
102. "Oil and Meal," *Pacific Fisherman* (Yearbook 1946): 409, 411.
103. "Herring Pack 1,200,000 Cases," *Pacific Fisherman* (April 1945): 96; Bower, *Alaska Fishery and Fur-Seal Industries: 1945*, 37–38.
104. Ward T. Bower, *Alaska Fishery and Fur-Seal Industries: 1946*, Statistical Digest No. 17 (Washington, DC: GPO, 1948), 40–41.
105. Kolloen, *The Decline and Rehabilitation of the Southeastern Alaska Herring Fishery*.
106. "Herring," *Pacific Fisherman* (Yearbook 1947): 221; James E. Hemming, Gordon S. Harrison, and Stephen R. Braund, *The Social and Economic Impacts of a Commercial Herring Fishery on the Coastal Villages of the Arctic/Yukon/Kuskokwim Area* (North Pacific Fishery Management Council Document No. 3, September 15, 1989), 71.
107. "Oil and Meal," *Pacific Fisherman* (Yearbook 1947): 305–319.
108. "Herring Belt Rebuilt for Kilt," *Pacific Fisherman* (April 1947): 85.
109. Seton H. Thompson, *Alaska Fisheries and Fur-Seal Industries: 1947*, Statistical Digest No. 20 (Washington, DC: GPO, 1950), 44–45; "20 Cents and 9-Men Basis for Herring Settlement," *Pacific Fisherman* (August 1947): 105; "Herring," *Pacific Fisherman* (Yearbook 1948): 235, 237.
110. "Oil & Meal," *Pacific Fisherman* (Yearbook 1948): 315–335.
111. "The WINNAH!—Max Jacobs," *Pacific Fisherman* (January 1948): 71.
112. Seton H. Thompson, *Alaska Fishery and Fur-Seal Industries: 1948*, Statistical Digest No. 23 (Washington, DC: GPO, 1952), 4, 35–37; "Drastic Herring Quota Cuts Forecast for 1949," *Pacific Fisherman* (March 1949): 39; "Oil & Meal," *Pacific Fisherman* (Yearbook 1949): 295.
113. "Herring," *Pacific Fisherman* (Yearbook 1949): 289.

CHAPTER 7: A CHRONICLE OF ALASKA'S HERRING INDUSTRY: 1949–1966

1. "Drastic Herring Quota Cuts Forecast for 1949," *Pacific Fisherman* (March 1949): 39.
2. "Fats Market Bust Leaves Herring Trade Up in the Air," *Pacific Fisherman* (May 1949): 83.
3. Chapter 3, *Session Laws of Alaska, 1949*.
4. Seton H. Thompson, *Alaska Fisheries and Fur-Seal Industries: 1949*, Statistical Digest No. 26 (Washington, DC: GPO, 1952), 46–48; "Oil and Meal," *Pacific Fisherman* (Yearbook 1950): 327, 329.
5. Chapter 68, *Session Laws of Alaska, 1949*.
6. Alaska Department of Fisheries, *Annual Report No. 1 for the Year 1949*, 12.
7. "Alaska Fishery Board," *Pacific Fisherman* (August 1949): 47.
8. Chapter 68, *Session Laws of Alaska, 1949*.
9. Seton H. Thompson, *Alaska Fisheries and Fur-Seal Industries: 1950*, Statistical Digest No. 29 (Washington, DC: GPO, 1953), 42–44.
10. Alaska Department of Fisheries, *Annual Report No. 1 for the Year 1949*, 16.
11. "12 Alaska Herring Plants Run on 3 Distinct Deals," *Pacific Fisherman* (July 1950): 58.
12. "Alaska Gains Greatly," *Pacific Fisherman* (Yearbook 1951): 301–317; Thompson, *Alaska Fisheries and Fur-Seal Industries: 1950*, 42–44.
13. "Alaska Herring Fishing Presents a Dizzy Picture," *Pacific Fisherman* (October 1951): 37.
14. "S.E. Alaska Herring Quota 100,000 Bbls.," *Pacific Fisherman* (April 1951): 69.
15. "Alaska Herring Floor Forced Up One-Third by Fishermen," *Pacific Fisherman* (July 1951): 48; "Alaska Herring Fishing Presents a Dizzy Picture," 37.
16. Seton H. Thompson, *Alaska Fishery and Fur-Seal Industries: 1951*, Statistical Digest No. 31 (Washington, DC: GPO, 1954), 45–46.
17. "Oil and Meal," *Pacific Fisherman* (Yearbook 1952): 283–299.
18. Thompson, *Alaska Fishery and Fur-Seal Industries: 1951*, 46.
19. "Ketchikan Makes Bait Early from Tongass," *Pacific Fisherman* (March 1952): 29.
20. "Herring Prospects Bleak with Product Prices Low," *Pacific Fisherman* (June 1952): 79.
21. "Oil and Meal," *Pacific Fisherman* (Yearbook 1953): 269–279.
22. "Six Alaska Herring Operations Underway," *Pacific Fisherman* (August 1952): 113.
23. "Oil and Meal," *Pacific Fisherman* (Yearbook 1953): 269–279.
24. "Alaska Winter Herring Fishery Is Tested," *Pacific Fisherman* (December 1952): 60; Lawrence N. Kolloen and Keith A. Smith, "Southeastern Alaska Exploratory Herring Fishing Operations, Winter 1952/53," *Commercial Fisheries Review* (November 1953): 1–24; E. J. Huizer, "History of Alaska Herring Fishery," *Alaska Department of Fisheries Annual Report for 1952*, 63–76.
25. International Convention for the High Seas Fisheries of the North Pacific Ocean, signed May 9, 1952; effective June 12, 1953, Annex 1b.
26. "Japanese Catch Herring Readily in Bering Sea," *Pacific Fisherman* (September 1960): 5.

27. "Jack Storfold Victim of Poisonous Fumes," *Pacific Fisherman* (November 1952): 8.

28. Alaska Senate Joint Memorial No. 9, February 25, 1953, *Session Laws of Alaska, 1953*.

29. Seton H. Thompson, *Alaska Fisheries and Fur-Seal Industries: 1953*, Statistical Digest No. 35 (Washington, DC: GPO, 1955), 53–55; "Early Herring Fishing Proves Unusually Good," *Pacific Fisherman* (August 1953): 65; "Oil and Meal," *Pacific Fisherman* (Yearbook 1954): 247–254.

30. "Sea Lion Fishery," *Pacific Fisherman* (April 1953): 42.

31. Gerald M. Reid, *Fishery Facts—2, Alaska's Fishery Resources: The Pacific Herring*, National Marine Fisheries Service Extension Publication (June 1972), 16.

32. "Winter Herring Studies on Prince William Sound," *Pacific Fisherman* (November 1953): 71.

33. Seton H. Thompson, *Alaska Fishery and Fur-Seal Industries: 1954*, Statistical Digest No. 37 (Washington, DC: GPO, 1956), 46–47; "By-Products," *Pacific Fisherman* (Yearbook 1955): 237–243.

34. Senate Joint Memorial No. 13, March 18, 1955, *Session Laws of Alaska, 1955*.

35. *Alaska Department of Fisheries Annual Report for 1955*, 20.

36. "By-Products," *Pacific Fisherman* (January 25, 1956): 219–225.

37. "Alaska Herring Outlook for 1956 Season," *Pacific Fisherman* (January 25, 1956): 224.

38. Seton H. Thompson and Donald W. Erickson, *Alaska Fishery and Fur-Seal Industries: 1956*, Statistical Digest No. 45 (Washington, DC: GPO, 1960), 62–63.

39. "Alaska Herring," *Pacific Fisherman* (January 25, 1957): 221.

40. "By-Products," *Pacific Fisherman* (January 25, 1957): 219–227.

41. "By-Products Industry Set Back by Strikes," *Pacific Fisherman* (January 25, 1958): 219–224.

42. "Extra Quota in Fishless Area Arouses Herringmen," *Pacific Fisherman* (September 1957): 47.

43. "By-Products," *Pacific Fisherman* (January 25, 1958): 219–224.

44. "Pacific Herring: Aerial Spawning Survey in Alaska," *Commercial Fisheries Review* (November 1959): 47; "By-Products Bounce Back," *Pacific Fisherman* (January 25, 1959): 225–235.

45. "John N. Cobb to Study Herring Offshore," *Pacific Fisherman* (August 1957): 25.

46. Chapter 63, *Session Laws of Alaska, 1957*.

47. §17, Chapter 64, *Session Laws of Alaska, 1959; 1959 Annual Report, Alaska Board of Fish and Game and Alaska Department of Fish and Game*, ADF&G Rept. No. 11, 3.

48. Gerald M. Reid, *Age Composition, Weight, Length, and Sex of Herring*, Clupea pallasii, *Used for Reduction in Alaska, 1929–66* (NOAA Tech. Rept. NMFS SSRF 634, July 1971), 4; Larry Matheney (former Storfold & Grondahl Packing Company employee), personal communication with author, March 2, 2020.

49. "By-Products Bounce Back," *Pacific Fisherman* (January 25, 1959): 225–235; "Radioactive Fish Tag Use Licensed by Atomic Energy Commission," *Commercial Fisheries Review* (April 1960): 29; Statement of Dr. Allyn Seymour, associate director, Laboratory of Radiation Biology, University of Washington, in *Application of Radioisotopes and Radiation in the Life Sciences, Hearings of the Subcommittee on Research Development, and Radiation of the Joint Committee on Atomic Energy*, 87th Cong., 1st

sess., *March 27–30, 1961* (Washington, DC: GPO, 1961), 273; "Tagging," *Commercial Fisheries Review* (January 1963): 54–55.

50. Alaska Department of Fish and Game, *Annual Report for 1959, 1959 Annual Report, Alaska Board of Fish and Game and Alaska Department of Fish and Game*, ADF&G Rept. No. 11, 70.

51. Alaska Department of Fish and Game, *Alaska Catch and Production, Commercial Fisheries Statistics, 1966*, 21.

52. "S.E. Alaska Quota Filled Early," *Pacific Fisherman* (August 1959): 31; "By-Products," *Pacific Fisherman* (January 25, 1960): 245–246.

53. A. T. Pruter, "Soviet Fisheries for Bottomfish and Herring off the Pacific and Bering Sea Coasts of the United States," MFR Paper No. 1225, *Marine Fisheries Review* (December 1976): 1–14; Yu. I. Dudnik and E. A. Usol'tsev, "The Herrings of the Eastern Part of the Bering Sea," translated from Russian, printed in Ronald I. Regnart and Carl M. Yanagawa, *Bering Sea Herring Fisheries, 1959–1969* (ADF&G A-Y-K Region Herring Rept. No. 2, circa 1972).

54. "By-Products," *Pacific Fisherman* (January 25, 1961): 205–210.

55. Alaska Board of Fish & Game and Alaska Department of Fish and Game, *Progress Report for the Years 1960-1961-1962*, Rept. No. 12, 60.

56. "Peru's Problem of Leadership," *Pacific Fisherman* (September 1960): 37–39; "By-Products," *Pacific Fisherman* (January 25, 1961): 205–210.

57. "Peru's Anchoveta Fishery," *Commercial Fisheries Review* (September 1971): 57–60.

58. International North Pacific Fisheries Commission, *Annual Report 1960*, 2; "Japan's Bering Herring Target 8,700 Tons," *Pacific Fisherman* (June 1960): 48; "Japanese Catch Herring Readily in Bering Sea," *Pacific Fisherman* (September 1960): 5.

59. "By-Products," *Pacific Fisherman* (January 25, 1962): 203–209; Alaska Department of Fish and Game, *Commercial Operators, 1961*, Statistical Leaflet No. 2.

60. "C. L. Anderson Retires with Work Well Done," *Pacific Fisherman* (December 1961): 37.

61. House Concurrent Resolution No. 34, *Session Laws of Alaska, 1962*.

62. "Herring Fishery," *Commercial Fisheries Review* (July 1962): 12.

63. "Herring in Good Abundance in Southeast Alaska," *Commercial Fisheries Review* (February 1963): 15; "Herring Fishery," *Commercial Fisheries Review* (September 1962): 13.

64. "By-Products," *Pacific Fisherman* (January 25, 1963): 173–174.

65. "'Egged Herring' Target of Japanese in Shelikof," *Pacific Fisherman* (May 1962): 16; "Herring Roe Industry Gets a Start in Norway," *Commercial Fisheries Review* (May 1962): 10; Affidavit of William A. Egan (Governor of Alaska), May 12, 1972, in Committee on Commerce, US Senate, 92nd Cong., 2d sess., hearing on *Provisional U.S. Charts Delimiting Alaskan Territorial Boundaries*, Serial No. 92-69, May 15, 1972 (Washington, DC: GPO, 1972), 34–35; Diplomatic Note from the Government of Japan to the United States of America, May 3, 1962, in Committee on Commerce, US Senate, 92nd Cong., 2d sess., hearing on *Provisional U.S. Charts Delimiting Alaskan Territorial Boundaries*, Serial No. 92–69, May 15, 1972 (Washington, DC: GPO, 1972), 38–39; "40-Mile Fishing Zone Sought," *Pacific Fisherman* (May 1962): 16; "Government to Protest Seizure of Fishing Vessels off Alaska," *Commercial Fisheries Review* (July 1962): 83.

66. "Japanese Fishermen Not Extradited for Violation of Fishing Regulations," *Commercial Fisheries Review* (June 1963): 20.

Notes

67. "Joint Roe Herring Fishery Planned," *Commercial Fisheries Review* (May 1963): 16; "Herring Roe Studies Needed for Japan Trade," *Pacific Fisherman* (May 1963): 6.

68. "Revised Fishing Regulations Issued," *Commercial Fisheries Review* (March 1963): 18; "By-Products," *Pacific Fisherman* (January 25, 1964): 172.

69. "Japanese Firm Interested in Alaska Herring Products," *Commercial Fisheries Review* (May 1964): 12.

70. "By-Products," *Pacific Fisherman* (January 25, 1965): 175; "Herring Fishery," *Commercial Fisheries Review* (February 1965): 15; Stephen H. Rogers, "Herring Fishery in Southeastern Alaska," *Commercial Fisheries Review* (August 1965): 1–6.

71. Alaska Department of Fish and Game, *Progress Report for the Years 1963 and 1964* (ADF&G Rept. No. 13), 70–71.

72. "Herring Fishery," *Commercial Fisheries Review* (May 1965): 4.

73. "By-Products," *Pacific Fisherman* (January 25, 1966): 158–159; Alaska Department of Fish and Game, *Petersburg-Wrangell Management Area Annual Report, 1965*, 26; Alaska Department of Fish and Game, *Petersburg-Wrangell Management Area Annual Report, 1967*, 21; Charles Burkey Jr. and Joan Ried, *Statistics of the Commercial Fishery for Pacific Herring from the Kodiak Area, Alaska* (ADF&G Tech. Fishery Rept. No. 88-11, August 1988), 13.

74. Alaska Department of Fish and Game, *Petersburg-Wrangell Management Area, Annual Report, 1966*, 24; "Herring Catch Lowest on Record," *Commercial Fisheries Review* (January 1967), 10; Alaska Department of Fish and Game, *Alaska Catch and Production, Commercial Fisheries Statistics, 1966*, Statistics Leaflet No. 13.

75. Alaska Department of Fish and Game, *Sitka Area Commercial Fisheries Management Annual Report, 1967*, 22.

76. Alaska Department of Fish and Game, *Commercial Operators, 1966*, Statistical Leaflet No. 12; Alaska Department of Fish and Game, *Alaska Catch and Production, Commercial Fisheries Statistics, 1966*, 13.

77. Alaska Department of Fish and Game, *Petersburg-Wrangell Management Area, Annual Report, 1966*, 25; Alaska Department of Fish and Game, *Commercial Operators, 1967*, Statistical Leaflet No. 14; Alaska Department of Fish and Game, *Commercial Operators, 1971*, Statistical Leaflet No. 22.

CHAPTER 8: BAIT HERRING

1. Alaska Department of Fish and Game, *Sitka Area Commercial Fisheries Management Annual Report, 1967*, 21.

2. Alaska Department of Fish and Game, "2020 Norton Sound Commercial Herring Bait Fishery to Open," Advisory Announcement, May 18, 2020; Gabe Colombo, "Norton Sound Herring Bait Harvest Breaks 10-Year Record," *KNOM Radio* (Nome, Alaska), May 23, 2018, https://www.knom.org/wp/blog/2018/05/23/norton-sound-herring-harvest-breaks-10-year-record/.

3. "Combination Herring and Halibut Boat," *Pacific Fisherman* (January 1911): 24.

4. Millard C. Marsh and John N. Cobb, *Fisheries of Alaska in 1907*, Bureau of Fisheries Doc. No. 632 (Washington, DC: GPO, 1908), 51.

5. Barton W. Evermann and E. L. Goldsborough, *The Fishes of Alaska*, US Bureau of Fisheries Doc. No. 624 (Washington, DC: GPO, 1907), 233.

6. Millard C. Marsh and John N. Cobb, *The Fisheries of Alaska in 1909*, Bureau of Fisheries Doc. No. 730 (Washington, DC: GPO, 1910), 48–50.

7. John N. Cobb, *The Commercial Fisheries of Alaska in 1905*, Bureau of Fisheries Doc. No. 603 (Washington, DC: GPO, 1906), 15–16; John N. Cobb, "Report on the Fisheries of Alaska [in 1906]," Bureau of Fisheries Doc. No. 618, in *Report of the Commissioner of Fisheries for the Year 1906 and Special Papers* (Washington, DC: GPO, 1907), 53.

8. Cobb, "Report on the Fisheries of Alaska [in 1906]," 49, 53.

9. Millard C. Marsh and John N. Cobb, *Fisheries of Alaska in 1910*, Bureau of Fisheries Doc. No. 746 (Washington, DC: GPO, 1911), 45–51.

10. "Review of the Halibut Industry for the Year 1910," *Pacific Fisherman* (February 1911): 20–22.

11. Barton Warren Evermann, *Alaska Fisheries and Fur Industries in 1911*, Bureau of Fisheries Doc. No. 766 (Washington, DC: GPO, 1912), 59–62.

12. "Alaska," *Pacific Fisherman* (November 1911): 19–20.

13. Evermann, *Alaska Fisheries and Fur Industries in 1911*, 61.

14. Ward T. Bower, *Alaska Fishery and Fur-Seal Industries in 1922*, Bureau of Fisheries Doc. No. 951 (Washington, DC: GPO, 1923), 75; Ward T. Bower, "Alaska Fisheries and Fur-Seal Industries in 1924," Bureau of Fisheries Doc. No. 992, Appendix IV to the *Report of the United States Commissioner of Fisheries for the Fiscal Year 1925* (Washington, DC: GPO, 1926), 136–137; Ward T. Bower, "Alaska Fishery and Fur-Seal Industries in 1926," Bureau of Fisheries Doc. No. 1023, Appendix IV to the *Report of the United States Commissioner of Fisheries for the Fiscal Year 1927* (Washington, DC: GPO, 1928), 294–297.

15. "New Alaska Fishery Regulations," *Fisheries Service Bulletin* (January 2, 1925): 3–4.

16. Ward T. Bower, "Alaska Fishery and Fur-Seal Industries in 1933," in *Report of the United States Commissioner of Fisheries for the Fiscal Year 1934* (Washington, DC: GPO, 1936), 283–286; New England Fish Company, "A Pictorial Story of the Salmon and Halibut Operation of the New England Fish Co. at Noyes Island and Ketchikan, Alaska," circa 1931, unpublished, Tongass Historical Society Collection, Ketchikan Museums.

17. E. J. Huizer, "History of Alaska Herring Fishery," *Alaska Department of Fisheries Annual Report for 1952*, 63–76.

18. Alaska Department of Fish and Game, *Sitka Area Commercial Fisheries Management Annual Report, 1966*, 21; Alaska Department of Fish and Game, *1983 Management Report for Southeast Alaska, Region 1*, 41–50; Kyle Hebert and Marc Pritchett, *Southeast Alaska–Yakutat Herring Fisheries*, ADF&G Regional Information Report No. 1J02-43 (November 2002), 5.

19. Evermann, *Alaska Fisheries and Fur Industries in 1911*, 61.

20. Alaska Department of Fish and Game, *Sitka Area Commercial Fisheries Management Annual Report, 1965*, 25; Alaska Department of Fish and Game, *Sitka Area Commercial Fisheries Management Annual Report, 1966*, 21; Alaska Department of Fish and Game, *Sitka Area Commercial Fisheries Management Annual Report, 1967*, 21.

21. Alaska Department of Fish and Game, *1963–64 Progress Report*, 70–71.

22. Seton H. Thompson and Donald W. Erickson, *Alaska Fishery and Fur-Seal Industries: 1956*, Statistical Digest No. 45 (Washington, DC: GPO, 1960), 62–63.

23. Kyle Hebert and Marc Pritchett, *Southeast Alaska–Yakutat Herring Fisheries*, ADF&G Regional Information Report No. 1J02-43 (November 2002), 5.

Notes

24. Thomas R. Schroeder, *A Summary of Historical Data for the Lower Cook Inlet, Alaska, Pacific Herring Sac Roe Fishery*, ADF&G Fishery Research Bulletin No. 89–04 (December 1989), 4.

25. Kyle Hebert and Marc Pritchett, *Report to the Board of Fisheries, Southeast Alaska–Yakutat Herring Fisheries, 2003*, ADF&G Regional Information Rept. No. 1J02-43 (November 2002), 10.

26. Fritz Funk, *Herring Bait Fisheries in Alaska*, ADF&G Regional Information Rept. No. 5J92-03 (March 1992), 2; Alaska Department of Fish and Game, *Statewide Herring Wholesale Value and Pounds Processed*, accessed September 25, 2020, https://www.adfg.alaska.gov/index.cfm?adfg=fishlicense.coar_herringproduction.

27. Alaska Department of Fish and Game, *Prince William Sound Area Annual Management Report, 1978*, 6, 13, 61–63; Alaska Department of Fish and Game, *Bristol Bay Area Annual Management Report, 1979*, 106; Alaska Department of Fish and Game, *Prince William Sound Area Annual Finfish Management Report, 1989*, Regional Information Rept. No. 2A9C-07 (February 1991), 32; Alaska Department of Fish and Game, *Prince William Sound Management Area, Herring Report to the Alaska Board of Fisheries*, Alaska Department of Fish and Game Regional Information Rept. No. 2A96-40 (December 1996), 13; Alaska Department of Fish and Game, *Prince William Sound Area 1993 Annual Finfish Management Report*, Regional Information Rept. No. 2A95-XX (April 1995), H.4.

28. Ralph B. Pirtle (Alaska Department of Fish and Game), *Prince William Sound Area Annual Finfish Management Report, 1979* (August 1980), 55; Alaska Department of Fish and Game, *Prince William Sound Area Annual Finfish Management Report, 1986* (November 1987), 22; Alaska Department of Fish and Game, *Prince William Sound Area Annual Finfish Management Report, 1989*, Regional Information Rept. No. 2C90-07 (February 1991), 32; Alaska Department of Fish and Game, *Prince William Sound Area 1990 Annual Finfish Management Report*, Regional Information Rept. No. 2C91-14 (November 1991), 16–17; Alaska Department of Fish and Game, *Prince William Sound Area 1991 Annual Finfish Management Report*, Regional Information Rept. No. 2A92-09 (May 1992), 19; Alaska Department of Fish and Game, *Prince William Sound Area 1992 Annual Finfish Management Report*, Regional Information Rept. No. 2A93-12 (June 1993), 22–23; Beaver Nelson (former herring seiner, Homer, Alaska), personal communication, August 28, 2020; Alaska Department of Fish and Game, *Prince William Sound Management Area, 2000 Annual Finfish Management Report*, Regional Information Rept. No. 2A02-02 (February 2002), 158.

29. Charles Burkey Jr. and Joan Ried, *Statistics of the Commercial Fishery for Pacific Herring from the Kodiak Area, Alaska*, ADF&G Tech. Fishery Rept. No. 88-11 (August 1988), 4; Dennis Gretsch, *Kodiak Management Area 1992 Commercial Herring Sac Roe and Food/Bait Fisheries, Report to the Board of Fisheries*, ADF&G Regional Information Rept. No. 4K92-40 (January 1992), 34; Dennis Gretsch, *Kodiak Management Area, Annual Herring Fisheries Management Report, 2001*, ADF&G Regional Information Rept. No. 4K02-25 (April 2002), 15–16; Alaska Commercial Fisheries Entry Commission, *Fishery Statistics—Permits & Permit Holders* (2020), accessed October 27, 2020, https://www.cfec.state.ak.us/pstatus/14052020.htm; Geoff Spalinger, *Kodiak Management Area Commercial Herring Food and Bait Fishery Harvest Strategy, 2014*, ADF&G Fishery Management Rept. No. 14-48 (November 2014), 1; Geoff Spalinger

(Alaska Department of Fish and Game), *Kodiak Management Area Herring Fisheries, Report to the Alaska Board of Fisheries* (January 2020), 19, 25.

30. Jim Paulin, "Dutch Harbor Herring Quota Set," *Bristol Bay Times*, July 14, 2017, http://www.thebristolbaytimes.com/article/1728dutch_harbor_herring_quota_set.

31. Alan Quimby (Alaska Department of Fish and Game), *Eastern Aleutian Island "Dutch Harbor" Food and Bait Herring Fishery*, Regional Information Rept. No. 4K90-34 (December 1990), 2–5; Lucas K. Stumpf, *Alaska Peninsula–Aleutian Islands Herring Sac Roe and Food and Bait Fisheries Annual Management Report*, 2017, ADF&G Regional Information Rept. No. 4K18-06 (May 2018), 17.

32. Cassandra Whiteside (Alaska Department of Fish and Game), personal communication, July 9, 2020; Alaska Administrative Code, 5 AAC 27.865, April 1988.

33. Alaska Department of Fish and Game, *Annual Management Report, 2000, Bristol Bay Area*, Regional Informational Report No. 2A01-10, 136; Alaska Department of Fish and Game, *2018 Bristol Bay Area Annual Management Report*, Fishery Management Rept. No. 19-12 (May 2019), 101; Lucas K. Stumpf, *Alaska Peninsula–Aleutian Islands Herring Sac Roe and Food and Bait Fisheries Annual Management Report*, 2017, ADF&G Regional Information Rept. No. 4K18-06 (May 2018), 15.

34. For the years 1960–1984, the data sources are ADF&G's annual catch and production reports. The exception is 1974 and 1975, for which the catch listed in the reports is obviously incorrect. For these years, in which there was no food/bait fishery in Prince William Sound, total production represents the sum of the harvests from Southeast Alaska and from the Kodiak area. The data source for the years 1984–1989 is ADF&G's *Statewide Herring COAR Production*, accessed November 2, 2020, http://www.adfg.alaska.gov/index.cfm?adfg=fishlicense.coar_herringproduction.

35. Stumpf, *Alaska Peninsula–Aleutian Islands Herring Sac Roe and Food and Bait Fisheries Annual Management Report*, 2017, 6; Jim Paulin, "Dutch Harbor Food-and-Bait Herring Fishery Finished," *Anchorage Daily News*, August 3, 2013.

36. Laine Welch, "Herring and Smelt Fisheries at Cook Inlet Fetch High Prices for Fishermen; Both Are Open Access," *Alaska Fish Radio*, April 18, 2018, http://www.alaskafishradio.com/herring-and-smelt-fisheries-at-cook-inlet-fetch-high-prices-for-fishermen-both-are-open-access/; Laine Welch, "Cook Inlet Bait Herring and Smelt Net Big Payouts for Few Participants," *National Fisherman*, April 24, 2018, https://www.nationalfisherman.com/alaska/cook-inlet-bait-herring-smelt-net-big-payouts-participants; Brian Marston and Alyssa Frothingham (Alaska Department of Fish and Game), *DRAFT: Upper Cook Inlet Commercial Fisheries Annual Management Report* (2019), 27, 134.

CHAPTER 9: GENESIS AND MANAGEMENT OF ALASKA'S ROE-HERRING FISHERY

1. Laine Welch, "Herring Ho-Hum? Alaska Season Opens with Limited Interest," *National Fisherman*, March 29, 2021, https://www.nationalfisherman.com/alaska/herring-ho-hum-alaska-season-opens-with-limited-interest.

2. Terry Johnson, "Roe Eruption," *Pacific Fishing* (April 1993): 45–51.

3. "Herring Roe Industry Gets a Start in Norway," *Commercial Fisheries Review* (May 1962): 10.

4. "Herring Roe Market Still Depressed," *Bering Sea Fisherman*, January 1982.

5. D. Douglas Coughenower, "Measuring Sac-Roe: Accuracy Means Bigger Profits," *Bering Sea Fisherman*, May 1984.

6. "Herring," *Pacific Fishing* (Yearbook 1990): 64–66; Alaska Department of Fish & Game, Statewide Herring Pounds and Estimated Ex-vessel Value, accessed March 25, 2022, https://www.adfg.alaska.gov/index.cfm?adfg=commercialbyfisheryherring.herring_grossearnings.

7. Tom Asakawa, "Japan: Salted Herring Roe Sales Rise After Five Years at Six Wholesale Markets," *Seafood News*, March 16, 2021.

8. "Mayor Answers Query on Herring Processing," *Daily Sitka Sentinel*, July 6, 1971; Howard O. Ness, "The Southeast Alaska Herring Fishery," MFR Paper No. 1239, *Marine Fisheries Review* (March 1977): 10–14; Howard O. Ness, "The Recent Development of the Southeast Alaska Herring Fishery," MFR Paper No. 1240, *Marine Fisheries Review* (March 1977): 15–18.

9. Chapter 9, *Session Laws of Alaska, 1977*; Chapter 27, *Session Laws of Alaska, 1980*; Chapter 14, *Session Laws of Alaska, 1983*.

10. Edward O. Otis and Lee F. Hammarstrom (Alaska Department of Fish and Game), *Overview of the Lower Cook Inlet Area Commercial Herring Fishery and Recent Stock Status*, Special Publication No. 04-13 (November 2004), 16; Charles Burkey Jr. and Joan Ried, *Statistics of the Commercial Fishery for Pacific Herring from the Kodiak Area, Alaska*, ADF&G Tech. Fishery Rept. No. 88-11 (August 1988), 16; Alaska Department of Fish and Game, *1964 Alaska Commercial Fishery Operators*, 21; Alaska Department of Fish and Game, *1964 Commercial Fisheries Catch and Production Statistics*, Statistical Leaflet No. 9, 18; Mike Geiger, "Norton Sound Herring Catch Sampling Program, 1964, Alaska Department of Fish and Game, AYK Herring Rept. No. 1," in *AYK Area 1964 Annual Management Report*, 110–117; Stanley Howitt, "Alaska," in "Part III, Committee Reports of International Law Division," American Bar Association, Section of International and Comparative Law, *Proceedings 1964* (1964): 121–130; "Taiyo Gyogyo Negotiating: Egan Reports Tokyo Firm Seeks to Buy Raw Salmon," *Daily Sitka Sentinel*, July 14, 1964; Western Alaska Enterprises to Alaska Department of Fish and Game, undated, Alaska Department of Fish and Game files, Nome, Alaska.

11. "Herring Roe Shortage Pushes Prices to Record High," *Commercial Fisheries Review* (March 1966): 59.

12. Alaska Department of Fish and Game, *Alaska Catch and Production, Commercial Fisheries Statistics, 1965*.

13. Alaska Department of Fish and Game, *Commercial Operators, 1965*, Statistical Leaflet No. 10.

14. Alaska Department of Fish and Game, *Commercial Operators, 1966*, Statistical Leaflet No. 12; Alaska Department of Fish and Game, *Alaska Catch and Production, Commercial Fisheries Statistics, 1966*, Statistics Leaflet No. 13.

15. Alaska Statehood Act, July 7, 1958 (72 Stat. 339), §6(e).

16. Constitution of the State of Alaska, 1959, Article 8, §4.

17. Chap. 206, *Session Laws of Alaska, 1975*; Alaska Department of Fish and Game, "Welcome to the Board of Fisheries," accessed January 19, 2016, http://www.adfg

.alaska.gov/index.cfm?adfg=fisheriesboard.main; Karl Ohls, "Alaska Fisheries Board Faces Mounting Pressures," *National Fisherman* (March 1984), 14.

18. Thomas F. Thornton et al., *Herring Synthesis: Documenting and Modeling Herring Spawning Areas within Socio-Ecological Systems over Time in the Southeastern Gulf of Alaska*, North Pacific Research Board Project No. 728 (June 2010), 290.

19. Alaska Department of Fish and Game, *Annual Management Report, 1986, Norton Sound—Port Clarence—Kotzebue*, 81.

20. Alaska Department of Fish and Game, *1982 Management Report for Southeast Alaska, Region I* (July 1986), 41–44; Belinda Chase, "Murky Waters: What's Next for Southeast Herring Fisheries?" *Ketchikan Daily News*, March 20–21, 1993.

21. Alaska Department of Fish and Game, *Southeast Alaska Roe Herring Fishery, 2004 Management Plan*, ADF&G Regional Information Rept. No. 1J04-06 (February 2004), 3; Kyle Hebert, *Southeast Alaska 2018 Herring Stock Assessment Surveys*, ADF&G Fishery Data Series No. 19-12 (April 2019), 1.

22. An Act Relating to the Regulation of Entry into Alaska Commercial Fisheries; and Providing for an Effective Date (AS16.43), April 26, 1973, *Session Laws and Resolutions of Alaska, 1973*.

23. Bruce Twomley (former commissioner, Alaska Commercial Fisheries Entry Commission), personal communication, April 4, 2020.

24. Howard O. Ness, "The Southeast Alaska Herring Fishery," MFR Paper No. 1239, *Commercial Fisheries Review* (March 1977): 10–14.

25. Beaver Nelson (former herring seiner, Homer, Alaska), personal communication, March 15, 2020.

26. Charles J. Stovall to Prince William Sound and Cook Inlet herring purse seine fishermen, December 23, 1976, in Alaska Commercial Fisheries Entry Commission, *Proposed Limited Entry Regulations for the Prince William Sound and Cook Inlet Herring Purse Seine Fisheries*, December 27, 1976.

27. Alaska Department of Fish and Game, *Prince William Sound Area Annual Management Report, 1978*, 61–63.

28. Burkey and Ried, *Statistics of the Commercial Fishery for Pacific Herring*, 13–14; Alaska Commercial Fisheries Entry Commission, *Fishery Statistics—Permits & Permit Holders* (1984), accessed August 21, 2020, https://www.cfec.state.ak.us/pstatus/14051984.htm; Alaska Commercial Fisheries Entry Commission, *Fishery Statistics—Permits & Permit Holders* (2009), accessed August 21, 2020, https://www.cfec.state.ak.us/pstatus/14052009.htm.

29. Dennis Gretsch, *Kodiak Management Area Annual Herring Management Report, 1997*, ADF&G Regional Information Rept. No. 4K98-41 (August 1998), 3–4; Dennis Gretsch, *Kodiak Management Area Herring Report to the Alaska Board of Fisheries, January 2005*, ADF&G Fishery Management Report No. 04-11 (December 2004), 2.

30. Alaska Commercial Fisheries Entry Commission, "Basic Information Table, Herring Roe, Purse Seine, Southeast," accessed March 10, 2020, https://www.cfec.state.ak.us/bit/X_G01A.htm.

31. Alaska Commercial Fisheries Entry Commission, "Basic Information Table, Herring Roe, Purse Seine, Prince William Sound," accessed March 10, 2020, https://www.cfec.state.ak.us/bit/X_G01E.htm.

32. Alaska Commercial Fisheries Entry Commission, "Basic Information Table, Herring Roe, Purse Seine, Cook Inlet," accessed March 10, 2020, https://www.cfec

.state.ak.us/bit/X_G01H.htm; Glenn Hollowell, Edward O. Otis, and Ethan Ford, *2011 Lower Cook Inlet Area Finfish Management Report*, Alaska Department of Fish and Game Fishery Management Rept. No. 12-30 (July 2012), 26; Glenn Hollowell, Edward O. Otis, and Ethan Ford, *2015 Lower Cook Inlet Area Finfish Management Report*, Alaska Department of Fish and Game Fishery Management Rept. No. 16-19 (May 2016), 187.

33. Alaska Commercial Fisheries Entry Commission, "Historical Fishery Codes Description Table," accessed March 11, 2020, https://www.cfec.state.ak.us/misc/FshyDesH.htm.

34. Alaska Commercial Fisheries Entry Commission, "Current Fishery Codes Description Table," accessed March 10, 2020, https://www.cfec.state.ak.us/misc/FshyDesC.htm.

35. Bruce Twomley (Chairman, Alaska Commercial Fisheries Entry Commission), *License Limitation in Alaska's Commercial Fisheries*, July 2003, accessed March 21, 2020, https://dlc.dlib.indiana.edu/dlc/bitstream/handle/10535/2203/Twomley%2CBruce.pdf?sequence=1&isAllowed=y.

36. Rachel Waldholz, "Co-op Herring Fishery Means Fewer Boats, Quiet Year in Sitka," *KCAW Radio* (Sitka, Alaska), March 18, 2015.

CHAPTER 10: SITKA SOUND ROE-HERRING FISHERY

1. John Grossmann, "The Thorniest Catch," *Gastronomica*, April 24, 2015, https://gastronomica.org/2015/04/24/thorniest-catch/.

2. Will Swagel, "Decade of the 80s a Time of Slow, Steady Growth for Sitka," *Daily Sitka Sentinel*, December 29, 1989.

3. Susan Froetschel, "Seiners Primed for Speedy Herring Fishery," *Daily Sitka Sentinel*, March 28, 1986.

4. Shannon Haugland, "Herring Fishery Wrapped Up," *Daily Sentinel*, March 23, 2000.

5. Alaska Department of Fish and Game, *1964 Alaska Commercial Fishery Operators*, 21; Alaska Department of Fish and Game, *1964 Alaska Catch and Production Statistics*, Statistical Leaflet No. 9, 15.

6. Alaska Department of Fish and Game, *Alaska Catch and Production, Commercial Fisheries Statistics, 1965*, ADF&G Statistical Leaflet No. 11, 17; Alaska Department of Fish and Game, *Alaska Fisheries, 1966 Commercial Operators*, ADF&G Statistical Leaflet No. 12, 20–21; Alaska Department of Fish and Game, *Alaska Catch and Production, Commercial Fisheries Statistics, 1966*, ADF&G Statistical Leaflet No. 13, 19; Alaska Department of Fish and Game, *Alaska Fisheries, 1967 Commercial Operators*, ADF&G Statistical Leaflet No. 14, 22; Alaska Department of Fish and Game, *Alaska Catch and Production, Commercial Fisheries Statistics, 1967*, ADF&G Statistical Leaflet No. 15, 18; Alaska Department of Fish and Game, *Commercial Operators, 1968*, ADF&G Statistical Leaflet No. 16, 20; Alaska Department of Fish and Game, *1968 Alaska Catch and Production, Commercial Fisheries Statistics*, ADF&G Statistical Leaflet No. 17, 18.

7. Alaska Department of Fish and Game, *2009 Southeast Alaska Commercial Herring Fishery Annual Management Report*, Fishery Management Rept. No. 10-39

(November 2010), 12, 36; Fortch Wayne, "On the Docks," *Daily Sitka Sentinel*, October 31, 1969.

8. "Fish & Game Board Changes Herring Rules," *Daily Sitka Sentinel*, March 5, 1970; "Herring Fishery Local Quotas Set," *Daily Sitka Sentinel*, March 20, 1970.

9. Alaska Department of Fish and Game, *2009 Southeast Alaska Commercial Herring Fishery Annual Management Report*, Fishery Management Rept. No. 10-39 (November 2010), 36; "Herring Season Closed," *Daily Sitka Sentinel*, April 8, 1970; Sitka Camp No. 1, Alaska Native Brotherhood, "Letter to Editor," *Daily Sitka Sentinel*, April 7, 1970.

10. Elvin Rottluff, Chairman, Southeast Alaska Trollers Assn., to Keith H. Miller, Governor of Alaska, "Letter to Editor," *Daily Sitka Sentinel*, April 7, 1970.

11. Alaska Department of Fish and Game, *1971 Commercial Operators*, Statistical Leaflet No. 22, 5; Ann Matthews, "Herring Season Is Over, But Cries Continue," *Daily Sitka Sentinel*, April 9, 1971; Ed Wyman, personal communication, July 11, 2020; Alaska Department of Fish and Game, *1971 Alaska Catch and Production, Commercial Fisheries Statistics*, Statistical Leaflet No. 23, 30.

12. "Sitka Sound Herring Harvest," *Petersburg Press*, April 5, 1973.

13. "Herring Harvest Nets Cash in Sitka Sound," *Daily Sitka Sentinel*, April 15, 1974.

14. "Sitka Herring Fishery Nets 8,000 Barrels," *Daily Sitka Sentinel*, April 19, 1976.

15. Alaska Commercial Fisheries Entry Commission, "Basic Information Table, Herring Roe, Purse Seine, Southeast," accessed March 10, 2020, https://www.cfec.state.ak.us/bit/X_G01A.htm.

16. "Active Harbor" (photo caption), *Daily Sitka Sentinel*, April 7, 1977; "Roe Herring Fishery Called Off for Now," *Daily Sitka Sentinel*, April 11, 1977.

17. "Herring Sac Roe Fishery in Sitka Nets 250 Tons," *Daily Sitka Sentinel*, April 19, 1978.

18. "Rich Herring Harvest," *Daily Sitka Sentinel*, April 6, 1979; "Cooperation Fine, But . . . Once Was Enough, Roe Seiners Agree," *Daily Sitka Sentinel*, April 26, 1979; "Herring Fishery Nets $1.6 Million," *Daily Sitka Sentinel*, April 7, 1980; Alaska Department of Fish and Game, "Sitka Seine Sac Roe, a Summary of the Sitka Sound Herring Sac Roe Fishery, 1979–2019," accessed September 28, 2020, https://www.adfg.alaska.gov/index.cfm?adfg=commercialbyareasoutheast.herring#harvest; Aaron Dupois (ADF&G, Sitka), personal communication, September 28, 2020.

19. Ron Rau, "The Great Sitka Herring Rush," *Alaskafest* (June 1984): 46–54.

20. James W. Parker, "Parker Outlines Results of Herring Fishery," *Daily Sitka Sentinel*, April 26, 1979.

21. "Fast Fish," *Daily Sitka Sentinel*, April 8, 1980; Alaska Department of Fish and Game, "Sitka Seine Sac Roe."

22. Sean Cockerham, "Sitka Sound Herring Fishery Finally Ends," *Daily Sitka Sentinel*, April 9, 1996; Alaska Department of Fish and Game, "Sitka Seine Sac Roe."

23. Will Swagel, "Seiners Finally Get Chance at Herring," *Daily Sitka Sentinel*, April 4, 1988; "Sitka Sound Herring Fishery Climbs Over Half-Way Mark," *Daily Sitka Sentinel*, April 6, 1988; Will Swagel, "Sitka Sound Herring Harvest Near Quota," *Daily Sitka Sentinel*, April 7, 1988; "Fishery Ending," *Daily Sitka Sentinel*, April 13, 1988; "Herring Fishery Ends; Work Starts for 1989," *Daily Sitka Sentinel*, April 15, 1988;

"Sitka Herring Spawn Hits a Record Level," *Daily Sitka Sentinel*, April 22, 1988; Alaska Department of Fish and Game, "Sitka Seine Sac Roe."

24. Bill Swagel, "Herring Fishery Called This Afternoon," *Daily Sitka Sentinel*, March 31, 1989; Alaska Department of Fish and Game, "Sitka Seine Sac Roe."

25. "Herring Fishermen Make Second Run," *Daily Sitka Sentinel*, April 6, 1990; Alaska Department of Fish and Game, "Sitka Seine Sac Roe."

26. "Fish & Game Holds Off Calling Herring Fishery," *Daily Sitka Sentinel*, April 1, 1991.

27. "Sitka Sound Herring Fishery Remains in the Waiting Stage," *Daily Sitka Sentinel*, April 4, 1991.

28. "Sitka Herring Fishery Under Way as Co-Op," *Daily Sitka Sentinel*, April 10, 1991; "Co-Op Herring Fishery Extended 2 More Days," *Daily Sitka Sentinel*, April 11, 1991; "Herring Fishery Closed Before Quota Reached," *Daily Sitka Sentinel*, April 15, 1991; Alaska Department of Fish and Game, "Sitka Seine Sac Roe."

29. Heather MacLean, "Hot Springs Hot Spot for Herring Fishery," *Daily Sitka Sentinel*, April 6, 1992; Shannon Haugland, "Sitka Herring Roe Fishery Fastest Yet," *Daily Sitka Sentinel*, April 7, 1992; Alaska Department of Fish and Game, "Sitka Seine Sac Roe."

30. Alaska Department of Fish and Game, *Southeast Alaska Sac Roe Herring Fishery, 1993 Management Plan*, ADF&G Regional Information Rept. No. 1J93-03 (March 1993), 6–7; Eben Punderson, "Steady Pace for Herring Fishery," *Daily Sitka Sentinel*, March 29, 1993.

31. Eben Punderson, "Herring Fishery Moved South of Sitka," *Daily Sitka Sentinel*, March 30, 1993.

32. "Herring Season a Slow One," *Anchorage Daily News*, April 30, 1993; Eben Punderson, "F&G Says No Go Yet for Herring Harvest," *Daily Sitka Sentinel*, March 26, 1993; Eben Punderson, "Herring Fishery Moved South of Sitka," *Daily Sitka Sentinel*, March 30, 1993; Eben Punderson, "Herring Roe Fishing Enters Last Stages," *Daily Sitka Sentinel*, March 31, 1993.

33. Eben Punderson, "Herring Roe Fishing Ends After a Week," *Daily Sitka Sentinel*, April 5, 1993.

34. Kyle Hebert, *2012 Report to the Alaska Board of Fisheries: Southeast Alaska—Yakutat Herring Fisheries*, ADF&G Fishery Management Rept. No. 11-74 (December 2011), 18.

35. Clyde Curry, personal communication, November 5, 2020; Beaver Nelson, "A Fisherman's Perspective and History of Alaska's Herring Fisheries," American Fisheries Society, Alaska Chapter, annual meeting, Girdwood, Alaska, November 16–18, 2011.

36. "Fishery Updated," *Daily Sitka Sentinel*, March 2, 1994; Steve Will, "Seiners Wrapping Up Herring Harvest," *Daily Sitka Sentinel*, March 31, 1994; Steve Will, "Herring Harvest Ends; Quality Good," *Daily Sitka Sentinel*, April 1, 1994.

37. Hebert, *2012 Report to the Alaska Board of Fisheries*, 18.

38. Shannon Haugland, "Herring Fishery on 2-Hour Notice Soon," *Daily Sitka Sentinel*, March 21, 1995; Shannon Haugland, "F&G: Herring Not Ripe for Fishing Yet," *Daily Sitka Sentinel*, March 24, 1995; Shannon Haugland, "Seiners Wind Up Herring Fishery," *Daily Sitka Sentinel*, March 27, 1995; Shannon Haugland, "Sitka Roe Herring Fishery Right On Target," *Daily Sitka Sentinel*, March 28, 1995.

39. Hebert, *2012 Report to the Alaska Board of Fisheries*, 18.

40. Sean Cockerham, "Sac Roe Herring Harvest Half Over," *Daily Sitka Sentinel*, March 25, 1996; Sean Cockerham, "F&G Holding Off on 2nd Herring Opening," *Daily Sitka Sentinel*, March 26, 1996; Sean Cockerham, "Nets Again Out in Sitka Herring Fishery," *Daily Sitka Sentinel*, April 1, 1996; Shannon Haugland, "Nothing Yet In Herring Roe Fishery," *Daily Sitka Sentinel*, April 3, 1996; Sean Cockerham, "Sitka Herring Harvest Picks Up," *Daily Sitka Sentinel*, April 8, 1996; Sean Cockerham, "Sitka Sound Herring Fishery Finally Ends," *Daily Sitka Sentinel*, April 9, 1996; Shannon Haugland, "Fishers Net a Fourth of Herring Quota in Quick First Opening," *Daily Sitka Sentinel*, March 19, 1997; Alaska Department of Fish and Game, *2009 Southeast Alaska Commercial Herring Fishery Annual Management Report*, Fishery Management Rept. No. 10-39 (November 2010), 37.

41. Shannon Haugland, "Fishers Net a Fourth of Herring Quota in Quick First Opening," *Daily Sitka Sentinel*, March 19, 1997; Shannon Haugland, "Fishers Out Again in Herring Roe Fishery," *Daily Sitka Sentinel*, March 20, 1997; Shannon Haugland, "Herring Fishing Ends After Five Openings," *Daily Sitka Sentinel*, March 24, 1997; Hebert, *2012 Report to the Alaska Board of Fisheries*, 18.

42. Shannon Haugland, "Sitka Sound Herring Fishery Opened Today," *Daily Sitka Sentinel*, March 16, 1998.

43. Shannon Haugland, "Sitka Herring Fishing Starts," *Daily Sitka Sentinel*, March 17, 1998; Shannon Haugland, "Herring Harvest Continues," March 18, 1998; Shannon Haugland, "Opening Three Today in Herring Fishery," *Daily Sitka Sentinel*, March 19, 1998; Shannon Haugland, "Seiners Wrap Up Sitka Herring Harvest," *Daily Sitka Sentinel*, March 20, 1998.

44. Hebert, *2012 Report to the Alaska Board of Fisheries*, 18.

45. Troy Etulain, "Seiners Haul in Large Number of Viewers," *Daily Sitka Sentinel*, March 23, 1999.

46. Troy Etulain, "First Opening Held in Sitka Herring Fishery," *Daily Sitka Sentinel*, March 22, 1999; Troy Etulain, "Sitka Herring Fishery Opening Day a Winner," *Daily Sitka Sentinel*, March 23, 1999; Shannon Haugland, "Herring Fishing to Go Co-Op," *Daily Sitka Sentinel*, March 25, 1999; Shannon Haugland, "Seiners Fish at Leisure for Remaining Herring Quota," *Daily Sitka Sentinel*, March 26, 1999.

47. Shannon Haugland, "Herring Quota Reached," *Daily Sitka Sentinel*, March 29, 1999.

48. Shannon Haugland, "Herring Seiners Net Over Half of Quota," *Daily Sitka Sentinel*, March 20, 2000; Shannon Haugland, "Seiners Get Second Run At Herring," *Daily Sitka Sentinel*, March 22, 2000; Shannon Haugland, "Herring Fishery Wrapped Up," *Daily Sitka Sentinel*, March 23, 2000; Alaska Department of Fish and Game, "Sitka Seine Sac Roe."

49. Shannon Haugland, "First Opening Held in Herring Harvest," *Daily Sitka Sentinel*, March 22, 2001; Shannon Haugland, "Herring Fishing Pauses After Full Opening Day," *Daily Sitka Sentinel*, March 23, 2001; Shannon Haugland, "Third Opening Called in Sitka's Herring Fishery," *Daily Sitka Sentinel*, March 26, 2001; Shannon Haugland, "Herring Harvesters Get Another Round," *Daily Sitka Sentinel*, March 27, 2001; Shannon Haugland, "Herring Fishery Ends With Bounteous Catch," *Daily Sitka Sentinel*, March 28, 2001.

50. Alaska Department of Fish and Game, "Sitka Seine Sac Roe."

51. Alaska Administrative Code, 5 AAC 27.195; Alaska Department of Fish and Game, *Southeast Alaska Sac Roe Herring Fishery, 2002 Management Plan*, Regional Information Rept. No. 1J02-11 (February 2002), 3–4; Shannon Haugland, "Sitkans Happy with New Herring Regs," *Daily Sitka Sentinel*, January 15, 2002; Alaska Department of Fish and Game, *Southeast Alaska Sac Roe Herring Fishery, 2006*, Fishery Management Rept. No. 06-07 (March 2006), 7.

52. Shannon Haugland, "First Roe Opening Takes Fifth of Quota," *Daily Sitka Sentinel*, March 23, 2009.

53. Chris Bernard, "Herring Harvest Opened in Silver Bay," *Daily Sitka Sentinel*, March 27, 2002; Chris Bernard, "Another Half-Hour of Herring Fishing Called," *Daily Sitka Sentinel*, March 29, 2002; Chris Bernard, "Sitka Herring Fishery: Three Down, More to Go," *Daily Sitka Sentinel*, April 1, 2002; Chris Bernard, "Herring Fishery: Fourth Down, Goal Still Unmet," *Daily Sitka Sentinel*, April 3; Chris Bernard, "Clock Ticking for Seiners; Herring Quota Still Unmet," *Daily Sitka Sentinel*, April 8, 2002; Chris Bernard, "Sac Roe Fishers Playing a Waiting Game," *Daily Sitka Sentinel*, April 9, 2002; Chris Bernard, "Seiners Study Options in Sitka Herring Fishery," *Daily Sitka Sentinel*, April 10, 2002; Chris Bernard, "Herring Fishers Ponder How to End the Season," *Daily Sitka Sentinel*, April 11, 2002; Chris Bernard, "Seiners Turn to Co-Op in Fifth Herring Fishery," *Daily Sitka Sentinel*, April 12, 2002; Chris Bernard, "Herring Fishers Wrap It Up," *Daily Sitka Sentinel*, April 15, 2002; Alaska Department of Fish and Game, "Sitka Seine Sac Roe."

54. Shannon Haugland, "Third Opening Ahead for Herring Fishery," *Daily Sitka Sentinel*, March 24, 2003; Shannon Haugland, "Final Herring Fishing Set for 10 Minutes," *Daily Sitka Sentinel*, March 26, 2003; Shannon Haugland, "10-Minute Opening Ends Herring Fishing," *Daily Sitka Sentinel*, March 27, 2003.

55. Hebert, *2012 Report to the Alaska Board of Fisheries*, 18.

56. Alaska Department of Fish and Game, *Southeast Alaska Roe Herring Fishery, 2004 Management Plan*, ADF&G Regional Information Rept. No. 1J04-06 (February 2004), 9; Shannon Haugland, "Forecast Good for '09 Herring Harvest," *Daily Sitka Sentinel*, December 5, 2008.

57. Shannon Haugland, "Fishing Opens for Sitka Sound Herring," *Daily Sitka Sentinel*, March 22, 2004; Shannon Haugland, "Second Herring Fishery Called at Redoubt Bay," *Daily Sitka Sentinel*, March 25, 2004; "Sac Roe Herring Fishery Getting Close to Cut-Off," *Daily Sitka Sentinel*, March 26, 2004; Shannon Haugland, "Third Time the Charm in Sitka's Herring Harvest," *Daily Sitka Sentinel*, March 29, 2004; "Unusual Fishing Style Confused Troopers?" *Daily Sitka Sentinel*, May 4, 2005; "Court Counts Down Seconds to Verdict," *Daily Sitka Sentinel*, May 5, 2005; John Barry, "Trial" (letter to editor), *Daily Sitka Sentinel*, May 9, 2005; Robert Woolsey, "Sitka Seiner Loses $170,000 of Herring in 28 Seconds," *Alaska Public Media*, August 15, 2007, https://www.alaskapublic.org/2007/08/15/sitka-seiner-loses-170000-of-herring-in-28-seconds/; Kyle Hebert, *2018 Report to the Alaska Board of Fisheries: Southeast Alaska–Yakutat Herring Fisheries*, ADF&G Fishery Management Rept. No. 17-58 (December 2017), 24.

58. Shannon Haugland, "One-Third Harvested in Sitka Herring Opening," *Daily Sitka Sentinel*, March 24, 2005; Shannon Haugland, "Fishers Get Second Run at Sitka Sound Herring," *Daily Sitka Sentinel*, March 25, 2005; Shannon Haugland,

"It's Over: Fishers Net Last of Herring Quota," *Daily Sitka Sentinel*, March 29, 2005; Hebert, *2018 Report to the Alaska Board of Fisheries*, 24.

59. Shannon Haugland, "Half of Quota to Go for Herring Fishers," *Daily Sitka Sentinel*, March 27, 2006; Shannon Haugland, "3 Down, 1 to Go in Herring Harvest?" *Daily Sitka Sentinel*, March 28, 2006; Shannon Haugland, "Co-Op Effort Ends Sitka Herring Fishing," *Daily Sitka Sentinel*, March 30, 2006; Hebert, *2018 Report to the Alaska Board of Fisheries*, 24.

60. "F&G Waiting for Wider Distribution of Herring," *Daily Sitka Sentinel*, March 28, 2007.

61. Shannon Haugland, "First Opening Called in Herring Sac Roe Fishery," *Daily Sitka Sentinel*, March 26, 2007; Shannon Haugland, "First Herring Opening Nets Third of Quota," *Daily Sitka Sentinel* March 27, 2007; "F&G Waiting for Wider Distribution of Herring," *Daily Sitka Sentinel*, March 28, 2007; Shannon Haugland, "Seiners Bearing Down on Herring Catch Limit," *Daily Sitka Sentinel*, April 2, 2007; "Seiners Get 4th Chance," *Daily Sitka Sentinel*, April 3, 2007.

62. Shannon Haugland, "'07 Herring Fishery: As Good As It Gets," *Daily Sitka Sentinel*, April 4, 2007; Hebert, *2018 Report to the Alaska Board of Fisheries*, 24.

63. Shannon Haugland, "Sitka Herring Quota Set at Record Level," *Daily Sitka Sentinel*, March 5, 2008; Shannon Haugland, "First Opening Called in Herring Fishery," *Daily Sitka Sentinel*, March 25, 2008; Craig Giammona, "Sitka Herring Fishing Off to Slow Start," *Daily Sitka Sentinel*, March 26, 2008; "Whale Gets Caught Up in Herring Fishery," *Daily Sitka Sentinel*, March 26, 2008.

64. Craig Giomonna, "Herring Quota Near in 2 Days of Fishing," *Daily Sitka Sentinel*, March 27, 2008; Craig Giammona, "Herring Fishing Ending with Record Haul," *Daily Sitka Sentinel*, March 31, 2008; Alaska Department of Fish and Game, "Sitka Seine Sac Roe."

65. Kathryn Shattuck, "What's on Today," *New York Times*, March 15, 2009; "Sitka Herring Fishery Nets Spot on TV," *Daily Sitka Sentinel*, March 13, 2009.

66. "Fish & Game: No Herring Seen Yet in Sitka Sound," *Daily Sitka Sentinel*, March 16, 2009; Shannon Haugland, "First Roe Opening Takes Fifth of Quota," *Daily Sitka Sentinel*, March 23, 2009; "Seiners Half-Way to Quota," *Daily Sitka Sentinel*, March 25, 2009; Shannon Haugland, "F&G Gets Set for Final Herring Opening," *Daily Sitka Sentinel*, March 20, 2009; Shannon Haugland, "Only 930 Tons to Go in Herring Fishing," *Daily Sitka Sentinel*, April 1, 2009; Shannon Haugland, "Silver Bay Catch Closes Herring Fishery," *Daily Sitka Sentinel*, April 3, 2009; Craig Giammona, "Chamber Gets Rundown on Sitka Herring," *Daily Sitka Sentinel*, April 9, 2009; Hebert, *2018 Report to the Alaska Board of Fisheries*, 24.

67. Shannon Haugland, "1st Herring Opening Nets Third of Catch," *Daily Sitka Sentinel*, March 25, 2010; "Catch of the Day," (photo caption), *Daily Sitka Sentinel*, March 25, 2010; "Herring Fleet on Standby," *Daily Sitka Sentinel*, March 26, 2010; Shannon Haugland, "2nd Herring Opening Yields 3,500 Tons," *Daily Sitka Sentinel*, March 29, 2010; Shannon Haugland, "Herring Fishing Ends Just Short of Target," *Daily Sitka Sentinel*, April 5, 2010; Hebert, *2018 Report to the Alaska Board of Fisheries*, 24.

68. Robert Woolsey, "At $12-million and counting, herring tops Sitka's fisheries," *KCAW Radio* (Sitka, Alaska), March 17, 2011, https://www.kcaw.org/2011/03/17/at

Notes

-12-million-and-counting-herring-tops-sitka039s-fisheries/; Alaska Department of Fish and Game, "Sitka Seine Sac Roe."

69. Alaska Department of Fish and Game, *2011 Southeast Alaska Sac Roe Herring Fishery Management Plan*, Regional Information Rept. No. 1J11-02 (March 2011), 10.

70. Alaska Commercial Fisheries Entry Commission, "Basic Information Table, Herring Roe, Purse Seine, Southeast," accessed March 10, 2020, https://www.cfec.state.ak.us/bit/X_G01A.htm.

71. Hebert, *2012 Report to the Alaska Board of Fisheries*, 5; Shannon Haugland, "Sitka Herring Quota Cut Sharply for 2013," *Daily Sitka Sentinel*, December 13, 2012; Hebert, *2018 Report to the Alaska Board of Fisheries*, 24; Alaska Department of Fish and Game, "Sitka Seine Sac Roe"; "Herring Fishing 2011," *Musings of Cate Morris*, April 4, 2011, https://catemorris.com/tag/alaska/.

72. "Herring Fishing Ends with Over Half to Go," *Daily Sitka Sentinel*, April 12, 2012; Shannon Haugland, "Sitka Herring Quota Cut Sharply for 2013," *Daily Sitka Sentinel*, December 13, 2012; Hebert, *2018 Report to the Alaska Board of Fisheries*, 24.

73. Shannon Haugland, "Sitka Herring Quota Cut Sharply for 2013," *Daily Sitka Sentinel*, December 13, 2012; Shannon Haugland, "More Herring Sought in Today's Opening," *Daily Sitka Sentinel*, March 28, 2013; "Harvest Half Over in Sitka Herring Fishery," *Daily Sitka Sentinel*, March 29, 2013; "Herring Fishery Winding Down," *Daily Sitka Sentinel*, April 3, 2013; Shannon Haugland, "Herring Fishery Ends With Half of Quota Left," *Daily Sitka Sentinel*, April 4, 2013; Hebert, *2018 Report to the Alaska Board of Fisheries*, 24.

74. Shannon Haugland, "3 Down, 1 to Go in Sitka Herring Fishery," *Daily Sitka Sentinel*, March 27, 2014; Shannon Haugland, "Sac Roe Season Ends with 3,935-Ton Day," *Daily Sitka Sentinel*, March 31, 2014; Hebert, *2018 Report to the Alaska Board of Fisheries*, 24.

75. "F&G Taking Stock as Herring Harvest Nears," *Daily Sitka Sentinel*, March 15, 2015; "Sitka Herring Fishery Launched," *Daily Sitka Sentinel*, March 19, 2015; Shannon Haugland, "Herring Fishery Slows as Harvest Nears the Quota," *Daily Sitka Sentinel*, March 25, 2015; "Seiners Close in on Herring Harvest Quota," *Daily Sitka Sentinel*, March 23, 2105; "Herring Harvest Is Over," *Daily Sitka Sentinel*, March 26, 2015; Hebert, *2018 Report to the Alaska Board of Fisheries*, 24.

76. "Herring GHL for 2016 Down From Last Year," *Daily Sitka Sentinel*, March 2, 2016; "Sitka Herring Fishery Put on 2-Hour Notice," *Daily Sitka Sentinel*, March 16, 2016; "Herring Fishery Opens in Area off Kruzof," *Daily Sitka Sentinel*, March 17, 2016; "Herring Fishing Opens; 4th of Quota Taken," *Daily Sitka Sentinel*, March 18, 2016; "F&G Looks for Another Spot for Herring Fishing," *Daily Sitka Sentinel*, March 25, 2016; Robert Woolsey, "Small Fish, Active Spawning Close Down Sitka's Sac Roe Harvest," *KCAW Radio* (Sitka, Alaska), March 28, 2016; Shannon Haugland, "One-Hour Opening Fails to Find Quality Herring," *Daily Sitka Sentinel*, March 23, 2016; "F&G Keeps Looking for More Spawning Herring," *Daily Sitka Sentinel*, March 28, 2016; Hebert, *2018 Report to the Alaska Board of Fisheries*, 24; Alaska Department of Fish and Game, "Sitka Seine Sac Roe."

77. Shannon Haugland, "Herring Fishery Closes with a Third Left to Go," *Daily Sitka Sentinel*, March 29, 2016.

78. "First Opening Lands 3,500 Tons of Herring," *Daily Sitka Sentinel*, March 20, 2017; "F&G Out All Day Seeking Seine Area," *Daily Sitka Sentinel*, March 21, 2017;

"2nd Opening Held in Sitka Herring Fishery," *Daily Sitka Sentinel*, March 22, 2017; Shannon Haugland, "Sitka Herring Fishery Hits Half-Way Mark," *Daily Sitka Sentinel*, March 23, 2017; Shannon Haugland, "Sitka Herring Fishing Down to Co-Op Effort," *Daily Sitka Sentinel*, March 27, 2017; Shannon Haugland, "Sitka Herring Harvest End; Numbers Good," *Daily Sitka Sentinel*, March 28, 2017; Hebert, *2018 Report to the Alaska Board of Fisheries*, 24.

79. Joe Viechnicki and Robert Woolsey, "On 2-Hour Notice: Sitka Herring Fleet Opts for 'Non-Competitive' Fishery," *KCAW Radio* (Sitka, Alaska), March 19, 2018; Klas Stolpe, "Sac Roe Fishery Opens North of Sitka," *Daily Sitka Sentinel*, March 26, 2018; "Second Round," *Daily Sitka Sentinel*, March 27, 2018; Klas Stolpe, "Herring Seiners Wait as 'Window' Shortens," *Daily Sitka Sentinel*, March 28, 2018; Klas Stolpe, "Herring Surveys Show Fishery May Be Over," *Daily Sitka Sentinel*, March 29, 2018; Klas Stolpe, "Low Herring Quality Puts Fishery in Doubt," *Daily Sitka Sentinel*, April 2, 2018; Klas Stolpe, "Herring Fishery Ends 8,330 Tons Short of GHL," *Daily Sitka Sentinel*, April 3, 2018; Charlie Ess, "Alaska Herring Quota Up, but Year Class Size May Not Make the Grade," *National Fisherman*, February 12, 2019, https://www.nationalfisherman.com/alaska/alaska-herring-quota-up-but-year-class-size-may-not-make-the-grade; Aaron Dupois (ADF&G, Sitka), personal communication, September 28, 2020.

80. Alaska Department of Fish and Game, *2019 Southeast Alaska Sac Roe Herring Fishery Management Plan*, ADF&G Regional Information Rept. No. 1J19-01 (March 2019), 6; Klas Stolpe, "Arial Survey Shows Active Herring Spawning," *Daily Sitka Sentinel*, April 1, 2019; Klas Stolpe, "Herring Spawning Widening," *Daily Sitka Sentinel*, April 2, 2019; "Aerial Survey Finds More Herring Spawning in Area," *Daily Sitka Sentinel*, April 3, 2019.

81. "Herring Sac Roe Harvest Nixed for Sitka," *Cordova Times*, April 11, 2019.

82. Katherine Rose, "Sitka Herring Roe Fishery Closed over Small Fish, Weak Markets and Coronavirus Uncertainty," *Alaska Public Media*, March 3, 2020, https://www.alaskapublic.org/2020/03/03/sitka-herring-roe-fishery-closed-over-small-fish-weak-markets-and-coronavirus-uncertainty/; Katherine Rose, "What 2019 Means for Sitka's Herring Future," *KCAW Radio* (Sitka, Alaska), June 25, 2019, https://www.kcaw.org/2019/06/25/what-2019-means-for-sitkas-herring-future/; "Sitka Sound Herring Fishery Not Likely in 2020," *Petersburg Pilot*, March 5, 2020.

83. Shannon Haugland, "Fish & Game Sums Up 2020 Herring Returns," *Daily Sitka Sentinel*, May 7, 2020.

84. Hebert, *2018 Report to the Alaska Board of Fisheries*, 18–19; Alaska Department of Fish and Game, *Commercial Herring Fisheries, Southeast Alaska & Yakutat*, accessed January 27, 2021, https://www.adfg.alaska.gov/index.cfm?adfg=commercialbyareasoutheast.herring#harvest.

85. "Big Herring Spawn," *Petersburg Press*, May 13, 1970; Alaska Department of Fish and Game, *1982 Management Report for Southeast Alaska, Region 1* (July 1986), 43, 115; Alaska Department of Fish and Game, *1989 Management Plan, Southeast Alaska Herring Roe Fishery*, ADF&G Regional Information Rept. No. 1J89-06 (March 1989), 6; Paul Johnson (formerly with Juneau Cold Storage), personal communication, July 17, 2020.

Notes

CHAPTER 11: RESURRECTION BAY AND PRINCE WILLIAM SOUND ROE-HERRING FISHERIES

1. Alaska Department of Fish and Game, *Cordova Area Annual Management Report, 1969*, 98.

2. Edward O. Otis and Lee F. Hammarstrom (Alaska Department of Fish and Game), *Overview of the Lower Cook Inlet Area Commercial Herring Fishery and Recent Stock Status*, Special Publication No. 04-13 (November 2004), 10; Alaska Department of Fish and Game, *Cook Inlet Area Annual Report, 1967*; Alaska Department of Fish and Game, *Cook Inlet Area Annual Report, 1968*.

3. Thomas R. Schroeder, *A Summary of Historical Data for the Lower Cook Inlet, Alaska, Pacific Herring Sac Roe Fishery*, ADF&G Fishery Research Bulletin No. 89-04 (December 1989), 1.

4. Beaver Nelson (former herring seiner, Homer, Alaska), personal communication, March 2020.

5. Alaska Department of Fish and Game, *Cook Inlet–Resurrection Bay Area Annual Management Report, 1969*; Otis and Hammarstrom, *Overview of the Lower Cook Inlet Area Commercial Herring Fishery*, 10.

6. Nelson, personal communication, March 2020.

7. Gretchen Bersch (Herring Northwest), personal communication, June 28, 2020; Ann Holmstrand (Herring Northwest), personal communication, June 28, 2020; Rhonda Hubbard (Seward Marine Services), personal communication, June 29, 2020; Alaska Department of Fish and Game, *Annual Management Report, 1970, Cook Inlet–Resurrection Bay Area*, 108–111.

8. Alaska Department of Fish and Game, *1970 Alaska Catch and Production, Commercial Fisheries Statistics*, Statistical Leaflet No. 21, 36; Alaska Department of Fish and Game, *Cook Inlet Annual Management Report, 1972*, 29–30.

9. Otis and Hammarstrom, *Overview of the Lower Cook Inlet Area Commercial Herring Fishery*, 2, 10.

10. Alaska Department of Fish and Game, *Cordova Area Annual Management Report, 1970*, 88; Alaska Department of Fish and Game, *Prince William Sound Management Area, 1990 Annual Finfish Management Report*, Regional Information Rept. No. 2C91-14 (November 1991), 137.

11. Alaska Department of Fish and Game, *Annual Management Report, Prince William Sound Area, 1971*, 2–6, 90; Alaska Department of Fish and Game, *1971 Commercial Operators*, ADF&G Statistical Leaflet No. 22, 11–13; Alaska Department of Fish and Game, *Prince William Sound Area Annual Management Report, 1975*, 79.

12. Alaska Department of Fish and Game, *Annual Report, 1971*, 15.

13. Alaska Department of Fish and Game, *Annual Report, 1973*, 14; Alaska Department of Fish and Game, *Prince William Sound Area Annual Management Report, 1972–1973*, 2–8, 138; Alaska Department of Fish and Game, *Prince William Sound Area Annual Management Report, 1975*, 79; Ward T. Bower, *Alaska Fisheries and Fur-Seal Industries in 1938*, Administrative Rept. No. 36, Appendix II to *Report of the United States Commissioner of Fisheries for Fiscal Year 1939* (Washington, DC: GPO, 1940), 137–140.

14. "9 Herring Plants to Run," *Pacific Fisherman* (June 1946): 97.

15. Stephen H. Rogers, "Herring Fishery in Southeastern Alaska," *Commercial Fisheries Review* (August 1965): 1–6; Nelson, personal communication, March 24,

2020; "One Last Season After Shaping Spotter Pilot Profession, Tom Parker Was Ready to Move On," *Anchorage Daily News*, April 11, 1991; Joel Gay, "Collision Mars PWS Herring Season," *Pacific Fishing* (June 1991): 21–22; Susan Froetschel, "Seiners Primed for Speedy Herring Fishery," *Daily Sitka Sentinel*, March 28, 1986.

16. Alaska Department of Fish and Game, *Prince William Sound Annual Management Report, 1978,* 61–63.

17. Alaska Department of Fish and Game, *Annual Report, Prince William Sound Area, 1974,* 65–68; Alaska Department of Fish and Game, *Prince William Sound Area Annual Management Report, 1975,* 77–79.

18. Alaska Department of Fish and Game, *Prince William Sound Area Annual Management Report, 1975,* 77–79; Alaska Department of Fish and Game, *Prince William Sound Management Area, 2000 Annual Finfish Management Report*, Regional Information Rept. No. 2A02-02 (February 2002), 158.

19. Beaver Nelson, "A Fisherman's Perspective and History of Alaska's Herring Fisheries," American Fisheries Society, Alaska Chapter, annual meeting, Girdwood, Alaska, November 16–18, 2011.

20. Doug McNair, "Alaska Roe Herring," *Pacific Fishing* (November 1983): 103–107.

21. Alaska Department of Fish and Game, *Prince William Sound Area Annual Management Report, 1976,* 71–73.

22. Anna Young, *The Lost Art of Alaska Fishing, Part One* (New York: Ruggedland Press, 2007), 63.

23. Young, *The Lost Art of Alaska Fishing,* 63–71.

24. Alaska Department of Fish and Game, *Prince William Sound Area Annual Management Report, 1976,* 71–73; Alaska Department of Fish and Game, *Prince William Sound Area Annual Management Report, 1977,* 17; Alaska Department of Fish and Game, *Prince William Sound Area Annual Management Report, 1978,* 13.

25. Alaska Department of Fish and Game, *Prince William Sound Area Annual Management Report, 1982,* 17.

26. Dennis Jones (former Icicle Seafoods employee), personal communication, April 2, 2020.

27. "The Recent Development of the Southeast Alaska Herring Fishery," MFR Paper No. 1240, *Marine Fisheries Review* (March 1977): 15–18.

28. Alaska Department of Fish and Game, *Prince William Sound Management Area, 1990 Annual Finfish Management Report*, ADF&G Regional Information Rept. No. 2C91-14 (November 1991), 12; Alaska Commercial Fisheries Entry Commission, Permit Status Report by Fishery Code, All Years, Fishery: G01E (Prince William Sound roe herring), accessed August 22, 2020, https://www.cfec.state.ak.us/pstatus/x_g01e.htm.

29. *Session Laws of Alaska, 1977,* Chapter 9, § 1.

30. Alaska Department of Fish and Game, *Prince William Sound Area Annual Management Report, 1977,* 15–17; Alaska Department of Fish and Game, *Prince William Sound Area 1993 Annual Finfish Management Report*, ADF&G Regional Information Rept. No. 2A95-XX (April 1995), H.4.

31. Alaska Department of Fish and Game, *Prince William Sound Area Annual Management Report, 1978,* 6, 13, 61–63; Daniel Sharp, Steve Morstad, and John Wilcock, *Prince William Sound Management Area, Herring Report to the Alaska Board of Fisheries*, ADF&G Regional Information Rept. No. 2A96-40 (December 1996), 13; Alaska Department of Fish and Game, *Prince William Sound Area 1993 Annual Finfish*

Management Report, ADF&G Regional Information Rept. No. 2A95-XX (April 1995), H.4.

32. Alaska Department of Fish and Game, *Prince William Sound Area Annual Management Report, 1979* (August 1980), 54–55, 57; Sharp, Morstad, and Wilcock, *Prince William Sound Management Area, Herring Report,* 13; Emilie Springer, "No Red Herring," *National Fisherman* (July 2017): 22–25.

33. Alaska Department of Fish and Game, *Prince William Sound Area Annual Management Report, 1980* (May 1981), 12–16; Sharp, Morstad, and Wilcock, *Prince William Sound Management Area, Herring Report,* 13–14.

34. Alaska Department of Fish and Game, *Prince William Sound Area Annual Management Report, 1981,* 12.

35. Alaska Department of Fish and Game, *Prince William Sound Area Annual Management Report, 1981,* 12–17; Alaska Department of Fish and Game, *Prince William Sound Area Annual Management Report, 1982,* 62; Alaska Department of Fish and Game, *Prince William Sound Annual Finfish Management Report, 1984,* 18; Alaska Department of Fish and Game, *Prince William Sound Annual Finfish Management Report, 1993,* ADF&G Regional Information Rept. No. 2A95-XX (April 1995), H.4, H.13.

36. Alaska Department of Fish and Game, *Prince William Sound Area Annual Management Report, 1982,* 12–17, 62.

37. Alaska Department of Fish and Game, *Prince William Sound Annual Finfish Management Report 1993,* H.4, H.13.

38. Alaska Department of Fish and Game, *Prince William Sound Annual Finfish Management Report, 1983,* 12–16, 72, 90; Nelson, personal communication, May 14, 2020; Alaska Department of Fish and Game, *Prince William Sound Annual Finfish Management Report, 1993,* H.4, H.13.

39. Alaska Department of Fish and Game, *Prince William Sound Annual Finfish Management Report, 1984,* 17–22; Alaska Department of Fish and Game, *Prince William Sound Annual Finfish Management Report, 1993,* H.4, H.13.

40. Alaska Department of Fish and Game, *Prince William Sound Annual Finfish Management Report, 1985,* 22–28, 96; Alaska Department of Fish and Game, *Prince William Sound Annual Finfish Management Report, 1993,* H.4, H.13.

41. Alaska Department of Fish and Game, *Prince William Sound Annual Finfish Management Report, 1985,* 22–28; Alaska Department of Fish and Game, *Prince William Sound Annual Finfish Management Report, 1993,* H.4, H.13.

42. Alaska Department of Fish and Game, *Prince William Sound Area Annual Finfish Management Report, 1986,* 23.

43. Alaska Department of Fish and Game, *Prince William Sound Area Annual Finfish Management Report, 1986,* 22.

44. Alaska Department of Fish and Game, *Prince William Sound Area Annual Finfish Management Report, 1986,* 20–26, 86; Alaska Department of Fish and Game, *Prince William Sound Annual Finfish Management Report, 1993,* ADF&G Regional Information Rept. No. 2A95-XX (April 1995), H.4, H.13.

45. Alaska Department of Fish and Game, *Prince William Sound Area Annual Finfish Management Report, 1986,* 25–26; Alaska Department of Fish and Game, *Prince William Sound Annual Finfish Management Report, 1993,* ADF&G Regional Information Rept. No. 2A95-XX (April 1995), H.4.

46. Alaska Department of Fish and Game, *Prince William Sound Area Annual Finfish Management Report, 1987* (March 1988), 13–15; Alaska Department of Fish and Game, *Prince William Sound Annual Finfish Management Report, 1993*, H.4, H.13.

47. Linda K. Brannian, *Forecast of the Pacific Herring Biomass in Prince William Sound, 1989*, ADF&G Regional Information Rept. No. 2A89-01 (January 1989), vii.

48. Alaska Department of Fish and Game, *Prince William Sound Area Annual Finfish Management Report, 1988*, ADF&G Regional Information Rept. No. 2C90-02 (March 1990), 22–32; Alaska Department of Fish and Game, *Prince William Sound Annual Finfish Management Report, 1993*, H.4, H.13.

49. Alaska Department of Fish and Game, *Prince William Sound Area Annual Finfish Management Report, 1988*, 22–32; Alaska Department of Fish and Game, *Prince William Sound Area Annual Finfish Management Report, 1989*, ADF&G Regional Information Rept. No. 2C90-07 (February 1991), 162; Alaska Department of Fish and Game, *Prince William Sound Annual Finfish Management Report, 1993*, H.4.

50. Dan Strickland, "Report From the Oil Slick in Alaska," *Pacific Fishing* (July 1989): 44–46.

51. Dan Strickland, "Report From the Oil Slick in Alaska," *Pacific Fishing* (July 1989): 44–46; Alaska Department of Fish and Game, *Prince William Sound Annual Finfish Management Report, 1989*, ADF&G Regional Information Rept. No. 2C90-07 (February 1991), 3–6; "Herring Roe Harvest in Prince William Sound Up," *Daily Sitka Sentinel*, May 1, 1991.

52. "Herring Fishermen Shut Out," *Anchorage Daily News*, April 4, 1989.

53. Alaska Department of Fish and Game, *Prince William Sound Annual Finfish Management Report, 1989*, 32.

54. Alaska Department of Fish and Game, *Prince William Sound Management Area 1990 Annual Finfish Management Report*, ADF&G Regional Information Rept. No. 2C91-14 (November 1991), 12–14, 137; Timothy T. Baker, John A. Wilcock, and Betsy W. McCracken, *Stock Assessment and Management of Pacific Herring In Prince William Sound, Alaska, 1990*, ADF&G Tech. Fishery Rept. No. 91-22 (December 1991), 11–12; Donna Parker, "Herring Roundup," *Pacific Fishing* (July 1990): 25–26; Alaska Department of Fish and Game, *Prince William Sound Annual Finfish Management Report, 1993*, H.4, H.13.

55. Alaska Department of Fish and Game, *Prince William Sound Area 1991 Annual Finfish Management Report*, ADF&G Regional Information Rept. No. 2A92-09 (May 1992), 15–21, 133; "Herring Roe Harvest in Prince William Sound Up," *Daily Sitka Sentinel*, May 1, 1991; Alaska Department of Fish and Game, *Prince William Sound Annual Finfish Management Report, 1993*, H.4, H.13.

56. "One Last Season After Shaping Spotter Pilot Profession, Tom Parker Was Ready to Move On," *Anchorage Daily News*, April 11, 1991.

57. Alaska Department of Fish and Game, *Prince William Sound Area 1991 Annual Finfish Management Report*, 21.

58. Alaska Department of Fish and Game, *Prince William Sound Area 1992 Annual Finfish Management Report*, ADF&G Regional Information Rept. No. 2A93-12 (June 1993), 16–19; Alaska Department of Fish and Game, *Prince William Sound Annual Finfish Management Report, 1993*, H.4, H.13.

59. Alaska Department of Fish and Game, *Prince William Sound Area 1992 Annual Finfish Management Report*, 23; Alaska Department of Fish and Game, *Prince William Sound Area 1993 Annual Finfish Management Report*, 19–27.

Notes

60. US Geological Survey, Western Fisheries Research Center, "Herring Disease Program," accessed May 25, 2020, https://www.usgs.gov/centers/wfrc/science/herring-disease-program?qt-science_center_objects=0#qt-science_center_objects; Alaska Department of Fish and Game, *Prince William Sound Area 1999 Annual Finfish Management Report*, Regional Information Rept. No. 2A00-32, 33.

61. Alaska Department of Fish and Game, *Prince William Sound Area 1993 Annual Finfish Management Report*, 19–27, H.4.

62. Alaska Department of Fish and Game, *Prince William Sound Area 1993 Annual Finfish Management Report*, 27.

63. Alaska Department of Fish and Game, *Prince William Sound Area 1994 Annual Finfish Management Report*, Regional Information Rept. No. 2A95-47 (December 1995), 18–20, 129.

64. Alaska Department of Fish and Game, *Prince William Sound Management Area 1996 Annual Finfish Management Report*, ADF&G Regional Information Rept. No. 2A97-17 (May 1997), 21.

65. Alaska Department of Fish and Game, *Prince William Sound Area 1997 Annual Finfish Management Report*, Regional Information Rept. No. 2A98-05 (May 1998), 23–31, 151; "2 Homer Men Killed in Crash—Herring Planes Collide," *Anchorage Daily News*, April 10, 1997; Alaska Department of Fish and Game, *Prince William Sound Area 2000 Annual Finfish Management Report*, Regional Information Rept. No. 2A02-02 (February 2002), H.4, H.13.

66. Alaska Department of Fish and Game, *Prince William Sound Area 1998 Annual Finfish Management Report*, Regional Information Rept. No. 2A99-20 (May 1999), 25–34; Alaska Department of Fish and Game, *Prince William Sound Area 2000 Annual Finfish Management Report*, H.4, H.13.

67. Alaska Department of Fish and Game, *Prince William Sound Area 1999 Annual Finfish Management Report*, Regional Information Rept. No. 2A00-32 (November 2000), 28.

68. Alaska Department of Fish and Game, *Prince William Sound Area 1999 Annual Finfish Management Report*, Regional Information Rept. No. 2A00-32 (November 2000), 27–36.

69. US Geological Survey, "Herring Disease Program"; *Exxon Valdez* Oil Spill Trustee Council, *Exxon Valdez Oil Spill Restoration Plan, 2014 Update Injured Resources and Services* (November 19, 2014), 26–28.

CHAPTER 12: LOWER COOK INLET AND KODIAK AREA ROE-HERRING FISHERIES

1. Alaska Department of Fish and Game, *1969 Commercial Operators*, Statistical Leaflet No. 18, 19; Alaska Department of Fish and Game, *Annual Report, 1975*, 23–24; Alaska Department of Fish and Game, *Annual Report, 1976*, 27; Edward O. Otis and Lee F. Hammarstrom, *Overview of the Lower Cook Inlet Area Commercial Herring Fishery and Recent Stock Status*, ADF&G Special Publication No. 04-13 (November 2004), 4.

2. Thomas R. Schroeder, *A Summary of Historical Data for the Lower Cook Inlet, Alaska, Pacific Herring Sac Roe Fishery*, ADF&G Fishery Research Bulletin No. 89-04

(December 1989), 1–4; Otis and Hammarstrom, *Overview of the Lower Cook Inlet Area Commercial Herring Fishery*, 2–3, 10–11, 14.

3. Alaska Commercial Fisheries Entry Commission, *Fishery Statistics—Permits & Permit Holders* (1977), accessed August 16, 2020, https://www.cfec.state.ak.us/pstatus/14051977.htm; Alaska Commercial Fisheries Entry Commission, *Fishery Statistics—Permits & Permit Holders* (2000), accessed August 17, 2020, https://www.cfec.state.ak.us/pstatus/14052000.htm.

4. Otis and Hammarstrom, *Overview of the Lower Cook Inlet Area Commercial Herring Fishery*, 2–3, 10–11, 14.

5. Dave Prokopowich, *Kodiak Management Area Commercial Herring Sac-Roe and Food/Bait Fisheries—1989, Report to the Board of Fisheries*, ADF&G Regional Information Rept. No. 4K89-30 (November 1989), 3.

6. Charles Burkey Jr. and Joan Ried, *Statistics of the Commercial Fishery for Pacific Herring from the Kodiak Area, Alaska*, ADF&G Tech. Fishery Rept. No. 88-11 (August 1988), 4, 16; Alaska Department of Fish and Game, *1964 Alaska Commercial Fisheries Operators*, Statistical Leaflet No. 8, 21; Alaska Department of Fish and Game, *1964 Commercial Fisheries Catch and Production Statistics*, Statistical Leaflet No. 9, 18; Prokopowich, *Kodiak Management Area Commercial Herring Sac-Roe and Food/Bait Fisheries—1989*, 17.

7. Burkey and Ried, *Statistics of the Commercial Fishery for Pacific Herring from the Kodiak Area*, 16.

8. "By-Products," *Pacific Fisherman* (January 25, 1966): 158–159; Alaska Department of Fish and Game, *Alaska Fisheries, 1965, Commercial Operators*, Statistical Leaflet No. 10, 16; Alaska Department of Fish and Game, *Alaska Catch and Production, Commercial Fisheries Statistics, 1965*, Statistical Leaflet No. 11, 17; Burkey and Ried, *Statistics of the Commercial Fishery for Pacific Herring from the Kodiak Area*, 13.

9. Burkey and Ried, *Statistics of the Commercial Fishery for Pacific Herring from the Kodiak Area*, 16; Alaska Department of Fish and Game, *Commercial Operators, 1966*, Statistical Leaflet No. 12, 20–22; Prokopowich, *Kodiak Management Area Commercial Herring Sac-Roe and Food/Bait Fisheries—1989*, 1–2.

10. Burkey and Ried, *Statistics of the Commercial Fishery for Pacific Herring from the Kodiak Area*, 13–14; Dennis Gretsch, Dave Prokopowich, and Kevin Brennan, *Kodiak Management Area Annual Herring Management Report, 1996*, ADF&G Regional Information Rept. No. 4K97-37 (May 1997), 1.

11. Burkey and Ried, *Statistics of the Commercial Fishery for Pacific Herring from the Kodiak Area*, 13–14; Gretsch, Prokopowich, and Brennan, *Kodiak Management Area Annual Herring Management Report, 1996*, 1.

12. Burkey and Ried, *Statistics of the Commercial Fishery for Pacific Herring from the Kodiak Area*, 13–14.

13. Alaska Commercial Fisheries Entry Commission, *Fishery Statistics—Permits & Permit Holders* (1984), accessed August 21, 2020, https://www.cfec.state.ak.us/pstatus/14051984.htm; Alaska Commercial Fisheries Entry Commission, *Fishery Statistics—Permits & Permit Holders* (2009), accessed August 21, 2020, https://www.cfec.state.ak.us/pstatus/14052009.htm.

14. Dennis Gretsch, *Kodiak Management Area Annual Herring Management Report, 1997*, ADF&G Regional Information Rept. No. 4K98-41 (August 1998), 1; Dennis Gretsch, *Kodiak Management Area Herring Report to the Alaska Board of Fisheries,*

January 2005, ADF&G Fishery Management Rept. No. 04-11 (December 2004), 2; Geoff Spalinger, *Kodiak Management Area Herring Sac Roe Fishery Harvest Strategy for the 2019 Season*, ADF&G Regional Information Rept. No. 4K19-05 (March 2019), 13–14; Donna Parker, "Herring Roundup," *Pacific Fishing* (July 1990): 25–26; Shannon Haugland, "Herring Fishing Ends After Five Openings," *Daily Sitka Sentinel*, March 24, 1997.

15. Geoff Spalinger, *Kodiak Management Area Herring Sac Roe Fishery Harvest Strategy for the 2019 Season*, ADF&G Regional Information Rept. No. 4K19-05 (March 2019), 13–14.

CHAPTER 13: TOGIAK ROE-HERRING FISHERY

1. Doug McNair, "Alaska Roe Herring," *Pacific Fishing* (November 1983): 103–107.
2. Jeffrey Skrade (Alaska Department of Fish and Game), "Management of Herring in the Eastern Bering Sea," *Proceedings of the Alaska Herring Symposium*, Alaska Sea Grant Rept. 80-4 (October 1980), 5-26.
3. Albrecht Schumacher (Federal Research Centre for Fisheries, Federal Republic of Germany), "Management of the North Sea Herring Fisheries," *Proceedings of the Alaska Herring Symposium*. Alaska Sea Grant Rept. 80-4 (October 1980), 239–249.
4. Darwin A. Biwer Jr., *Herring Fishery in the Togiak District of Bristol Bay*, ADF&G Bristol Bay Rept. No. 17 (June 1969), 5.
5. Jeffrey Skrade (former ADF&G fisheries biologist), personal communication, November 15, 2020.
6. Fritz Funk, ed., *Preliminary Forecasts of Catch and Stock Abundance for 1993 Alaska Herring Fisheries*, ADF&G Regional Information Rept. No. 5J93–06 (May 1993), 18; Alaska Department of Fish and Game, "2001 Bristol Bay Herring Fishery," in *Bristol Bay Area 2001 Annual Management Report*, ADF&G Regional Information Rept. No. 2A02-18 (May 2002), 131; Alaska Department of Fish and Game, *2018 Bristol Bay Area Annual Management Report*, ADF&G Fishery Management Rept. No. 19-12 (May 2019), 101.
7. Alaska Department of Fish and Game, "Bristol Bay Herring Fishery," in *Bristol Bay Area 2002 Annual Management Report*, ADF&G Regional Information Rept. No. 2A03-18 (April 2003), 119; Alaska Department of Fish and Game, *2018 Bristol Bay Area Annual Management Report*, 101.
8. Alaska Department of Fish and Game, *Bristol Bay Area Annual Management Report, 1967*, 18; Alaska Department of Fish and Game, *Bristol Bay Area Annual Management Report, 1976*, 129.
9. Biwer, *Herring Fishery in the Togiak District of Bristol Bay*, 5–6; Alaska Department of Fish and Game, *Bristol Bay Area Annual Management Report, 1968*, 29–30; Alaska Department of Fish and Game, *Bristol Bay Area Annual Management Report, 1976*, 129; Alaska Department of Fish and Game, *Bristol Bay Area Annual Management Report, 1986* (September 1987), 228.
10. Alaska Department of Fish and Game, *Bristol Bay Area Annual Management Report, 1969*, 17.
11. Gretchen Bersch, crewmember aboard *Diver 1*, personal communication, June 28, 2020.

12. Alaska Department of Fish and Game, *Bristol Bay Area Annual Management Report, 1976* (January 1979), 129.

13. Louis H. Barton and Vidar G. Wespestad, "Distribution, Biology and Stock Assessment of Western Alaska's Herring Stocks," *Proceedings of the Alaska Herring Symposium*. Alaska Sea Grant Rept. 80-4 (October 1980), 27–53.

14. Flip Todd, "Agent Reviews Foreign Catch," *Anchorage Times*, January 21, 1975.

15. Alaska Department of Fish and Game, *Bristol Bay Area Annual Management Report, 1975* (May 1976), 27.

16. Magnuson-Stevens Fishery Management and Conservation Act, April 13, 1976 (90 Stat. 331).

17. Brad Matsen, "Old Captain K. and the Alaska Herring Boom," *Alaska Fisherman's Journal* (October 1982): 41–42; Beaver Nelson (former herring seiner, Homer, Alaska), personal communication, October 31, 2020.

18. Alaska Department of Fish and Game, *Bristol Bay Area Annual Management Report, 1981* (April 1982), 151–153.

19. Emilie Springer, "Togiak Sac Roe Herring: Catch the Story Before It's Gone," *Homer News*, August 3, 2017, https://www.homernews.com/news/togiak-sac-roe-herring-catch-the-story-before-its-gone/.

20. Beaver Nelson (former herring seiner, Homer, Alaska), personal communication, April 5, 2020.

21. Skrade, "Management of Herring in the Eastern Bering Sea," 5–26.

22. Alaska Department of Fish and Game, *Bristol Bay Area Annual Management Report, 1979*, 53–55; Springer, "Togiak Sac Roe Herring"; Nelson, personal communication, April 5, 2020.

23. Alaska Department of Fish and Game, *Bristol Bay Area Annual Management Report, 1977*, 26–27.

24. Nelson, personal communication, April 24, 2020.

25. Tom Swanson (Icicle Seafoods), personal communication, February 12, 2020; Nelson, personal communication, April 5, 2020.

26. Alaska Department of Fish and Game, *Bristol Bay Area Annual Management Report, 1977*, 26–27; Alaska Department of Fish and Game, *Bristol Bay Area 1990 Annual Management Report*, Regional Information Rept. No. 91-1 (April 1991), H32.

27. "Herring Roe Marketing Channels in Japan," *Bering Sea Fisherman*, April 1981.

28. Donna Parker, "Few Hurrahs on the Herring Horizon," *Pacific Fishing* (March 1990): 21, 38.

29. Alaska Department of Fish and Game, "Bristol Bay Herring Fishery," in *Bristol Bay Area 2002 Annual Management Report*, 119; Nelson, personal communication, July 3, 2020.

30. Alaska Department of Fish and Game, *Bristol Bay Area Annual Management Report, 1978*, 34–38; Alaska Department of Fish and Game, *Bristol Bay Area 1990 Annual Management Report*, H32; Jeffrey Skrade (former ADF&G fisheries biologist), personal communication, May 2, 2020.

31. Alaska Department of Fish and Game, *Commercial Herring Fisheries*, accessed April 16, 2020, https://www.adfg.alaska.gov/index.cfm?adfg=commercialbyareakodiak.main; Alaska Administrative Code, 5 AAC 27.865 (7).

32. Alaska Department of Fish and Game, *Bristol Bay Area Annual Management Report, 1979*, 106.

33. Skrade, personal communication, April 27, 2020.

34. "3 Killed in Mid-air Collision," *Daily Sitka Sentinel*, May 31, 1978.

35. Alaska Department of Fish and Game, *Bristol Bay Area Annual Management Report, 1979*, 53.

36. Alaska Department of Fish and Game, *Bristol Bay Area Annual Management Report, 1979*, 53–55; Alaska Department of Fish and Game, *Bristol Bay Area Annual Management Report, 1988*, 318; Alaska Department of Fish and Game, "2001 Bristol Bay Herring Fishery," 131, 153; Alaska Department of Fish and Game, *2018 Bristol Bay Area Annual Management Report*, 101.

37. Jim Edenso, "Welcome," *Proceedings of the Alaska Herring Symposium*, Alaska Sea Grant Rept. 80-4 (October 1980), 1–2.

38. Jim H. Branson to Jay Hammond, February 28, 1980.

39. Alaska Department of Fish and Game, *Bristol Bay Area Annual Management Report, 1980* (September 1981), 37–44, 158; Alaska Department of Fish and Game, *Bristol Bay Area 1990 Annual Management Report*, H32.

40. Skrade, personal communication, May 2, 2020.

41. Nelson, personal communication, May 14, 2020.

42. Prepared Statement of Paul D. Kelly, Attorney for Western Alaska Cooperative Marketing Association and Bristol Bay Herring Marketing Cooperative, in *200-Mile Fishery—ICCAT, Hearings Before the Subcommittee on Fisheries and Wildlife Conservation and the Environment of the Committee on Merchant Marine and Fisheries, House of Representatives, on Magnuson Fishery Conservation and Management Act Oversight, September 24–25, October 14, 1981*, 97th Cong., 1st sess., Serial No. 97-18 (Washington, DC: GPO, 1982), 464–466.

43. Swanson, personal communication, April 29, 2020.

44. Bruce Twomley (former chairman, Alaska Commercial Fisheries Entry Commission), personal communication with author, October 2, 2020.

45. "In-state Processing of Herring Approved," *Daily Sitka Sentinel*, April 23, 1980; Chapter 27, *Session Laws of Alaska, 1980*; "Judge Oks Selling Herring," *Daily Sitka Sentinel*, April 29, 1981.

46. Alaska Department of Fish and Game, *Bristol Bay Area Annual Management Report, 1981*, 151–175; Skrade, personal communication, November 16, 2020.

47. Magnuson-Stevens Fishery Management and Conservation Act, April 13, 1976 (90 Stat. 331), § 306.

48. Foreign fish processing permits, Alaska Administrative Code, 5 AAC 39.198.

49. Removal of herring from state, Alaska Statute 16.10.175.

50. Prepared Statement of Paul D. Kelly, Attorney for Western Alaska Cooperative Marketing Association and Bristol Bay Herring Marketing Cooperative, 464–466.

51. Alaska Department of Fish and Game, *Bristol Bay Area Annual Management Report, 1982*, 153–156; Alaska Department of Fish and Game, *Bristol Bay Area 1990 Annual Management Report*, H32; Takaharu Ohyama (Japanese North Pacific Longline Association), personal communication, June 8, 2020; Paul Kelly (Bristol Bay Herring Marketing Cooperative), personal communication, June 4, 2020; Henry Mitchell (Bering Sea Fishermen's Association), personal communication, June 5, 2020; "Fishermen Get Okay to Sell Direct to Japan," *Daily Sitka Sentinel*, March 31, 1981; "Judge Oks Selling Herring," *Daily Sitka Sentinel*, April 29, 1981; *Bristol Bay*

Herring Marketing Cooperative et al. v. Ronald Skoog et al., No. A81-043 Civ. (D.Alaska 1982); Skrade, personal communication, November 14, 2020.

52. Alaska Department of Fish and Game, *Bristol Bay Area Annual Management Report, 1987*, 249–278; Kelly, personal communication, June 4, 2020.

53. Alaska Department of Fish and Game, *Bristol Bay Area Annual Management Report, 1981*, 151–175; Alaska Department of Fish and Game, *Bristol Bay Area 1990 Annual Management Report*, H32.

54. Alaska Department of Fish and Game, *Bristol Bay Area Annual Management Report, 1982*, 181–188; Alaska Department of Fish and Game, *Bristol Bay Area 1990 Annual Management Report*, H32.

55. Charles Meachan, "Herring Fishery Gears Up for 1983," *Bristol Bay Fisherman*, April 1983.

56. Alaska Department of Fish and Game, *Bristol Bay Area Annual Management Report, 1983* (March 1984), 173–186; Alaska Department of Fish and Game, *Bristol Bay Area 1990 Annual Management Report*, H32, H38.

57. Alaska Department of Fish and Game, *Bristol Bay Area Annual Management Report, 1984*, 198–219; Alaska Department of Fish and Game, *Bristol Bay Area 1990 Annual Management Report*, H32, H38; Kris Freeman, "Roe Herring," *Pacific Fishing* (Yearbook 1985): 109–115; "Pilots Killed in Collision," *Daily Sitka Sentinel*, May 22, 1984.

58. Skrade, personal communication, April 27, 2020; "Artifact Looters Rob Graves and History," *Anchorage Daily News*, November 27, 1994.

59. "Village Wants to Rid Fishery of Liquor Trade," *Anchorage Daily News*, April 11, 1988.

60. Alaska Department of Fish and Game, *Bristol Bay Area Annual Management Report, 1985*, 142–161; Freeman, "Herring Roe," 59–61; Alaska Department of Fish and Game, *Bristol Bay Area 1990 Annual Management Report*, H32.

61. Alaska Department of Fish and Game, *Bristol Bay Area Annual Management Report, 1986*, 225–255; Alaska Department of Fish and Game, *Bristol Bay Area 1990 Annual Management Report*, H32.

62. Alaska Department of Fish and Game, *Bristol Bay Area Annual Management Report, 1987*, 249–278; Alaska Department of Fish and Game, *Bristol Bay Area 1990 Annual Management Report*, H32; "Early Herring Have Fishermen in a Frenzy," *Daily Sitka Sentinel*, April 28, 1987.

63. 5 AAC 27.831.

64. Alaska Department of Fish and Game, "Bristol Bay Herring Fishery," in *Bristol Bay Area 1995 Annual Management Report*, ADF&G Regional Information Rept. No. 2A96-06 (April 1996), 2; Nelson, personal communication, November 16, 2020.

65. Alaska Department of Fish and Game, *Bristol Bay Area Annual Management Report, 1988*, 291–324; "Harvest Starts Off Strong," *Anchorage Daily News*, May 19, 1988; "Seward Fisherman Makes a Huge Haul," *Daily Sitka Sentinel*, June 7, 1988.

66. "Herring Roe," *Pacific Fishing* (Yearbook 1989): 131.

67. Alaska Department of Fish and Game, *Bristol Bay Area 1989 Annual Management Report*, ADF&G Regional Information Rept. No. 2K90-03 (June 1990), 186–223.

68. "Spill Cleanup Robs Herring Fishery of Boats," *Anchorage Daily News*, May 9, 1989.

Notes

69. "Herring Roe," *Pacific Fishing* (Yearbook 1989): 131.

70. Alaska Department of Fish and Game, *Bristol Bay Area 1989 Annual Management Report*, 199.

71. Alaska Department of Fish and Game, *Bristol Bay Area 1989 Annual Management Report*, 186–223; "Herring Roe," *Pacific Fishing* (Yearbook 1990): 140.

72. Alaska Department of Fish and Game, *Bristol Bay Area 1990 Annual Management Report*, H1–H38.

73. "Boat Trash Pickup Leaves Beach Clean," *Daily Sitka Sentinel*, July 18, 1990.

74. Magnuson-Stevens Fishery Management and Conservation Act, as amended, § 3(2). Amended by the Sustainable Fisheries Act, October 11, 1996 (110 Stat. 3561).

75. Fritz Funk, Leslie Watson, and Richard Berning, *Revised Estimates of Bycatch of Herring in 1989 Bering Sea Trawl Fisheries*, ADF&G Regional Information Rept. No. 5590-01 (January 1990), Executive Summary; North Pacific Fishery Management Council, *Fishery Management Plan for the Groundfish Fishery of the Bering Sea/Aleutian Islands*, Amendment 16(a), September 1990; North Pacific Fishery Management Council, *Bering Sea/Aleutian Islands Groundfish Fishery Management Plan, Amendment Action Summaries* (May 2016), 27; 56 *Fed. Reg.* 32984–32990, July 18, 1991.

76. Alaska Department of Fish and Game, "Bristol Bay Herring Fishery," in *Bristol Bay Area 1991 Annual Management Report*, ADF&G Regional Information Rept. No. 2A92-08 (April 1992), 1–37.

77. "Herring Season a Slow One," *Anchorage Daily News*, April 30, 1993.

78. Alaska Department of Fish and Game, "Bristol Bay Herring Fishery," in *Bristol Bay Area 1992 Annual Management Report*, ADF&G Regional Information Rept. No. 2A93-32 (July 1993), 31, 33; Springer, "Togiak Sac Roe Herring."

79. Alaska Department of Fish and Game, "Bristol Bay Herring Fishery," in *Bristol Bay Area 1993 Annual Management Report*, ADF&G Regional Information Rept. No. 2A94-02 (February 1994), 3–10, 20; Alaska Department of Fish and Game, "Bristol Bay Herring Fishery," in *Bristol Bay Area 1997 Annual Management Report*, ADF&G Regional Information Rept. No. 2A98-08 (April 1998), 136; "Record Herring Return Could Spawn Trouble," *Anchorage Daily News*, February 23, 1993; "Togiak Herring Arrive, Processors Don't," *Anchorage Daily News*, April 26, 1993; "Herring Season a Slow One," *Anchorage Daily News*, April 30, 1993; Alaska Department of Fish and Game, "Bristol Bay Herring Fishery," in *Bristol Bay Area 1998 Annual Management Report*, ADF&G Regional Information Rept. No. 2A99-18 (March 1999), 134.

80. "Low Togiak Herring Fishery Numbers May Boost Prices," *Daily Sitka Sentinel*, May 12, 1993.

81. Alaska Department of Fish and Game, "Bristol Bay Herring Fishery," in *Bristol Bay Area 1994 Annual Management Report*, ADF&G Regional Information Rept. No. 2A95-11 (March 1995), 4–8, 18–24; "Fishery Closes," *Daily Sitka Sentinel*, May 20, 1994.

82. "Unalaska Processing Plant Making Surimi from Herring," *Daily Sitka Sentinel*, May 31, 1994.

83. "Kodiak Plant Using Herring to Make Surimi, Not Fishmeal," *Daily Sitka Sentinel*, April 29, 1997.

84. "Artifact Looters Rob Graves and History," *Anchorage Daily News*, November 27, 1994.

85. National Marine Fisheries Service, Alaska Region, *Final Summary* of Tribal Consultation Teleconference to Discuss *Bering Sea Herring Bycatch Management, August 14, 2020*, https://media.fisheries.noaa.gov/dam-migration/bering_sea_herring_bycatch_tribal_consultation_2020_final.pdf, accessed December 20, 2021.

86. Alaska Department of Fish and Game, "Bristol Bay Herring Fishery," in *Bristol Bay Area 1995 Annual Management Report*, ADF&G Regional Information Rept. No. 2A96-06 (April 1996), 6–14, 27–33; Tim Sands, *Overview of the Togiak Herring District Sac Roe and Spawn-on-Kelp Fisheries of Bristol Bay, Alaska, 2009; A Report to the Alaska Board of Fisheries*, Alaska Department of Fish and Game Special Pub. No. 09-15 (November 2009), 11; "Togiak Fishery Doing Well," *Daily Sitka Sentinel*, May 11, 1995; Alaska Department of Fish and Game, *Statewide Herring Wholesale Value and Pounds Processed*, accessed September 25, 2020, https://www.adfg.alaska.gov/index.cfm?adfg=fishlicense.coar_herringproduction.

87. Alaska Department of Fish and Game, "Bristol Bay Herring Fishery," in *Bristol Bay Area 1996 Annual Management Report*, ADF&G Regional Information Rept. No. 2A97-14 (May 1997), 144–148, 153–164; Sands, *Overview of the Togiak Herring District Sac Roe and Spawn-on-Kelp Fisheries* 11; Alaska Commercial Fisheries Entry Commission, *Changes in Roe Herring Markets: A Review of Available Evidence*, CFEC Rept. No. 05-5N (December 2005), 2; Alaska Department of Fish & Game, Statewide Herring Pounds and Estimated Ex-vessel Value, accessed March 25, 2022, https://www.adfg.alaska.gov/index.cfm adfg=commercialbyfisheryherring.herring_grossearnings.

88. Alaska Department of Fish and Game, "Bristol Bay Herring Fishery," in *Bristol Bay Area 1997 Annual Management Report*, 138–144, 154; "Herring Fishery Poised to Open," *Anchorage Daily News*, May 16, 1997; Sands, *Overview of the Togiak Herring District Sac Roe and Spawn-on-Kelp Fisheries* 11; "Togiak Herring Fishery Ends," *Daily Sitka Sentinel*, May 8, 1997.

89. Alaska Department of Fish and Game, "Bristol Bay Herring Fishery," in *Bristol Bay Area 1998 Annual Management Report*, ADF&G Regional Information Rept. No. 2A99-18 (March 1999), 125–155.

90. Alaska Department of Fish and Game, "Bristol Bay Herring Fishery," in *Bristol Bay Area 1999 Annual Management Report*, Regional Information Rept. No. 2A00-20 (March 2000), 5–10, 14–24.

91. Alaska Department of Fish and Game, "2000 Bristol Bay Herring Fishery," in *Bristol Bay Area 2000 Annual Management Report*, ADF&G Regional Information Rept. No. 2A01-10 (April 2001), 124–140.

92. Alaska Department of Fish and Game, "2001 Bristol Bay Herring Fishery," in *Bristol Bay Area 2001 Annual Management Report*, ADF&G Regional Information Rept. No. 2A02-18 (May 2002), 137–158; "3 Accused of Wasting Herring in Togiak Fishery," *Daily Sitka Sentinel*, January 24, 2003; Alaska Commercial Fisheries Entry Commission, *CFEC Permits and Estimates of Gross Earnings in the Togiak Sac Roe Herring Purse Seine and Gillnet Fisheries, 1983–2017*, CFEC Report No.18-6N (November 2018), 20.

93. "Skid Roe: Japanese Shell out Less for Alaska Herring Eggs," *Anchorage Daily News*, January 27, 2002.

94. Alaska Department of Fish and Game, "2002 Bristol Bay Herring Fishery," 124–145.

Notes

95. Alaska Department of Fish and Game, "2003 Bristol Bay Herring Fishery," in *Bristol Bay Area 2003 Annual Management Report*, Regional Information Rept. No. 2A04-16 (April 2004), 38–45, 126–131.

96. Alaska Department of Fish and Game, "2004 Bristol Bay Herring Fishery," in *Bristol Bay Area 2004 Annual Management Report*, Regional Information Rept. No. 2A05-41 (June 2005), 33.

97. Alaska Department of Fish and Game, "2004 Bristol Bay Herring Fishery," 35–39, 124–129.

98. Alaska Department of Fish and Game, "2005 Bristol Bay Herring Fishery," in *Bristol Bay Area 2005 Annual Management Report*, Regional Information Rept. No. 06-37 (June 2006), 33.

99. Alaska Department of Fish and Game, *Bristol Bay Area 2018 Annual Management Report*, Regional Information Rept. No. 19-2 (May 2019), 100; Alaska Department of Fish and Game, "2019 Togiak Herring Season Summary" (news release), June 6, 2019.

100. 77 *Fed. Reg.* 60649, October 4, 2012.

101. Isabelle Ross, "Togiak Herring Fishermen Can Tap a Huge Quota in 2020, but Some Are Still Staying Home," *KDLG Radio* (Dillingham, Alaska), December 13, 2019, https://www.kdlg.org/post/togiak-herring-fishermen-can-tap-huge-quota-2020-some-are-still-staying-home#stream/0; Isabelle Ross, "Togiak's Herring Buyer Says Floating Processor Will Have 'Zero Impact' on Community," *KDLG Radio* (Dillingham, Alaska), April 8, 2020, https://www.alaskapublic.org/2020/04/08/togiaks-herring-buyer-says-floating-processor-will-have-zero-impact-on-community/; "Togiak Herring Harvesters Complete Season," *Cordova Times*, June 24, 2020, https://www.thecordovatimes.com/2020/06/24/togiak-herring-harvesters-complete-season/; Nelson, personal communication, May 26, 2020; Tim Sands (ADF&G), personal communication, April 2, 2021; KMXT Radio, Kodiak, Alaska, "Alaska Fisheries Report," April 1, 2021, https://kmxt.org/2021/04/alaska-fisheries-report-april-1-2021/.

102. 85 *Fed. Reg.* 13568, March 9, 2020; North Pacific Fishery Management Council, Agenda B-2: National Marine Fisheries Service Management Report, June 7, 2020, https://meetings.npfmc.org/CommentReview/DownloadFile?p=313e81d5-db42-42e0-a736-e9758acd4f35.pdf&fileName=B2%20%20NMFS%20Report%20.pdf; NOAA Fisheries, *NMFS Prohibits Directed Fishing for Pollock in the Summer and Winter Herring Savings Areas of the Bering Sea and Aleutian Islands*, Information Bulletin No. 20-42, June 4, 2020, https://www.fisheries.noaa.gov/bulletin/ib-20-42-nmfs-prohibits-directed-fishing-pollock-summer-and-winter-herring-savings; 85 *Fed. Reg.* 35381, June 10, 1991; Diana Stram and Michael Fey (staff, North Pacific Fishery Management Council), *E1 Supplemental Information on BSAI Herring PSC and Distribution*, June 10, 2020, https://meetings.npfmc.org/CommentReview/DownloadFile?p=52003716-9e53-4e8e-861a-6829f4d17f39.pdf&fileName=E1%20Presentation%20to%20Council%20-%20BSAI%20Herring%20Supplemental.pdf.

103. 85 *Fed. Reg.* 36509, June 17, 1991.

104. Alaska Department of Fish and Game, "2021 Togiak Herring Outlook," Advisory Announcement, March 19, 2021.

105. Tim Sands (Alaska Department of Fish and Game), personal communication, May 13, 2021; Beaver Nelson (former herring seiner, Homer, Alaska), personal

communication, May 23, 2021; "ADF&G's Togiak Herring Forecast Is Robust," *Fishermen's News*, December 15, 2021, https://fishermensnews.com/adfgs-togiak-herring-forecast-is-robust/.

106. Alaska Department of Fish and Game, "2022 Togiak Herring Forecast," Advisory Announcement, December 8, 2021.

107. "Alaska Fisheries Report," KMXT Radio (Dillingham, Alaska), April 1, 2021, https://kmxt.org/2021/04/alaska-fisheries-report-april-1-2021/.

CHAPTER 14: NORTON SOUND ROE-HERRING FISHERIES

1. "Golovin Herring Season," *Pacific Fisherman* (March 1925): 46.

2. Mike Geiger, "Norton Sound Herring Catch Sampling Program, 1964," Alaska Department of Fish and Game, AYK Herring Rept. No. 1, in Alaska Department of Fish and Game, *AYK Area 1964 Annual Management Report*, 110–117; Western Alaska Enterprises to Alaska Department of Fish and Game, undated, Alaska Department of Fish and Game files, Nome, Alaska.

3. Leon A. Verhoeven, ex. director, Pacific Marine Fisheries Commission, to Hon. Robert W. Packwood, March 11, 1970, in *Fisheries Legislation, 1969–70, Hearings Before the Subcommittee on Energy, Natural Resources, and the Environment of the Committee on Commerce, on S. 1151, S. 1889, S. 2230, S. 2396, S. 2825, S 3102, S. 3176, S. 3492, H.R. 1049, and H.R. 4813*, US Senate, 91st Cong., 1st and 2d Sess., Serial 91-57, October 23, 1969 and March 12 and 13, 1970 (Washington, DC: GPO, 1970), 238; Mortimer L. Henry, draft "Observations and Notes of the Norton Sound Herring Fishery, 1969," Alaska Department of Fish and Game files, Nome; "Japanese Captains Found Guilty, Fined," *Daily Sitka Sentinel*, June 13, 1969; Alaska Department of Fish and Game, *Annual Management Report, 1981, Norton Sound–Port Clarence–Kotzebue*, 125.

4. Alaska Department of Fish and Game, *Annual Management Report, 1979, Norton Sound–Port Clarence–Kotzebue*, 135–139.

5. James Mackovjak, *Icicle Seafoods, Inc., A Concise Report* (November 1979), unpublished.

6. Alaska Department of Fish and Game, *2015 Annual Management Report Norton Sound, Port Clarence, and Arctic, Kotzebue Areas*, Fishery Management Rept. No. 17-15, 24; Alaska Department of Fish and Game, *2018 Annual Management Report Norton Sound, Port Clarence, and Arctic, Kotzebue Areas*, Fishery Management Rept. No. 20-05, 24–25.

7. Alaska Department of Fish and Game, *Annual Management Report, 1980, Norton Sound–Port Clarence–Kotzebue*, 113–119.

8. Alaska Department of Fish and Game, *Annual Management Report, 1981, Norton Sound–Port Clarence–Kotzebue*, 83; Alaska Department of Fish and Game, *Annual Management Report, 1982, Norton Sound–Port Clarence–Kotzebue*, 90.

9. Beaver Nelson (former herring seiner, Homer, Alaska), personal communication, May 29, 2020.

10. "In-state Processing of Herring Approved," *Daily Sitka Sentinel*, April 23, 1980.

11. Chapter 27, *Session Laws of Alaska, 1980*.

Notes

12. Chapter 9, *Session Laws of Alaska*, 1977; Chapter 27, *Session Laws of Alaska*, 1980; Chapter 14, *Session Laws of Alaska*, 1983; "News from Juneau," *Bering Sea Fisherman*, May 1980; "Cape Romanzof Herring," *Bering Sea Fisherman*, June 1981; "Sheffield Signs Bering Sea Herring Dumping Bill," *Bering Sea Fisherman*, May 1983.

13. Alaska Department of Fish and Game, *Annual Management Report, 1981, Norton Sound–Port Clarence–Kotzebue*, 83–84; "Western Alaska Herring Harvest," *Bering Sea Fisherman*, July 1988.

14. Alaska Department of Fish and Game, *Annual Management Report, 1984, Norton Sound–Port Clarence–Kotzebue*, 82–83; Charles Lean (former ADF&G management biologist), personal communication, August 6, 2020.

15. Charles F. Lean, Fredrick J. Bue, and Tracy L. Lingnau, *Annual Management Report, 1992, Norton Sound–Port Clarence–Kotzebue*, ADF&G Regional Information Rept. No. 3A93-15, 80–81

16. Charles F. Lean, Fredrick J. Bue, and Tracy L. Lingnau, *Annual Management Report, 1992, Norton Sound–Port Clarence–Kotzebue*, ADF&G Regional Information Rept. No. 3A93-15, 80–81; Thomas A. Morehouse (Institute of Social and Economic Research), *Native Participation in Alaska's Commercial Fisheries*, presented at Conference on Native Self-Reliance, Fisheries and Aquatic Resources Panel, August 20, 1984, https://pubs.iseralaska.org/media/f704c57e-a42b-4d3c-a06d-bef6f0e7faf9/1984_08_20-NativeCommercialFisheries.pdf; Alaska Administrative Code, 5 AAC 27.987; "Board Grants Exclusive Herring Areas," *Bering Sea Fisherman*, February 1983; "Exclusive-Use Herring Areas OK, Court Rules," *Daily Sitka Sentinel*, October 12, 1987; Alaska Department of Fish and Game, *Annual Management Report, 1984, Norton Sound–Port Clarence–Kotzebue*, 81–82; Alaska Department of Fish and Game, *Annual Management Report, 1985, Norton Sound–Port Clarence–Kotzebue*, 81–82; "Western Alaska Herring Limited Entry Studied," *Bering Sea Fisherman* (August 1987); "Herring Limited Entry for Western Alaska," *Bering Sea Fisherman* (November 1987).

17. Charles Lean, *The Development of the Norton Sound Commercial Herring Fishery, 1979–1988*, ADF&G Regional Information Rept. No. 3N89-04 (January 1989), 6.

18. Alaska Department of Fish and Game, *Annual Management Report, 1989, 1990, 1991, Norton–Sound Port Clarence–Kotzebue*, Regional Information Rept. No. 3A92-12; Alaska Department of Fish and Game, *Annual Management Report, 2002, Norton Sound–Port Clarence–Kotzebue*, Regional Information Rept. No. 3A03-30; Alaska Department of Fish and Game, *Annual Management Report, 2004, Norton Sound–Port Clarence–Kotzebue*, Regional Information Rept. No. 3A05-04; Alaska Department of Fish and Game, *2005 Annual Management Report, Norton Sound, Port Clarence, and Kotzebue*, Regional Information Rept. No. 07-32.

19. Alaska Department of Fish and Game, *Annual Management Report, 1989, 1990, 1991, Norton–Sound Port Clarence–Kotzebue*; "Herring Roe," *Pacific Fishing* (Yearbook 1989): 131.

20. Alaska Department of Fish and Game, *Annual Management Report, 1989, 1990, 1991, Norton–Sound Port Clarence–Kotzebue*.

21. Alaska Department of Fish and Game, *Annual Management Report, 1992, Norton–Sound Port Clarence–Kotzebue*.

22. "Ice Keeps Processors from Herring Grounds," *Daily Sitka Sentinel*, June 19, 1992.

23. "Norton Sound Declared Economic Disaster Area," *Daily Sitka Sentinel*, July 29, 1992.

24. Alaska Department of Fish and Game, *Annual Management Report, 1993, Norton–Sound Port Clarence–Kotzebue*, Regional Information Rept. No. 3A95-06.

25. Alaska Department of Fish and Game, *Annual Management Report, 1994, Norton Sound–Port Clarence–Kotzebue*, Regional Information Rept. No. 3A96-02.

26. Alaska Department of Fish and Game, *Annual Management Report, 1995, Norton Sound–Port Clarence–Kotzebue*, Regional Information Rept. No. 3A96-30.

27. Alaska Department of Fish and Game, *Annual Management Report, 1996, Norton Sound–Port Clarence–Kotzebue*, Regional Information Rept. No. 3A97-30; Alaska Commercial Fisheries Entry Commission, *Changes in Roe Herring Markets: A Review of Available Evidence*, CFEC Rept. No. 05-5N (December 2005), 2.

28. Alaska Department of Fish and Game, *Annual Management Report, 1997, Norton Sound–Port Clarence–Kotzebue*, Regional Information Rept. No. 3A98-28.

29. Alaska Department of Fish and Game, *Annual Management Report, 1998, Norton Sound–Port Clarence–Kotzebue*, Regional Information Rept. No. 3A99-32.

30. Alaska Department of Fish and Game, *Annual Management Report, 1999, Norton Sound–Port Clarence–Kotzebue*, Regional Information Rept. No. 3A01-05.

31. Alaska Department of Fish and Game, *Annual Management Report, 2000, Norton Sound–Port Clarence–Kotzebue*, Regional Information Rept. No. 3A02-02, 36.

32. Alaska Department of Fish and Game, *Annual Management Report, 2000, Norton Sound–Port Clarence–Kotzebue*.

33. Alaska Department of Fish and Game, *Annual Management Report, 2001, Norton Sound–Port Clarence–Kotzebue*, Regional Information Rept. No. 3A03-04.

34. Alaska Department of Fish and Game, *Annual Management Report, 2002, Norton Sound–Port Clarence–Kotzebue*, Regional Information Rept. No. 3A03-30.

35. Alaska Department of Fish and Game, *Annual Management Report, 2003, Norton Sound–Port Clarence–Kotzebue*, Regional Information Rept. No. 3A04-19.

36. Alaska Department of Fish and Game, *Annual Management Report, 2004, Norton Sound–Port Clarence–Kotzebue*, Regional Information Rept. No. 3A05-04.

37. Alaska Department of Fish and Game, *2005 Annual Management Report, Norton Sound, Port Clarence, and Kotzebue*, Regional Information Rept. No. 07-32.

38. Alaska Department of Fish and Game, *2018 Annual Management Report, Norton Sound, Port Clarence, and Arctic, Kotzebue Areas*, Regional Information Rept. No. 20-05, 150; Alaska Department of Fish and Game, "2019 Norton Sound Commercial Herring Bait Fishery Opens," news release, May 9, 2019; Alaska Department of Fish and Game, "2020 Norton Sound Commercial Herring Bait Fishery Opens," news release, May 18, 2020.

CHAPTER 15: FOOD HERRING IN THE MODERN ERA

1. "Herring Fillet Machines for PFI," *Petersburg Press*, December 21, 1972; Alaska Department of Fish and Game, *Annual Report, 1973*, 13; "Filleting Machine Developed for Herring and Pilchard," *Commercial Fisheries Review* (February 1963): 63–64.

2. Alaska Legislature, "Alaska Canned Chum & Herring Demonstration Project," TPS Rept. No. 54827v1 (2010).

3. Alaska Legislature, "Alaska Seafood Marketing Institute–Canned Salmon, Herring and Protein Powder Project," TPS Rept. No. 61026v1 (2013).

4. Alaska Seafood Marketing Institute, "Alaska Herring Development Project," June 2016, https://www.alaskaseafood.org/wp-content/uploads/2016/06/Alaska-Herring-Development-Project-Final.pdf; Bruce Schactler (Alaska Seafood Marketing Institute), personal communication, December 18, 2020.

5. McDowell Group, *Analyses of Specialty Alaska Seafood Products* (November 2017), 64.

6. Laine Welch, "Seafood Lovers Get Another Taste of Herring," *Anchorage Daily News*, June 17, 2016; "Seattle Spotlights Alaska Herring with Alaska Herring Week," *FSR*, June 12, 2017, https://www.fsrmagazine.com/content/seattle-spotlights-alaska-herring-alaska-herring-week.

7. Suzanna Caldwell, "Is Having a Special Week in Seattle Enough to Make Alaska Herring Cool Again?" *Anchorage Daily News*, July 3, 2017, https://www.adn.com/business-economy/2017/07/03/alaska-herring-just-had-its-own-week-in-seattle-is-it-enough-to-make-herring-cool-again/.

8. McDowell Group, *Analyses of Specialty Alaska Seafood Products* (November 2017), 65.

9. Alaska Herring Week, "Alaska Herring Week 2018 Hiatus," accessed December 18, 2020, https://nwherringweek.com/alaska-herring-week-2018-hiatus/.

10. Adwoa Gyimah-Brempongk, "Taking the Bait," *Edible Seattle*, accessed December 23, 2020, https://www.edibleseattle.com/explore/features/taking-the-bait/; Adwoa Gyimah-Brempongk, "Herring Gets No Respect. This Man Wants to Change That," *Northwest News Network*, May 3, 2017, https://www.nwnewsnetwork.org/post/herring-gets-no-respect-man-wants-change; Kara Elder, "Shopping Cart: Deckhand's Daughter Smoked Herring," *Washington Post*, June 14, 2017.

CHAPTER 16: GENESIS OF ALASKA'S HERRING SPAWN-ON-KELP FISHERY

1. Alaska Department of Fish and Game, *Prince William Sound Area Annual Finfish Management Report, 1982* (March 1983), 15.

2. "USDI Inspection of Herring-Eggs-on-Kelp," *Commercial Fisheries Review* (June 1966): 6; Dan Strickland, "Herring Roe By the Pound," *Pacific Fishing* (March 1990): 43–51; "Allocates Import Quota for Herring Roe on Kelp," *Commercial Fisheries Review* (June 1968): 74.

3. Alaska Department of Fish and Game, *2020 Southeast Alaska Herring Spawn-on-Kelp Pound Fishery Management Plan*, ADF&G Regional Information Rept. No. 1J20-03 (February 2020), 12.

4. Alaska Department of Fish and Game, *Annual Report for 1958*, 20.

5. Alaska Statehood Act, July 7, 1958 (72 Stat. 339).

6. Chapter 34, *Session Laws of Alaska, 1959*, March 11, 1959.

7. Alaska Department of Fish and Game, *Ketchikan Management Area Annual Report, 1969*, 30.

8. Alaska Department of Fish and Game, *Annual Report for 1959*, 112.

9. Anonymous Petersburg, Alaska, resident, personal communication, February 12, 2020.

10. Harry Yoshimura (Mutual Fish Company), personal communication, March 14, 2020.

11. Constitution of the State of Alaska, 1959, Article 8, §4.

12. Chapter 36, *Session Laws of Alaska*, 1960, March 16, 1960.

13. Alaska Department of Fish and Game, *Ketchikan Management Area Annual Report*, 1966, 30; "Kelp and Herring Spawn," *Commercial Fisheries Review* (October 1960): 14.

14. "Allocates Import Quota for Herring Roe on Kelp," 74; Alaska Department of Fish and Game, *1962 Catch and Production Statistics*, Statistical Leaflet No. 5.

15. Alaska Department of Fish and Game, *1963–64 Progress Report*, ADF&G Rept. No. 11, 70–71; Alaska Department of Fish and Game, *Ketchikan Management Area Annual Report*, 1965, 27; Alaska Department of Fish and Game, *Commercial Operators, 1963*, Statistical Leaflet No. 6; Dave Ohmer (Alaskan Glacier Seafoods), personal communication, March 11, 2020; "Craig Herring Harvest Nets 203,000 Lbs.," *Petersburg Press*, April 7, 1967.

16. "Herring-Roe-on-Kelp Harvest Is Held," *Commercial Fisheries Review* (June 1967): 9.

17. Alaska Department of Fish and Game, *1963–64 Progress Report*, 70–71; Alaska Department of Fish and Game, *Annual Report, 1966, Ketchikan Management Area*, 28–30.

18. Alaska Department of Fish and Game, *Sitka Area Commercial Fisheries Management Annual Report*, 1965, 25; Alaska Department of Fish and Game, *Sitka Area Commercial Fisheries Management Annual Report*, 1967, 23.

19. Alaska Department of Fish and Game, *Sitka Area Commercial Fisheries Management Annual Report*, 1965, 26.

20. Alaska Department of Fish and Game, *Sitka Area Commercial Fisheries Management Annual Report*, 1965, 25–26.

21. Alaska Department of Fish and Game, *Sitka Area Commercial Fisheries Management Annual Report*, 1965, 25.

22. Alaska Department of Fish and Game, *Annual Report, 1966, Ketchikan Management Area*, 28–30; "Herring-Eggs-on-Kelp Fishery at Craig," *Commercial Fisheries Review* (July 1966): 13.

23. Carl Vertrees, "Harvesters Charge Gross Mismanagement by F&G of Local Herring Spawn Fishery," *Daily Sitka Sentinel*, April 14, 1966; "50 Tons of Herring Spawn Harvested," *Daily Sitka Sentinel*, April 14, 1966; "Order New Restrictions in Herring Spawn Fisheries," *Daily Sitka Sentinel*, April 19, 1966; Ed Wyman, personal communication, March 13, 2020; "Hydaburg Herring Spawn Will Not Open," *Daily Sitka Sentinel*, May 1, 1967; Alaska Department of Fish and Game, *Alaska Catch and Production, Commercial Fisheries Statistics, 1966*, ADF&G Statistics Leaflet No. 13; Alaska Department of Fish and Game, *Sitka Area Commercial Fisheries Management Annual Report*, 1966, 23–25.

24. Ann Mathews, "About Town," *Daily Sitka Sentinel*, April 8, 1969; Wyman, personal communication, March 13, 2020.

25. "Craig Herring Harvest Nets 203,000 Lbs."

Notes

26. "Herring-Roe-on-Kelp Harvest Is Held," 9; Alaska Department of Fish and Game, *Sitka Area Commercial Fisheries Management Annual Report, 1967*, 23; NOAA/National Marine Fisheries Service, Alaska Region *Status Review of Southeast Alaska Herring (Clupea pallasi), Threats Evaluation and Extinction Risk Analysis* (March 2014), 40–41.

27. Alaska Department of Fish and Game, *Sitka Area Commercial Fisheries Management Annual Report, 1967*, 23–25.

28. "Set Quotas for Herring-Spawn-on-Kelp Fishery," *Commercial Fisheries Review* (May 1968): 20.

29. "No Eggs-on-Kelp Season to Be Held Here," *Daily Sitka Sentinel*, May 1, 1968; Alaska Department of Fish and Game, *1968 Alaska Catch and Production*, 18.

30. "Permits to Harvest Hair Kelp Granted," *Daily Sitka Sentinel*, April 10, 1968; "Hair Kelp Harvest to Be Ended," *Daily Sitka Sentinel*, April 12, 1968; Alaska Department of Fish and Game, *1968 Alaska Catch and Production*, 18.

31. Alaska Department of Fish and Game, *Cook Inlet Area Annual Report, 1967*, 31.

32. Alaska Department of Fish and Game, *Annual Management Report, Prince William Sound Area, 1970*, 88.

33. Alaska Department of Fish and Game, *Annual Management Report, Prince William Sound Area, 1974*, 67; Alaska Department of Fish and Game, *Annual Management Report, Prince William Sound Area, 1977*, 15–16; Alaska Department of Fish and Game, *Annual Report, Prince William Sound Area, 1978*, 73; Daniel Sharp, Steve Morstad, and John Wilcock, *Prince William Sound Management Area Herring Report to the Alaska Board of Fisheries*, ADF&G Regional Information Rept. No. 2A96-40 (December 1996), 16; D. B. Pleschner, "Alaska Roe-on-Kelp, Part I," *Pacific Fishing* (September 1986): 34–1.

34. Bill Davidson and Dave Gordon, *Hoonah Sound Herring Spawn-on-Kelp Pound Fishery, 2001 Management Plan*, ADF&G Regional Information Rept. No. 1J01-04 (January 2001), 5.

35. Alaska Department of Fish and Game, *Cook Inlet Annual Management Report, 1972*, 31.

36. Alaska Department of Fish and Game, *Prince William Sound Area Annual Finfish Management Report, 1982* (March 1983), 15.

37. D. B. Pleschner, "Alaska Roe-on-Kelp, Part II," *Pacific Fishing* (October 1986): 30–39.

38. Alaska Department of Fish and Game, *2007 Southeast Alaska Herring Spawn-On-Kelp Pound Fishery Management Plan*, ADF&G Regional Information Rept. No. 1J07-02 (March 2007), 6.

39. Alaska Department of Fish and Game, *Prince William Sound Area 1990 Annual Finfish Management Report*, Regional Information Rept. No. 2C91-14 (November 1991), 139.

40. Shannon Haugland, "Hoonah's Herring Fishery Hitting Snags," *Daily Sitka Sentinel*, April 20, 2006.

41. Ralph B. Pirtle (ADF&G), *[Prince William Sound Area] Annual Finfish Management Report, 1979* (August 1980), 56; Gene J. Sandone, Samuel Sharr, and James A. Brady, *Prince William Sound, Commercial Harvest of Pacific Herring 1984–1987*, ADF&G Fishery Research Bulletin No. 88-08 (December 1988), 3; D. B. Pleschner,

"Alaska Roe-on-Kelp, Part I," *Pacific Fishing* (September 1986): 34–41; D. B. Pleschner, "Alaska Roe-on-Kelp, Part II," *Pacific Fishing* (October 1986): 30–39.

42. Alaska Department of Fish and Game, *Prince William Sound Area Annual Finfish Management Report, 1980* (May 1981), 10–12; Alaska Department of Fish and Game, *Prince William Sound Area 1990 Annual Finfish Management Report*, Regional Information Rept. No. 2C91-14 (November 1991), 139–140.

CHAPTER 17: PRINCE WILLIAM SOUND HERRING SPAWN-ON-KELP FISHERIES, 1981–1999

1. Alaska Department of Fish and Game, *Prince William Sound Area Annual Finfish Management Report, 1981* (February 1982), 2–3, 14–16.

2. Alaska Department of Fish and Game, *Prince William Sound Area Annual Finfish Management Report, 1982*, 12–17, 64.

3. Alaska Department of Fish and Game, *Prince William Sound Area Annual Finfish Management Report, 1983* (March 1984), 12–15.

4. Alaska Department of Fish and Game, *Prince William Sound Area Annual Finfish Management Report, 1984* (April 1985), 20–22.

5. Alaska Department of Fish and Game, *Prince William Sound Area Annual Finfish Management Report, 1985* (June 1986), 26–27.

6. Alaska Department of Fish and Game, *Prince William Sound Area Annual Finfish Management Report, 1986*, 24–25, 86; Steve Morstad, Timothy T. Baker, and James A. Brady, *Prince William Sound Area 1990 Annual Finfish Management Report*, ADF&G Regional Information Rept. No. 2A92-02 (November 1991), Abstract; Alaska Department of Fish and Game, *Prince William Sound Area 1991 Annual Finfish Management Report*, Regional Information Rept. No. 2A92-09 (May 1992), 15–16; D. B. Pleschner, "Alaska Roe-on-Kelp, Part I," *Pacific Fishing* (September 1986): 34–1; D. B. Pleschner, "Alaska Roe-on-Kelp, Part II," *Pacific Fishing* (October 1986): 30–39; Wayne Donaldson, Steve Morstad, and John Wilcock, *Prince William Sound Management Area Herring Report to the Board of Fisheries*, ADF&G Regional Information Rept. No. 2A93-36 (December 1993), 5; Alaska Department of Fish and Game, *Finfish Fisheries, Southeast Alaska–Yakutat Region, 1991*, ADF&G Regional Information Rept. No. 1J93-10 (May 1993), 5.22; Ed Wyman (former Prince William Sound herring-pound group member), personal communication, March 13, 2020, and August 31, 2020.

7. Alaska Department of Fish and Game, *Prince William Sound Area Annual Finfish Management Report, 1987* (March 1988), 15–16.

8. Alaska Department of Fish and Game, *Prince William Sound Area Annual Finfish Management Report, 1988*, ADF&G Regional Information Rept. No. 2C90-02 (March 1990), 26–29.

9. Alaska Department of Fish and Game, *Prince William Sound Area 1990 Annual Finfish Management Report*, ADF&G Regional Information Rept. No. 2C91-14 (November 1991), 14–16.

10. Alaska Department of Fish and Game, *Prince William Sound Management Area 1991 Annual Finfish Management Report*, ADF&G Regional Information Rept. No. 2A92-09 (May 1992), 15–19; Alaska Department of Fish and Game, *Prince William*

Sound Management Area 1992 Annual Finfish Management Report, ADF&G Regional Information Rept. No. 2A93-12 (June 1993), H7.

11. Alaska Department of Fish and Game, *Prince William Sound Management Area 1992 Annual Finfish Management Report*, 19–22.

12. Alaska Department of Fish and Game, *Prince William Sound Management Area 1992 Annual Finfish Management Report*, 23–25, H7.

13. Alaska Department of Fish and Game, *Prince William Sound Management Area 1996 Annual Finfish Management Report*, ADF&G Regional Information Rept. No. 2A95-47 (December 1995), 19.

14. Alaska Department of Fish and Game, *Prince William Sound Management Area 1995 Annual Finfish Management Report*, ADF&G Regional Information Rept. No. 2A96-25 (June 1996), 14.

15. Alaska Department of Fish and Game, *Prince William Sound Management Area 1996 Annual Finfish Management Report*, 21.

16. Alaska Department of Fish and Game, *Prince William Sound Management Area 1997 Annual Finfish Management Report*, ADF&G Regional Information Rept. No. 2A98-05 (May 1998), 29–30, 154.

17. Alaska Department of Fish and Game, *Prince William Sound Management Area 1998 Annual Finfish Management Report*, ADF&G Regional Information Rept. No. 2A99-20 (May 1999), 31–33, 151–152.

18. Alaska Department of Fish and Game, *Prince William Sound Area 1999 Annual Finfish Management Report*, Regional Information Rept. No. 2A00-32 (November 2000), 27–36.

CHAPTER 18: SOUTHEAST ALASKA HERRING SPAWN-ON-KELP POUND FISHERIES

1. Will Swagel, "Herring Fishery OK's for Peril Straits Area," *Daily Sitka Sentinel*, February 7, 1989; Alaska Department of Fish and Game, *Hoonah Sound Herring Spawn-on-Kelp Pound Fishery, 1990 Management Plan*, ADF&G Regional Information Rept. No. 1J89-20 (November 1989), 3; Bill Davidson and Dave Gordon, *Hoonah Sound Herring Spawn-on-Kelp Pound Fishery 2000 Management Plan*, ADF&G Regional Information Rept. No. 1J00-06 (January 2000), 6; Alaska Department of Fish and Game, "Memorandum: Hoonah Sound Spawn-on-Kelp Pound Fishery, 1994 Post-Season Evaluation and Re-Cap," May 17, 1994.

2. Alaska Department of Fish and Game, *Hoonah Sound Herring Spawn-on-Kelp Pound Fishery 1990 Management Plan*, 5; Alaska Department of Fish and Game, *Hoonah Sound Herring Spawn-on-Kelp Pound Fishery 1993 Management Plan*, Regional Information Rept. No. 1J 92-16 (December 1992), 11; Alaska Department of Fish and Game, *Craig/Klawock Herring Spawn-on-Kelp Pound Fishery Management Plan, 1992*, ADF&G Regional Information Rept. No. 1592-06 (January 1992), 3, 7.

3. Alaska Department of Fish and Game, *Hoonah Sound Herring Spawn-on-Kelp Fishery 1991 Management Plan*, ADF&G Regional Information Rept. No. 1 J90-36 (December 1990), 2; Will Swagel, "Herring Pound Fishery Set in Hoonah Sound," *Daily Sitka Sentinel*, March 23, 1990; Will Swagel, "Herring Roe Pound Fishery at

Hoonah Sound Going Well," *Daily Sitka Sentinel*, April 24, 1992; Greg Streveler, personal communication, October 31, 2020.

4. "Fish & Game Assesses Herring Pound Fishery," *Daily Sitka Sentinel*, May 4, 1990.

5. Alaska Department of Fish and Game, *Finfish Fisheries, Southeast Alaska–Yakutat Region, 1991*, ADF&G Regional Information Rept. No. 1J93-10 (May 1993), 5, 10; Bill Davidson and Dave Gordon, *Hoonah Sound Herring Spawn-on-Kelp Pound Fishery, 2002 Management Plan*, ADF&G Regional Information Rept. No. 1J02-04 (January 2002) 14; Alaska Department of Fish and Game, *2007 Southeast Alaska Herring Spawn-on-Kelp Pound Fishery Management Plan*, ADF&G Regional Information Rept. No. 1J07-02 (March 2007), 23.

6. Alaska Department of Fish and Game, *Craig/Klawock Herring Spawn-on-Kelp Pound Fishery Management Plan, 1992*, ADF&G Regional Information Rept. No. 1592-06 (January 1992), 6; Davidson and Gordon, *Hoonah Sound Herring Spawn-on-Kelp Pound Fishery 2000 Management Plan*, 8.

7. Alaska Department of Fish and Game, Memorandum: "Hoonah Sound Spawn-on-Kelp Pound Fishery, 1994 Post-Season Evaluation and Re-Cap, May 17, 1994."

8. Alaska Department of Fish and Game, Memorandum: "Hoonah Sound Spawn-on-Kelp Pound Fishery, 1995 Post-Season Evaluation and Re-Cap," May 25, 1995; Alaska Department of Fish and Game, Memorandum: "1997 Hoonah Sound Spawn-on-Kelp Pound Fishery Summary," October 6, 1997; Shannon Haugland, "Open Pounds OK'd in Hoonah Fishery," *Daily Sitka Sentinel*, February 4, 1997.

9. Alaska Department of Fish and Game, *Craig/Klawock Herring Spawn-on-Kelp Fishery, 1996 Management Plan*, ADF&G Regional Information Rept. No. 1J95-27 (December 1995), 6; Alaska Administrative Code, 20 AAC 05.320(f).

10. Alaska Department of Fish and Game, "Hoonah Sound Herring Pound Fishery," news release, October 16, 1995; Davidson and Gordon, *Hoonah Sound Herring Spawn-on-Kelp Pound Fishery 2000 Management Plan*, 6, 13; Bill Davidson and Dave Gordon, *Hoonah Sound Herring Spawn-on-Kelp Pound Fishery, 2001 Management Plan*, Regional Information Rept. No. 1J01-04 (January 2001), 5, 13.

11. Alaska Department of Fish and Game, *2007 Southeast Alaska Herring Spawn-on-Kelp Pound Fishery Management Plan*, ADF&G Regional Information Rept. No. 1J07-02 (March 2007), 23; Kyle Hebert, *Southeast Alaska 2013 Herring Stock Assessment Surveys*, ADF&G Fishery Data Series No. 14-13 (February 2014), 15; Alaska Department of Fish and Game, *2015 Southeast Alaska Herring Spawn-on-Kelp Pound Fishery Management Plan*, ADF&G Regional Information Rept. No. 1J15-01 (March 2015), 27; Alaska Department of Fish and Game, *2013 Southeast Alaska Herring Spawn-on-Kelp Pound Fishery Management Plan*, ADF&G Regional Information Rept. No. 1J13-01 (March 2013), 5; Alaska Department of Fish and Game, *2019 Southeast Alaska Herring Spawn-on-Kelp Pound Fishery Management Plan*, Regional Information Rept. No. 1J19-02 (March 2019), 4, 22; Alaska Department of Fish and Game, *2020 Southeast Alaska Herring Spawn-on-Kelp Pound Fishery Management Plan*, ADF&G Regional Information Rept. No. 1J20-03 (February 2020), 4; Alaska Department of Fish and Game, *2021 Southeast Alaska Herring Summary*, Advisory Announcement, May 28, 2021.

Notes

12. Delores Garza (Klawock Cooperative Association), *Department of Interior and Related Agencies Appropriations for 1992, Hearing Before a Subcommittee of the Committee on Appropriations*, 102nd Cong., 1st sess., March 12, 1991 (Washington, DC: GPO, 1991), 47–50.

13. Paul R. Larson, *Legislative Report, Prince of Wales Island Proposal for Developing Commercial Utilization of Local Herring Roe-on-Kelp Resources*, ADF&G Regional Information Rept. No. 1J88-35 (November 1988).

14. Klawock Heenya Corporation, "Alaska Board of Fisheries 1990/1991 Finfish Proposals Cook Inlet/Prince William Sound/Southeast (including Shellfish)" (January 1991), proposal no. 151; "Board Decides to Allow Roe Fishery," *Daily Sitka Sentinel*, January 28, 1991.

15. Alaska Department of Fish and Game, *Craig/Klawock Herring Spawn-on-Kelp Fishery 1993 Fishery Management Plan*, ADF&G Regional Information Rept. No. 1J92-17 (December 1992), 2, 13; Brian Kandoll, Craig/Klawock pound operator, personal communication, October 19, 2020; Fritz Funk, ed., *Preliminary Forecasts of Catch and Stock Abundance for 1993 Alaska Herring Fisheries*, ADF&G Regional Information Rept. No. 5J93-06 (May 1993), 10–11.

16. Alaska Department of Fish and Game, *Craig/Klawock Herring Spawn-on-Kelp Pound Fishery 1994 Management Plan*, ADF&G Regional Information Rept. No. 1J94-05 (January 1994), 14; Joel Gay, "Alaska Herring Roundup," *Pacific Fishing* (July 1993): 20, 65.

17. Alaska Department of Fish and Game, *Craig/Klawock Herring Spawn-on-Kelp Pound Fishery 1996 Management Plan*, ADF&G Regional Information Rept. No. 1J95-27 (December 1995), 16.

18. Alaska Department of Fish and Game, *Craig/Klawock Herring Spawn-on-Kelp Fishery, 1996 Management Plan*, 6; Alaska Administrative Code, 20 AAC 05.320(f).

19. Alaska Department of Fish and Game, *Craig/Klawock Herring Spawn-on-Kelp Pound Fishery 2000 Management Plan*, ADF&G Regional Information Rept. No.1 1J00-02 (January 2000), 5, 15–16; Klawock Heenya Corporation, "Alaska Board of Fisheries 1990/1991 Finfish Proposals."

20. Alaska Department of Fish and Game, *Craig/Klawock Herring Spawn-on-Kelp Pound Fishery 2000 Management Plan*, 5; Alaska Department of Fish and Game, *Southern Southeast Herring Spawn-on-Kelp Pound Fishery, 2004 Management Plan*, Regional Information Rept. No. 1J04-05 (February 2004), 16; Alaska Department of Fish and Game, *2020 Southeast Alaska Herring Spawn-on-Kelp Pound Fishery Management Plan*, ADF&G Regional Information Rept. No. 1J20-03 (February 2020), 12.

21. Alaska Department of Fish and Game, *Craig/Klawock Herring Spawn-on-Kelp Pound Fishery, 2001 Management Plan*, ADF&G Regional Information Rept. No. 1J01-03 (January 2001), 11–12; Bo Meredith (ADF&G), personal communication, September 18, 2020.

22. Alaska Department of Fish and Game, *Craig/Klawock Herring Spawn-on-Kelp Pound Fishery, 2003 Management Plan*, ADF&G Regional Information Rept. No. 1J03-17 (March 2003), 12.

23. Alaska Department of Fish and Game, *2013 Southeast Alaska Herring Spawn-on-Kelp Pound Fishery Management Plan*, ADF&G Regional Information Rept. No. 1J13-01 (March 2013), 7; Alaska Department of Fish and Game, *2020 Southeast Alaska Herring Spawn-on-Kelp Pound Fishery Management Plan*, 12; Tom Swanson

(permit holder, Craig/Klawock herring spawn-on-kelp pound fishery), personal communication, September 21, 2020.

24. Alaska Department of Fish and Game, *2017 Southeast Alaska Herring Spawn-on-Kelp Pound Fishery Management Plan*, ADF&G Regional Information Rept. No. 1J17-01 (March 2017), 7, 15.

25. Angela Denning, "Fishermen Forced to Share Pounds in Herring Fishery," *KFSK Radio* (Petersburg, Alaska), February 1, 2017, https://www.kfsk.org/2017/02/01/fishermen-forced-to-share-pounds-in-herring-fishery/.

26. Alaska Department of Fish and Game, *2020 Southeast Alaska Herring Spawn-on-Kelp Pound Fishery Management Plan*, 12; Angela Denning, "Regulations Liberalized in the Craig Spawn-on-Kelp Fishery," *KFSK Radio* (Petersburg, Alaska), April 3, 2017, https://www.ktoo.org/2017/04/03/regulations-liberalized-craig-spawn-kelp-fishery/; Meredith, personal communication, September 22, 2020.

27. Alaska Department of Fish and Game, *Craig/Klawock Herring Spawn-on-Kelp Pound Fishery, 2003 Management Plan*, 12; Alaska Department of Fish and Game, *2020 Southeast Alaska Herring Spawn-on-Kelp Pound Fishery Management Plan*, 12; Riley Woodford, "Record Herring Event Highlights Roe on Kelp Fishery," *Alaska Fish & Wildlife News* (May 2020), http://www.adfg.alaska.gov/index.cfm?adfg=wildlifenews.view_article&articles_id=953.

28. Alaska Department of Fish and Game, *2007 Southeast Alaska Herring Spawn-on-Kelp Pound Fishery Management Plan*, 3–4; Alaska Department of Fish and Game, *2020 Southeast Alaska Herring Spawn-on-Kelp Pound Fishery Management Plan*, 3–4, 19.

29. Alaska Department of Fish and Game, *2007 Southeast Alaska Herring Spawn-on-Kelp Pound Fishery Management Plan*, 3; Alaska Department of Fish and Game, *2020 Southeast Alaska Herring Spawn-on-Kelp Pound Fishery Management Plan*, 3, 17.

CHAPTER 19: TOGIAK AND NORTON SOUND HERRING SPAWN-ON-KELP FISHERIES

1. Alaska Department of Fish and Game, *Bristol Bay Area Annual Management Report, 1968* (May 1969), 29–30; Alaska Department of Fish and Game, *Annual Management Report—1980–Bristol Bay Area* (September 1981), 159.

2. Alaska Department of Fish and Game, *Annual Management Report—1981–Bristol Bay Area* (April 1982), 173; Alaska Department of Fish and Game, *Bristol Bay Area Annual Management Report, 1970* (October 1971), 20.

3. Alaska Department of Fish and Game, *Annual Management Report—1974–Bristol Bay Area* (February 1976), 18.

4. Jeffrey Skrade (former ADF&G fisheries biologist), personal communication, November 23, 2020; Alaska Department of Fish and Game, *Annual Management Report—1978–Bristol Bay Area* (December 1981), 35–36.

5. Alaska Department of Fish and Game, *Annual Management Report—1979–Bristol Bay Area* (February 1982), 56.

6. Alaska Department of Fish and Game, *Annual Management Report—1980–Bristol Bay Area*, 39–40.

Notes

7. Alaska Department of Fish and Game, *Annual Management Report–1981–Bristol Bay Area*, 156–157.

8. Alaska Department of Fish and Game, *Annual Management Report–1982–Bristol Bay Area* (April 1983), 189–191, 208.

9. Alaska Department of Fish and Game, *Annual Management Report–1983–Bristol Bay Area* (March 1984), 181–182.

10. Alaska Department of Fish and Game, *Annual Management Report–1984–Bristol Bay Area* (April 1985), 203–204, 222, 225.

11. Alaska Department of Fish and Game, *Annual Management Report–1985–Bristol Bay Area* (May 1986), 148–151.

12. Alaska Department of Fish and Game, *Annual Management Report–1986–Bristol Bay Area* (September 1987), 233–236, 255.

13. Alaska Department of Fish and Game, *Annual Management Report–1987–Bristol Bay Area* (February 1988), 258–259, 269, 275, 278; UPI, "Gunfight leaves 3 Alaskans Dead," May 4, 1987, https://www.upi.com/Archives/1987/05/04/Gunfight-leaves-3-Alaskans-dead/3173547099200/.

14. Alaska Department of Fish and Game, *Annual Management Report–1988–Bristol Bay* ADF&G Regional Information Rept. No. 4089-09 (September 1989), 303–324.

15. Alaska Department of Fish and Game, *Annual Management Report–1989–Bristol Bay* ADF&G Regional Information Rept. No. 2K90-03 (June 1990), 200–201, 211, 220.

16. Alaska Department of Fish and Game, *Annual Management Report–1990–Bristol Bay Area*, ADF&G Regional Information Rept. No. 91-1 (April 1991), H2, H16–H17, H35; Alaska Commercial Fisheries Entry Commission, 1990 and 1996 Permit Status, accessed October 3, 2020, https://www.cfec.state.ak.us/pstatus/14051990.htm and https://www.cfec.state.ak.us/pstatus/14051996.htm; Alaska Department of Fish and Game, *Overview of the Togiak Sac Roe and Spawn-on-Kelp Fisheries of Bristol Bay, Alaska; A Report to the Alaska Board of Fisheries*, ADF&G Special Publication No. 06-25 (November 2006), 8.

17. Alaska Department of Fish and Game, "Bristol Bay Herring Fishery," in *Annual Management Report–1991–Bristol Bay Area*, ADF&G Regional Information Rept. No. 2A92-08 (April 1992), 14–15; Alaska Department of Fish and Game, *Overview of the Togiak Sac Roe and Spawn-on-Kelp Fisheries of Bristol Bay*, 8.

18. Alaska Department of Fish and Game, "Bristol Bay Herring Fishery," in *Annual Management Report–1992–Bristol Bay Area*, ADF&G Regional Information Rept. 2A93-32 (July 1993), 21–23, 36.

19. Alaska Department of Fish and Game, "Bristol Bay Herring Fishery," in *Annual Management Report–1993–Bristol Bay Area*, ADF&G Regional Information Rept. No. 2A94-02 (February 1994), 7–9.

20. Alaska Department of Fish and Game, "Bristol Bay Herring Fishery," in *Annual Management Report–1994–Bristol Bay Area*, ADF&G Regional Information Rept. No. 2A95-11 (March 1995), 7–8, 21, 24.

21. Alaska Department of Fish and Game, "Bristol Bay Herring Fishery," in *Annual Management Report–1995–Bristol Bay Area*, ADF&G Regional Information Rept. No. 2A96-06 (April 1996), 15–16.

22. Alaska Department of Fish and Game, "Bristol Bay Herring Fishery," in *Annual Management Report–1996–Bristol Bay Area*, ADF&G Regional Information Rept. No. 2A97-14 (May 1997), 148–149, 161, 164.

23. Alaska Department of Fish and Game, "Bristol Bay Herring Fishery," in *Annual Management Report–1997–Bristol Bay Area*, ADF&G Regional Information Rept. No. 2A98-08 (April 1998), 141–142.

24. Alaska Department of Fish and Game, "Bristol Bay Herring Fishery," in *Annual Management Report–1998–Bristol Bay Area*, ADF&G Regional Information Rept. No. 2A99-18 (March 1999), 139–140.

25. Alaska Department of Fish and Game, "Bristol Bay Herring Fishery," in *Annual Management Report–1999–Bristol Bay Area*, ADF&G Regional Information Rept. No. 2A00-20 (March 2000), 5–23.

26. Alaska Department of Fish and Game, *Annual Management Report–2000–Bristol Bay Area*, ADF&G Regional Information Rept. No. 2A01-10 (April 2001), 128; Alaska Department of Fish and Game, *Annual Management Report–2001–Bristol Bay Area*, ADF&G Regional Information Rept. No. 2A02-18 (May 2002), 143.

27. Alaska Department of Fish and Game, *Annual Management Report–2002–Bristol Bay Area*, ADF&G Regional Information Rept. No. 2A03-18 (April 2003), 131.

28. Susan E. Merkouris and Charles Lean, *Annual Management Report, 1987, Norton–Sound Port Clarence–Kotzebue*, ADF&G Regional Information Rept. No. 3N88-27 (August 1988), 85; Charles Lean, *The Development of the Norton Sound Commercial Herring Fishery, 1979–1988*, ADF&G Regional Information Rept. No. 3N89-04 (January 1989), 2; Charles F. Lean, Fredrick J. Bue, and Tracy L. Lingnau, *Annual Management Report, 1989, 1990, 1991, Norton–Sound Port Clarence–Kotzebue*, ADF&G Regional Information Rept. No. 3A92-12 (May 1982), 133.

29. Alaska Department of Fish and Game, *Annual Management Report, 1998, Norton Sound–Port Clarence–Kotzebue*, Regional Information Rept. No. 3A99-3233 (September 1999), 35–36.

30. Alaska Department of Fish and Game, *Annual Management Report, 2000, Norton Sound–Port Clarence–Kotzebue*, ADF&G Regional Information Rept. No. 3A02-02 (January 2002), 36; Alaska Department of Fish and Game, *Annual Management Report, 1999, Norton Sound–Port Clarence–Kotzebue*, ADF&G Regional Information Rept. No. 3A01-05 (March 2001), 82.

31. Alaska Department of Fish and Game, *2013 Annual Management Report, Norton Sound–Port Clarence Area and Arctic–Kotzebue Area*, ADF&G Fishery Management Report No. 15-09 (March 2015), 22, 137.

EPILOGUE

1. Yogi Berra, *The Yogi Book* (New York: Workman Publishing Company, 2010), 154.

2. Laine Welch, "Herring Ho-Hum? Alaska Season Opens with Limited Interest," *National Fisherman*, March 29, 2021, https://www.nationalfisherman.com/alaska/herring-ho-hum-alaska-season-opens-with-limited-interest.

3. Alaska Department of Fish and Game, "Sitka Sound Herring Fishery Announcement," Advisory Announcement, April 30, 2021; Tim Sands (ADF&G), personal communication, May 13, 2021.

4. NOAA Fisheries, *NMFS Prohibits Directed Fishing for Pollock in the Summer and Winter Herring Savings Areas of the Bering Sea and Aleutian Islands*, Information Bulletin No. 20-42, June 4, 2020, https://www.fisheries.noaa.gov/bulletin/ib-20-42-nmfs-prohibits-directed-fishing-pollock-summer-and-winter-herring-savings; Diana Stram and Michael Fey (staff, North Pacific Fishery Management Council), *E1 Supplemental Information on BSAI Herring PSC and Distribution*, June 10, 2020, https://meetings.npfmc.org/CommentReview/DownloadFile?p=52003716-9e53-4e8e-861a-6829f4d17f39.pdf&fileName=E1%20Presentation%20to%20Council%20-%20BSAI%20Herring%20Supplemental.pdf.

5. Alaska Department of Fish and Game, "Sitka Sound Herring Fishery Announcement," Advisory Announcement, January 11, 2021; Alaska Department of Fish and Game, "Sitka Sound Herring Fishery Announcement," Advisory Announcement, April 30, 2021.

6. Alaska Department of Fish and Game, "2021 Togiak Herring Forecast," Advisory Announcement, December 14, 2020.

7. Sands (ADF&G), personal communication, May 24, 2021.

8. Geoff Spalinger (ADF&G), *Kodiak Management Area Herring Sac Roe Fishery Harvest Strategy for the 2021 Season*, Regional Information Rept. No. 4K21-02 (March 2021), 2.

9. Jeremy Botz (ADF&G), personal communication, May 13, 2021.

10. Marine Stewardship Council, "AS [Atlanto-Scandian] herring quota-sharing," news release, October 5, 2020, https://www.msc.org/media-centre/press-releases/as-herring-quota-sharing.

11. "H.R. 2236—116th Congress (2019–2020): Forage Fish Conservation Act," *Congress.Gov*, accessed May 23, 2021, https://www.congress.gov/search?q={%22source%22:%22legislation%22,%22search%22:%22\%22forage%20fish%20conservation%20act\%22%22}&searchResultViewType=expanded; "S.5053—116th Congress (2019–2020) Forage Fish Conservation Act," *Congress.Gov*, accessed May 23, 2021, https://www.congress.gov/search?q={%22source%22:%22legislation%22,%22search%22:%22\%22forage%20fish%20conservation%20act\%22%22}&searchResultViewType=expanded; "Blunt, Blumenthal Introduce Measure to Protect Forage Fish," Senator Roy Blunt press release, April 29, 2021, https://www.blunt.senate.gov/news/press-releases/blunt-blumenthal-introduce-measure-to-protect-forage-fish.

12. Office of Representative Jared Huffman, "Huffman, Case Introduce the Sustaining America's Fisheries for the Future Act, Legislation to Update Federal Fisheries Management," July 26, 2021, https://huffman.house.gov/media-center/press-releases/huffman-case-introduce-the-sustaining-americas-fisheries-for-the-future-act-legislation-to-update-federal-fisheries-management.

13. "The WINNAH!—Max Jacobs," *Pacific Fisherman* (January 1948): 71. Image of Pacific herring (figure 20.1) is from the *Report of the Commissioner of Fish and Fisheries for the Fiscal Year Ending June 30, 1901*, 57th Cong. 1st sess., Ho. Doc. No. 705 (Washington, DC: GPO, 1902), 534.

SUGGESTED READINGS

Franklin, H. Bruce. *The Most Important Fish in the Sea: Menhaden and America.* Seattle: Island Press, 2007.

Funk, Fritz, James Blackburn, Douglas Hay, A.J. Paul, Robert Stephenson, Reidar Toresen, and David Witherell, eds. *Proceedings of the Symposium, Herring 2000: Expectations for a New Millennium, Anchorage, Alaska, February 23–26, 2000.* University of Alaska Sea Grant, 2001.

Melteff, Brenda R., and Vidar G. Wespestad, compilers and eds. *Proceedings of the Alaska Herring Symposium, Anchorage, Alaska, February 19–21, 1980.* University of Alaska Sea Grant, 1980.

Reid, Gerald M. *Fishery Facts-2, Alaska's Fishery Resources—The Pacific Herring.* National Marine Fisheries Service Extension Publication (June 1972).

Thornton, Thomas F., and Madonna L. Moss. *Herring and People of the North Pacific: Sustaining a Keystone Species.* Seattle: University of Washington Press, 2021.

Wespestad, Vidar, Jeremy Collie, and Elizabeth Collie, eds. *Proceedings of the International Herring Symposium, Anchorage, Alaska, USA, October 23–25, 1990.* University of Alaska Sea Grant, 1991.

INDEX

abundance, herring, 13, 41, 68, 98, 105, 125, 127, 151, 155, 158, 199, 201, 211, 241, 285, 289, 304
Act for the Protection and Regulation of the Fisheries of Alaska (1906), 30, 32
Act for the Protection of the Fisheries of Alaska (1924). *See* White Act
Afognak Island, 60, 67, 75, 77, 80, 213
age, herring, 9–10, 145, 153, 207, 209, 219
Alaska: curers, 58–59, 91; Department of Commerce, Community, and Economic Development, 267; Department of Environmental Conservation, 232; Department of Fisheries, 27, 44, 50, 112–13, 119, 121, 124; fish species in, 3; gaining statehood, 122–23; Herring Bays in, 3; herring industry of, 4–5; herring oil, 25, 90; herring production, 57, 210, 239; Herring Week, 268–69; officials, 125–26; packers, 57, 59, 61, 86, 91, 99; roe-herring fisheries, 146, 150, 155, 218, 234; waters, 19, 40, 42, 120, 125, 133
Alaska Board of Fish & Game, 277
Alaska Commercial Fisheries Entry Commission, 137, 149, 151, 194, 196, 202, 212, 214, 229, 259, 298, 303
Alaska Department of Fish and Game (ADF&G), 12, 17, 42, 121, 124, 126, 133, 144, 148
Alaska Fishermen's Cooperative, 112, 116
Alaska Fish Salting and By-Products Company, 33
Alaska Global Food Aid Program, 266
Alaska Herring & Sardine Company, 47–50
Alaska Herring Corporation, 231, 235
Alaska Herring Development Project, 268
Alaska Herring Marketing Association, 86
Alaska Herring Packers Association, 78
Alaska Legislature, 79, 256–57; accepting reduction plants, 33; and Canned Salmon, Herring and Protein Powder Project, 267; on cost of fish processing plants, 112; creating Department of Fisheries, 112–13; dividing Board of Fish and Game, 147–48; on herring quota, 118; in-state herring processing bill, 256; levy taxes, 32; maintaining "liberal attitude," 123; memorializing Congress, 92; and nobbing machine, 267; outlawing roe stripping, 20, 146; passing in-state herring processing bill, 230; passing Limited Entry Act, 149; passing memorial, 117; passing 1962 resolution, 124; replacing boards/commissions, 121; using taxation authority, 32–33; on wasting commercially taken herring, 196
Alaska Native Claims Settlement Act (1971), 301
Alaskan Glacier Seafoods, 276
Alaska Oil and Guano Company, 16, 23–26, 31–33, 43, 54, 82
Alaska-Pacific Herring Company, 48–50, 52
Alaska Pulp Company, 160, 277
Alaska Seafood Marketing Institute, 266–69, 267
Alaska Statehood Act, 147, 275
Albatross (vessel), 14, 25
Anchorage Daily News, 268
anchovetas (*Engraulis ringens*), 123
Ancon (vessel), 22
Anderson, Clarence, 50, 67, 113, 124
Anderson, Dean, 236
Anderson, Ray, 176, 181, 183, 226. *See also* Seward Marine Services
Andrews, Roy Chapman, 41
Angoon (Kootznoowoo), 21–22
Apex Fish Company, 102, 104
Arctic Whaling & Fishing Company, 63
Arden Salt Company, 56
Arensten & Company, 27–28, 98
Atlantic herring (*Clupea harengus*), 9

391

bait herring, 18–20, 45, 47, 58, 73, 77, 96, 104, 128, 130; Dutch Harbor, 137–39; Kodiak, 136–37; Prince William Sound, 135–36; Southeast Alaska, 130–35; Upper Cook Inlet, 139
Baranof Packing Company, 78
Barnes, Katherine, 95
barrels, 21, 23, 43–44, 54–55
beach seines, 14–15, 18, 62, 144, 254, 256, 258–59, 263–64
Bell, Frank, 96
Bersch, Peter, 181–82
Bianca (vessel), 75
Big Port Walter, 20, 28, 33–34, 48, 50, 52, 96, 98, 121, 127–28
biomass: kelp, 308–9; mature herring, 226, 252, 299; peak biomass, 293; spawning, 10, 138, 148, 153, 203, 206–7, 217, 226, 238–39, 300
Blumenthal, Richard, 310
Bosch, Carl, 24
Bower, Ward, 17, 38, 96, 97, 107
Brady, John, 44
brailed, term, 27
Branson, Jim, 228
Bristol Bay, 10–11, 18, 64, 138, 143, 145; and modern-era food herring, 267–68; and spawn-on-kelp fisheries, 273, 278, 280; and Togiak roe-herring fishery, 217–53
Bristol Bay Fisherman, 143
Bristol Bay Herring Marketing Cooperative, 231
Brophy, Bob, 221–22
Browning, Robert, 51
Buchan, Andrew, 86–87, 108
Buchan & Heinen Packing Company, 36, 41, 84, 86–87, 89, 104, 108, 115
Bureau of Commercial Fisheries, 125, 276
Bureau of Fisheries, US, 16–19; bait herring, 130–33; chronicle of herring industry, 86, 88, 90, 94–99, 120; fishery regulation, 65–70; herring industry expansion, 73, 78; and herring industry development, 30–32, 34, 36, 41; and salted herring, 46–47, 49–50, 53, 56, 58, 61
Buschmann, Leif, 61
Buschmann, Peter, 44

Canned Salmon, Herring and Protein Powder Project, 267
canning, 30, 43, 47–49, 267–68

Cape Ommaney, 41, 47, 93, 97, 106–7, 110, 117, 125
Case, Ed, 320
Chatham Strait Fish Company, 36, 77, 82, 96, 102, 104–5, 109
Chichagof (vessel), 181–82, 193, 223, 296, 306
Chilkat Inlet, 22
circuit seiners, 150, 212, 215
Clark, Walter, 32
closed pound, 274, 282–84; fisheries, 288, 290, 296, 300; in Prince William Sound, 285–90, 293–95; spawn-on-kelp fishery, 279, 289, 296, 299–300, 304
Clupeidae, family, 3–4
Cobb, John, 16, 21, 29, 44, 52, 53, 132
Coei Maru 11 (vessel), 254–55
combat fishing, 218
Commercial Fisheries Entry Commission, 137, 149–51, 155, 196–98, 212, 214–15, 229, 291, 298, 303, 313
Commercial Fisheries of Alaska, The (Cobb), 21
competitive fishery, 137, 156, 158, 161–62, 164–65, 169
controversy, herring fishery, 19–20
Cook Inlet. *See* Lower Cook Inlet
Coolidge, Calvin, 65
cooperative: fishery, 137, 151, 157, 162, 166; fishing, 151, 168, 176; hybrid competitive-cooperative system, 159
COVID-19, 177
Craig/Klawock, spawn-on-kelp fisheries in, 300–306
Cullen-Harrison Act, 85
cured herring, 61–63; 1932 production estimate, 84; 1934 season production, 87; 1937 production of, 91–93; European availability, 90; industry, 19, 48–49, 53, 57, 59–60, 72, 93, 99; market, 10, 82, 90; pack, 56–57, 60, 77, 86, 99; production, 80, 84, 87, 103, 106, 109. *See also* Scotch-cured herring
curing: demise of industry of, 84–110; operations, 5, 33, 54; postwar years, 56–61; Scotch method of, 49–54; stations for, 57
Curry, Clyde, xiii, 356

Dahlgren, Edwin, 99–100, 105
Daily Sitka Sentinel, 152, 158–59, 163
Davidson, Bill, 152, 162, 164–65, 167
Deckhand Seafoods, 269

Index

decomposition method, 145, 154. *See also* stripping, roe
DeJong, Bob, 12, 297
Delegate Act, 30
diet, herring, 11
Dingell, Debbie, 319
Diver 1 (vessel), 181–83, 220
Dolphin (vessel), 28
Donna Lane (vessel), 75–77, 80, 83, 90
double-permit pound, 299, 304
Dutch Harbor, 81, 106, 151; bait herring in, 137–39

early development, herring industry: catching herring, 27–28; herring use expansion, 29–35; modern purse seining, 35–39; reducing herring, 21–26; whaling, 40–42
early Native uses, herring, 13–17
Edgar C (vessel), 36
Edible Herring Products Company, 121, 127–28
eggs, herring, 12–13
Emlach, W. J., 57
Endangered Species Act, 42
Ernest Sound, spawn-on-kelp pound fisheries in, 307
Esther (vessel), 75
Evermann, Barton Warren, 30, 40, 47
ex-vessel value, 136, 138, 144; Kodiak fishery, 214–16; Lower Cook Inlet fishery, 212; Norton Sound fishery, 255–56, 261–65; Prince William Sound fishery, 197–209; Sitka Sound fishery, 156–76; spawn-on-kelp fisheries, 278–79, 287, 290, 292, 302, 306–7; Togiak fishery, 227–28, 232, 234–36, 238, 244–49
expansion, herring industry, 72–74; floating salteries, 75–78; packers, 78–83
Exxon Valdez (oil tanker), 136, 150, 179, 203–4, 206, 210, 237, 282, 290

Favorite (vessel), 28
feeding, herring, 11
fertilizer. *See* meal/fertilizer, herring
Fish and Wildlife Service, US, 18, 97–98, 101–2, 104–6, 109, 111–13, 116–20
Fish Egg Island, 3, 119, 134, 301
fish meal, 18–20, 22–24, 33, 104, 109, 120, 123–24, 126, 195
Fish Reduction and Saltery Workers' Union, 89–90

fisheries: early development of, 21–42; early regulation/research, 64–71; expansion of, 72–83; herring basics, 3–20; in Kodiak area, 211–16; in Lower Cook Inlet, 211–16; management of, 147–51; in Norton Sound, 254–65, 308–17; in Prince William Sound, 179–210, 285–95; regulation of, 64–71; in Resurrection Bay, 179–210; salted herring, 43–63; in Sitka Sound, 152–78; in Southeast Alaska, 296–307; in Togiak, 217–53, 308–17
floating salteries, 75–78, 81, 83
Floe, Olaf, 78–79, 91, 93
food herring, 20–21, 57, 266–69
Forage Fish Conservation Act, 319–20
Fordney-McCumber Tariff, 60
Franklin Packing Company, 1, 56, 58, 67, 75, 79
Fred D. Parr (vessel), 277
freighter, 76, 87, 179, 224–25, 230, 233, 254, 256

giant kelp (*Macrocystis pyrifera*), 273, 276, 278–80, 283–89, 291, 296–97, 316–17
gibbing, 50–51, 76, 107
gillnets, 18, 52, 65, 98, 102, 118, 121, 207, 226, 232–33, 241–42, 254–56, 260, 264
Glacier Kelp Group, 288
Goldsborough, E. L., 30
Golovin, community, 61–63
Gordon, Dave, 151, 169, 171, 175
Great Depression, 82–83, 91
Gregory, Homer, 95
Gribble, Ron, 208
growth, herring, 9–10
guano, market for, 23–24
guideline harvest level (GHL), 135–36, 148, 298
gypsy fleet, 257

Haber, Fritz, 24
hair kelp (*Desmarestia* sp.), 273
Halibut Cove, 52, 61, 64–66, 78, 81, 86
Hammond, Jay, 228
Harding, Warren G., 64–65
Herring Bay, 3
Herring Cove, 3
Herring Fisherman's Union of the Pacific, 88–89, 94

herring industry, chronicle of: cured-herring industry demise (1932–48), 84–110; reduction operations (1949–66), 111–39. *See also* early development, herring industry; expansion, herring industry; reduction operations, herring industry
Herring Marketing Cooperative, 231
Herring Northwest, 181–83
herring-catch books, 68
Higgins, Elmer, 73
"High the Herring," 85
Hirohito, emperor, 260
Homstrand, Bob, 181–82
Hoonah Sound, spawn-on-kelp fisheries in, 296–300
Huffman, Jared, 320
Huizer, E. J., 27, 44

Ichthyophonus hoferi (fungus), 207
Ickes, Harold, 102–3, 103, 106, 112
Icy Strait Packing Company, 44
Imlach, W. J., 66
Imlach Packing Company, 66
impounding, 46–47
in-state herring processing bill. *See* Togiak, roe-herring fishery in
industry, herring: early development of, 21–42; expansion of, 72–83; reduction operations in (1949–66), 111–29
Infinite Glory (vessel), 170
International Convention for the High Seas Fisheries, 116–17, 124

Japan, 10, 18–20, 177–78, 197–98, 318; amending International Convention, 124; booming economy, 222; death of emperor of, 237; fishing abstinence, 116–17; freezing herring for, 224–25; herring roe shortage in, 147; herring spawn on kelp in, 273, 276, 281; *kazunoko* in, 125, 143; March 11 earthquake, 172–73; and Pearl Harbor, 101–2; roe herring importance in, 143–45; surplus in, 162, 215; and Togiak roe-herring fishery, 217–25; wea yen in, 244
John N. Cobb (vessel), 116, 118, 120–21
Jones, E. Lester, 16–17, 32
Jones, Wesley, 31
Juneau Canning Company, 29
Juneau Cold Storage, 146, 178

Kaitlin Packing Company, 278
Kandoll, Brian, 301
Katlian Packing Company, 133
kazunoko, 19, 125, 143–45, 143, 162, 172–74, 215–16, 237, 244, 258–52. *See also* roe herring
Kazunoko (vessel), 211
kazunoko kombu (herring spawn-on-kelp), 19, 272–74
kelp (*kazunoko kombu*), spawning on, 273–81
Kenai Peninsula, 3, 67, 106, 146, 254
Killisnoo, reduction plant at, 21–28, 31–32, 43, 54, 66
king salmon, 4, 19, 30, 131, 133
kippering, 48
Kitano, Shigeyoshi, 220–21, 257
Klawock Cooperative Association, 300
Klawock Heenya Corporation, 300–301
Klie, August H. D., 49–50
Kodiak, 60; bait herring in, 136–37; establishing reduction plants on, 88; 1941–42 herring quota, 101–3; 1943 herring quota, 103–4; roe-herring fishery in, 212–16
Kodiak-Afognak, 10, 36, 64
Kodiak Operators, 102
Kolloen, Lawrence, 105, 107
Kootznahoo Inlet, 66
Kuiu Island, 36, 107, 120
Kulukak Bay, 217, 219, 225–26, 236, 308
Kyokko Suisan Alaska, 221, 257

Larsen Bay, 213
Lean, Charles, 259
LeConte Glacier, 131
legislation, attempted regulation through, 70–71. *See also* Alaska Legislature
Libby, McNeill & Libby Company, 76
Limited Entry Act, 149–50
limited-entry system, 151, 299
liquor, problem with, 233
Louise (vessel), 28
Lower Cook Inlet, roe-herring fishery in, 211–12
Lynn Canal, 70, 178

Macrocystis pyrifera (giant kelp), 15, 273
Magnuson-Stevens Fishery Conservation and Management Act, 220, 230, 255, 319
Maple Leaf C (vessel), 36

on Storfold & Grondahl Company, 79–80; summarizing 1923 season, 61; summarizing 1928 herring fishery, 80; on Unalaska herring fishery, 81; Unalaska herring fishery, 81; whale slaughter, 40–41

Pacific herring (*Clupea pallasii*), 3–5; abundance, 13; age/growth of, 9–10; bait herring, 130–39; catching, 27–28; commercial fishery of, 17–18; commercial harvest, 18; controversy involving, 19–20; diet/feeding, 11; early Native uses of, 13–17; eggs, 12–13; expansion of uses for, 29–35; general description of, 9; range/migration, 10; reducing, 21–26; spawning, 11–12; spawning on kelp, 273–81

Pacific Herring Packers Association, 24, 70, 79, 97, 102

Pacific Rim (floating reduction plant), 112, 114, 116

packers, organizing, 78–83

packing, herring, 24, 50, 61–62, 66, 74, 79, 87

Paoli, Mike, 208

Parker, James, 154, 156

Parker, Tom, 205

Pearl Harbor, 101–2

Pelican Cold Storage, 153

permit holders, 137, 157, 159, 161, 164, 166, 168, 174, 176, 247, 291, 293, 299, 305, 314

Peru, 123–24, 126–28

Petersburg Cold Storage, 147, 153, 276

Petersburg Fisheries, 154, 181, 195, 266

Petroff, Ivan, 14, 15

pickle (brine), 51–52

Pioneer Mining & Ditch Company, 62

pollock, 220, 238, 242–43, 251–52, 318

Port Ashton Packing Company, 114–15

pounds: closed, 282–84; open, 281–82

Pribilof Islands, 10, 122, 127, 220, 238

Prince William Sound, 3, 18, 34, 42, 44–45; areas, 65, 80, 109, 113, 119, 121; bait herring in, 135–36; 1941–42 herring quota, 101–3; packers, 90, 109; purse seining in, 35–36; spawning in, 11–12

Prince William Sound, roe-herring fishery in: districts, 185; doubled catch (1972), 185; emergency-opening debacle (1979), 197–98; establishing new fishing district (1980), 198; fleet growth (1976), 194–95; fungus infection (1994), 207; GHL reduction (1987), 202; hazards (1973), 186–90; herring abundance (1986), 201–2; herring population decline (1999), 209–10; highest projection forecast (1992), 206; limiting catch (1974), 191; management change (1975), 191–93; modifying management strategy (1982–85), 199–201; oil spill (1989), 203–4; plane crash (1991), 205; producing economic boost (1971), 184–85; seine fleet increase (1981), 198–99; seiner decline (1998), 208–9; seiner management (1990), 204–5; spotter plane tragedy (1995–96), 207–8; spring harvest (1988), 203; viral hemorrhagic septicemia (VHS), 206–7

Prince William Sound, spawn-on-kelp fisheries in: 1986 season, 287–88; 1993 season, 292; 1995 season, 293; 1996 season, 293; closing fisheries (1989), 290; general abundance (1984), 286–87; herring abundance decline (1987), 289; importing *Macrocystis* kelp (1981), 285; increasing biomass threshold (1994), 292–93; limited wild fishery entry (1992), 291–92; nonexistent season (1985), 287; oil spill response (1990), 290; pound operation management plan (1991), 290–91; pound operation options (1997–99), 293–95; protecting harvest areas (1982), 285–86; restricting closed-pound fishery (1988), 290; use of open-pounded kelp (1983), 286

Prohibition, 58

Puget Sound Reduction Company, 77

purse seining, 35–39, 222, 227, 256; boats, 69, 125; cooperatives, 247; fleets, 201, 237, 240, 247–49; openings, 236, 240, 247; permits, 150, 215

Quigmy River, 242

quotas, herring, 77, 96, 99, 101–3, 107, 109–13, 119, 123, 134, 136, 214, 309; 1948, 109–10

range, herring, 10

raw herring, reduction of, 19

Index

Maren 1 (vessel), 223–25, 230
Marine Stewardship Council, 318
Marsh, Millard, 30
Marubeni Corporation, 221
McDowell Group, 268
McNair, Doug, 192
McNeil, Dan, 73–74
meal/fertilizer, herring, 33, 57, 69, 73, 78, 80–82, 86, 130
menhaden (*Brevoortia tyrannus*), 4, 22
Metervik Bay, 3, 226, 308
migration, herring, 10
modern era, food herring in, 266–69
Montague Island, 12, 18, 186, 203, 206, 208–9, 294
Moore, H. F., 53, 223–24
Moore, Ken, xiii, 223
Moser, Jefferson, 14–15, 25–26, 28
Muriel (vessel), 75–76
Murkowski, Frank, 266
mustard sardines, 29
Mutual Fish Company, 275

Nagel, Charles, 31
Nassau Fish Company, 75, 77, 81
National Atmospheric and Oceanic Administration, 224
National Marine Fisheries Service, 196, 220, 228, 238, 251, 318
Natives: early herring uses, 13–17; eating herring eggs, 15; employment, 28; herring roe and, 153; kelp and, 280–81, 300, 308; primitive rakes of, 14; reducing herring, 21–22
Nelson, Beaver, 179–81, 184, 192, 222, 224–25
Nelson, Edward, 12
New England Fish Company, 132–33
North American Fisheries Company, 76
North Pacific Fishery Management Council, 228
North Pacific Longline Association, 231
North Pacific Seafoods, 268–69
Northwestern Herring Company, 24, 78, 88, 93, 96
Northwest Trading Company, 21–23
Norton Sound, 146, 150; commercial fishery in, 18; cured herring in, 61–63; herring migration, 10; roe-herring fisheries, 254–59; spawn-on-kelp fishery in, 316–17; spawning in, 11–12

Norton Sound, roe-herring fisheries in, 254–59; clearing ice early (1995–96), 262; declaring economic disaster (1992), 260–61; diminishing interest (2001–2), 264; and death of Japanese emperor (1989), 260; 1988 season, 259; 1990 herring record, 260; 1991 average price, 260; 1999 market, 263; projecting fewer herring (1997), 262–63; projecting herring abundance (1998), 263; scratch fishery (1994), 261–62; smallest fishing effort (1993), 261; 2000 season, 264; 2003–5 harvests, 265; 2006–18 harvest, 265

O'Malley, Henry, 34, 60, 64, 70–71
OBI Seafoods, 255
Ocean Beauty Seafoods (OBI), 255, 267
Oceanic Fisheries Company, 90, 102, 105, 109, 115
oil spill. See *Exxon Valdez*
oil, herring, 22–26, 72–73, 81, 86–88, 90–94, 108–9, 112, 115–16, 118, 120–22, 124–27
open pound, 281–82; in spawn-on-kelp fisheries, 285–87, 292–95, 316

Pacific Cured Fish Association, 55, 58–60, 78
Pacific Fisherman, 58–59; on 1918 pack; on 1927–28 production, 80; 1933 Scotch-cured herring pack, 86; 1943 herring quota, 104; on Act for the Protection and Regulation of the Fisheries of Alaska, 30–31; and Alaska Herring Packers Association, 78; on Aleutian Islands herring, 82–83; Aleutian Islands herring, 83; barrel quality, 54; booming reduction industry, 72–73; celebrating legislation, 85; chronicle of herring industry (1932–48), 87–88, 91, 93–94, 99, 101–10; chronicle of herring industry (1949–66), 112–17, 120–21, 123–25, 129; describing salt operation, 56; on *Donna Lane*, 76; on freezing bait, 132; herring curing industry, 34; on herring meal, 24; herring pounds, 46; interviewing Clarence Anderson, 67–68; *Muriel* alterations, 76; on Port Walter plant, 50; Rounsefell writing in, 68; Scotch curing, 49–50;

395

red feed, 45–49
reduction: defense of, 31–32; demise of plants for, 111–29; early development of, 21–26; floating plants, 112, 114, 116; floating salteries and, 75–78; herring-reduction plant, 20, 24, 33–34, 77–78, 80, 88, 92, 96–97, 114–17, 121, 125, 186, 188; plant development (1935), 88; plants for, 75–78
reduction operations, herring industry: abundance constraint (1958), 121–22; amending fishing regulations (1952), 115–17; closing Edible Herring Products Company (1966), 128–29; debacle of 1949, 111–13; declining fish meal/herring oil production (1961), 124; establishing catch limits (1956), 119–20; limited operations (1954), 118; passing memorial (1953), 117–18; plant dismantling (1965), 127–28; prohibiting reduction (1963), 126; pursuing knowledge of herring stocks (1957), 120–21; quota increase (1955), 118–19; quota reduction (1951), 114–15; return to "normal" operations (1950), 113–14; Shelikof Strait seizure, 124–26; sustained-yield principle (1960), 123–24; US-Japan negotiations (1964), 126–27; young herring preponderance (1959), 122–23
regulation: ADF&G regulations, 288, 291, 301; of early herring fishery, 64–67; fisheries, 130, 133; of fishing time, 230; and research, 67–70; through legislation, 70–71
Report on the Population, Industries, and Resources of Alaska, 14
research, 67–70
Resurrection Bay, roe-herring fishery in, 179–84
retros, 205
revenues, 137, 151, 155, 174, 297
ribbon kelp (*Laminaria* sp.), 273
Robby (vessel), 179
rockweed (*Fucus* sp.), 273
roe herring, 143–45; fishery management, 147–51; genesis of industry of, 146–47; harvests, 181, 183–84, 201, 211, 213–14, 234, 239–41, 246, 248, 260, 263; Kodiak fishery, 212–16; Lower Cook Inlet fishery, 211–12; Norton Sound fisheries, 254–65; Prince William Sound fishery of, 184–210; production, 153, 206; Resurrection Bay fishery of, 179–84; seasons, 153–54, 158, 186, 192, 196–97, 200, 233, 246, 264, 288; Sitka Sound fishery of, 152–78; stripping, 145–46; Togiak fishery, 217–53
Rogers, Stephen, 37
Roosevelt, Franklin, 85, 94, 97
Rounsefell, George, 36–37, 46–47, 67–68
rousing, 50–51

sac roe. *See* roe herring
St. Matthew Island, 220
salt, 55–56
salted herring, 43–45; barrels, 54–55; cured herring, 61–63; feedy herring, 45–49; post–World War I years, 56–61; salt, 55–56; Scotch curing, 49–54
salteries, 34, 44–46, 73–74, 87, 90, 99, 132; floating, 75–78; and herring packers, 81, 83
sardines, 29, 35, 68
schooner, 75–77, 81
Scidmore, Eliza, 14, 22, 25
Scotch herring, 49–54
Scotch-cured herring, demise of industry of: doubled production, 88; 1932 production, 84; 1933 production, 84–86; 1936 production, 91; 1938 production, 93–95; 1939 production, 95–98; 1940 production, 98–100
Scow Bay, 24, 44, 131–33
Sea of Japan, 10
seasons, establishing, 106
Seattle Times, 181
Seward Marine Services, 181, 183
Shady Lady (vessel), 171
Sheakley, James, 25
sieve kelp (*Agarum* sp.), 273
Sitka Cold Storage, 134, 153
Sitka Sound, 12, 18, 42, 66, 133, 137, 139, 145, 148, 151
Sitka Sound, roe-herring fishery in, 152–54; 11 percent roe content (2007), 169; 1992 opening, 159; abundance decrease (1991), 158–59; addressing controversy (2002), 166–67; banner year (1988–89), 156–57; coming-of-age moment (1979), 155–56; competition (1981–87), 156; constrained

kazunoko market, 177; cooperative agreement (2013), 173–74; declining roe-herring market (2018), 176; exceeding quota (2006), 169; five/six openings (2011), 172–73; four openings (2009–11), 171–72; GHL increase (1980/2005), 156, 168; GHL reduction (1978), 155; GHL reduction (1990/2000), 157–58, 165; Halibut Point–Middle Island fishing (2003), 167; handling carcasses (1976), 154; harvestable size (2014), 174; highest spawning herring return (2004), 167–68; high-quality roe content (2001), 165; hybrid competitive-cooperative system (1993), 159–60; implementing limited entry (1977), 155; less marketable fish (2016), 175; limiting boat usage (1996), 161–62; low ex-vessel prices (2015), 174–75; managing multiple openings (1999), 163–64; marketable herring size (2017), 175–76; national publicity (2008), 170; permit holders, 156–57; poor herring volume (2012), 173; quota decrease (1998), 162–63; season cancellation (2020), 177; second-largest ever quota (1997), 162; six-year-old class (1994), 160–61; three-year-old class (1995), 161

Sitka Sound Seafoods, 66, 153–54, 160

Skrade, Jeffrey, 217

Southeast Alaska, 3–5, 14, 153–55, 179–80, 197, 254, 273; 1943 herring quota, 103–4; bait herring in, 130–35; commercial fisheries in, 17–18; commercial fishery in, 18; competition in, 149; ending herring fishing in, 98–100; fishery regulation/research, 66–70; herring cultural value in, 13–17; herring industry development, 21–22, 24, 28–30, 33–35; industry chronicle (1932–48), 84, 86–110; industry chronicle (1949–66), 111–28; industry expansion, 73, 77–83; limited-entry commission, 150–52; overfishing in, 95–98; reduction plant shuttered in, 20; roe-herring fishery genesis, 146–47; salted-herring early years, 43, 46–47, 50, 54, 60–61; spawn-on-kelp fishery, 273–80, 283–88, 296; spawning in, 11–12; whaling station established in, 40–42

Southeast Alaska, spawn-on-kelp pound fisheries in: Craig/Klawock, 300–306; Ernest Sound, 307; Hoonah Sound, 296–300; Tenakee Inlet, 306–7

Southeast Alaska Herring Championship, 86–87

Southwestern Alaska Fishery Reservation, 64, 65

Southwestern Fisheries Company, 104

Southwestern Herring, Inc., 102

spawn-on-kelp fisheries: genesis of, 273–81; in Norton Sound, 316–17; pounds and, 281–84; in Prince William Sound, 285–95; in Southeast Alaska, 296–307; in Togiak, 308–14

spawning: biomass, 10, 138, 148, 154, 203, 206–7, 226, 238–39, 300; herring, 11–12, 152, 153, 157, 163, 165–67, 175, 201, 203, 208, 261, 282, 285, 292; populations, 62, 148, 161, 206, 213–14, 217, 283, 296, 298

Special Subcommittee on Alaskan Fisheries, 34–35

Spiridon Bay, 146, 213, 254

Spuhn, Carl, 15–16, 21, 31–32

steamships, 23, 54, 77, 131

stink plants, 72

Storfold, Jack, 117

Storfold, Olaf, 3

Storfold & Grondahl Packing Company, 79–80, 93, 101–4, 115

Storis (vessel), 254–55

Stovall, Charles, 150

stripping, roe, 145–46, 153, 182–84, 196, 220, 257

Sustaining America's Fisheries for the Future Act, 320

Taiyo Gyogyo Ltd., 146

Tenakee Inlet, spawn-on-kelp fisheries in, 306–7

Territory of Alaska, establishing, 32. *See also* Alaska

Thomas, Ashton, 29, 45, 47–48, 56, 60

Thompson, Harold, 160

Thompson, Seton, 115

Thornton, Thomas, 3

Togiak, roe-herring fishery in: catching yesterday's fish (1996), 244; COVID-19 response (2020), 251–53; efficiency development (1986), 234–35; effort increase (1979), 226–27; environmental issues (1984), 232–33; final "normal" years (2005–19), 250–51; fishing time regulation (1981), 230–31; fog presence (1988), 235–36; genesis of, 217–25; herring abundance (1992), 239–40; herring symposium (1980), 228–30; last joint venture operation (1987), 235; limited capacity openings (2000–2001), 246–48; low ex-vessel revenue (1998), 245; modifying management plan (2004), 250; monitoring processing capacity (1999), 245–46; 1982–83 record, 231–32; 1991 participation, 238–39; opening duration change (1985), 234; participation decline (2002), 248–49; potential trawl closures (1990), 237–38; regulation adoption (1995), 243–44; seine fleet increase (1978), 225–26; strong preseason herring-abundance forecast (1994), 241–43; tender shortage (1989), 236–37; traditional "race for fish" decline (2003), 249–50; and weak yen (1997), 244–45; worldwide glut (1993), 240–41

Togiak, spawn-on-fish fisheries in, 308; 1981–83 harvest, 309–10; crewmember prohibition (1992), 313; decline of interest (1993), 313; establishing harvest quota (1984), 310–11; greatest effort recorded (1991), 313; gunfire incident (1987), 311; limiting fishery entry, 312–13; 1986 harvest, 311; 1988 harvest, 312; 1994–96 harvest, 314; production peak (1989), 312; weather problems (1985), 311

Totem (vessel), 223
tow pound, 283–84
Tyee Company, 40

Unalaska, 10, 46, 68, 80–83, 87, 92, 98, 101, 138, 242
unions, 88–90, 94
United States. *See* Alaska; Bureau of Fisheries, US; Fish and Wildlife Service, US
Upper Cook Inlet, bait herring in, 139
Utopian Fisheries Company, 75, 90
Uyak Bay, 215

Valencia (vessel), 35
Vanderbilt, J. M., 21
vessels. *See various entries by name*
viral hemorrhagic septicemia (VHS), 206–7

Wage Stabilization Board, 114
Wakefield, J. Howard, 112
Wakefield, Lee, 46, 47–48, 67, 79
Washing Meal & Reduction Company, 127
Washington Fish & Oyster Company, 213
Webster, John, 58
Western Alaska Canned Chum & Herring Demonstration Project, 267
Western Alaska Cooperative Marketing Association, 231
Western Alaska Enterprises, 146
Western Cooperage Company, 54
Western Fur and Trading Company, 21
Westward Seafoods, 242
whale, humpback (*Megaptera novaeangliae*), 41, 170
whaling, 40–42
White Act (Act for the Protection of the Fisheries of Alaska), 65
White, Wallace, 64–65
Whitney-Fidalgo Seafoods, 179
Wickersham, James, 31, 70–71
Wilde, W. W., 102
Wilson Fisheries Company, 33, 48–49
Winning Hand (vessel), 164
Winthers, John, 63
women, 51
World War I, 47, 49–50; postwar years, 56–61
World War II, 38, 63, 97, 145, 181, 194–95, 213, 220; need for fish oil/meal during, 104–6
wrack (*Saccharina* sp.), 273
Wrangell Narrows, 44, 130, 266
Wyman, Bob, 153, 277–78

Yakutat & Southern Railway Company, 44
Yardam Knot (vessel), 315
Yoshimura, Harry, 275

Zachar Bay, 96, 105, 109, 127–29, 213
Zenpu Maru 8 (vessel), 254–55

www.ingramcontent.com/pod-product-compliance
Lightning Source LLC
Chambersburg PA
CBHW020514080526
44583CB00013B/590